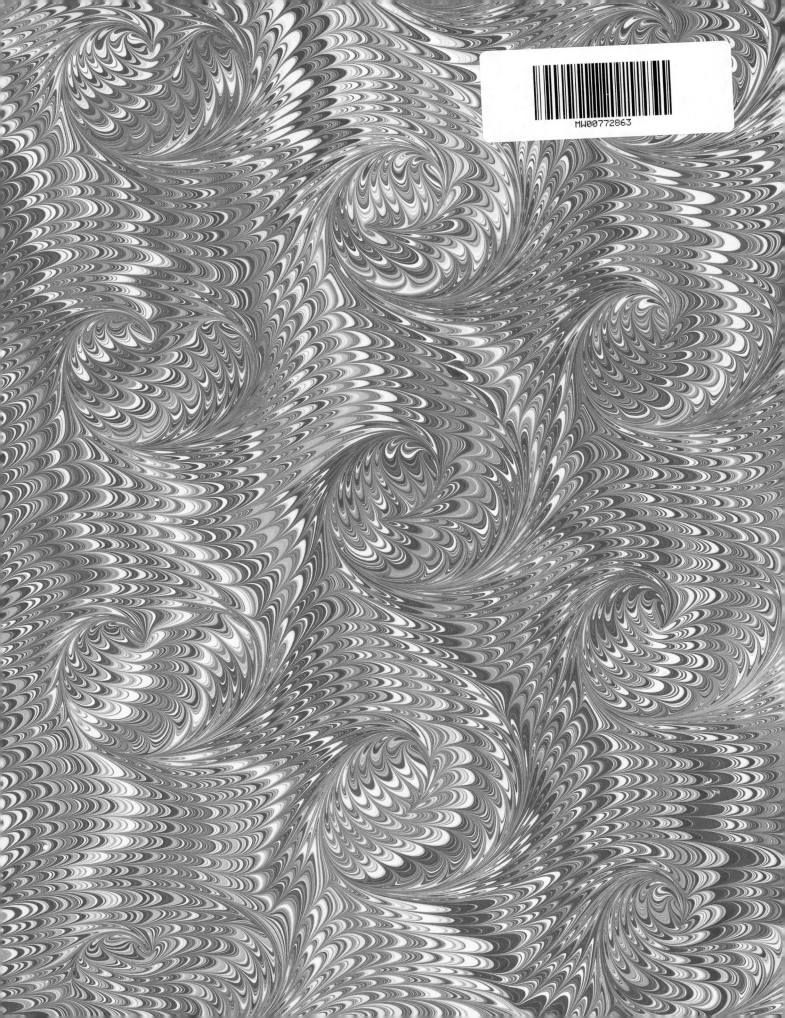

MW00772863

Gray Ghosts

Peter E. Davies

This book is dedicated to the memory of
Captain Foster Schuler Teague, USN.

Gray Ghosts

U.S. Navy and Marine Corps F-4 Phantoms

Peter E. Davies

Schiffer Military History
Atglen, PA

Acknowledgments

Thanks are due to the following for their generous assistance: Steven P. Albright, Ben Backes (Fox One Decals), CAPT George K. "Bullet" Baldry USN (Ret), CDR Stephen J. Barkley USN (Ret), LT Henry A. "Bart" Bartholomay USN (Ret), Mary Barr (McDonnell Douglas), LT Lynn R. Batterman USN (Ret), LT David Batson USN (Ret), LT Bill "Burner" Beardsley USN (Ret), LCDR Jerry Beaulier USN (Ret), James M. "Taco" Bell USN (Ret), CDR Don Bentley USN (Ret), Tom Bennington USN (Ret), RADM Peter B. Booth USN (Ret), LT Gene Blair USN (Ret), CDR Alan Bradford USN (Ret), CAPT John Brickner USN (Ret), LTJG Dick Brooker USN (Ret), James N. Butler USN (Ret), CDR James Carlton USN, CAPT Roger Carlquist USN (Ret), CAPT Roy "Outlaw" Cash, USN (Ret), CAPT A.W. "Hap" Chandler USN (Ret), CDR C. Doug Clower USN (Ret), LCDR Matthew J. Connelly III USN (Ret), COL John D. Cummings USMC (Ret), Frank Curcio (USN (Ret), Charles D'Ambrosia USN (Ret), CDR David "Skinny" Daniels USN (Ret), CDR Curtis R. "Dozo" Dosé USN (Ret), CAPT Hugh N. Dyer USN (Ret), Terry Edwards USN (Ret), CAPT Orville G. "Tex" Elliott USN (Ret), VADM Donald D.Engen USN (Ret), CDR Grover G. Erickson USN (Ret), COL Michael Fagan, USMC, Michael A. France, MGEN Paul A. Fratarangelo USMC (Ret), CDR William "Farkle" Freckleton USNR (Ret), CDR Guy H. Freeborn USN (Ret), CAPT Brian E. "Bulldog" Grant USN (Ret), CDR William E. Greer III USN (Ret), Doug Gresham USMC (Ret), CAPT Charles M. de Gruy USN (Ret), Eugene R. Hamamoto USMC (Ret), LT COL John Harty USAF (Ret)/McDonnell Douglas, CAPT Jerry B. "Devil" Houston, USN (Ret), CAPT William B. Haff, USN (Ret), CAPT Thomas H. Idema, USMC (Ret), CDR Jan C. Jacobs, USN (Ret), Wes Johnson USMC (Ret), LT W. Fritz Klumpp, USN (Ret)/ McDonnell Douglas, CAPT William D. Knutson, USN (Ret), LT COL John Jay Kuenzle, USMC (Ret), Marty Lachow USMC (Ret), CDR Scott Lamoreaux USN (Ret), COL Ed J. Love USMC (Ret), CAPT Eugene P. "Geno" Lund, USN (Ret), CAPT Lonnie K. McClung, USN (Ret), LCDR William J. Mayhew USN (Ret), CDR Peter B. Mersky USN, LGEN Thomas Miller, USMC (Ret), LT COL Charlie "Burner" Mitchell, USMC (Ret), CAPT John Nash USN (Ret), LT COL James O'Donnell, USMC, CWO J. J. O'Brien, USMC (Ret), David Parsons USN (Ret), LT COL Buck Peck USMC (Ret), AMS Rodney D. Preston USN (Ret), Chris M. Reed, COL Manfred A. "Fokker" Rietsch USMC, Angelo Romano, CDR Stephen A. Rudloff USN (Ret), James H. Ruliffson USN (Ret), COL Dayton Robinson USMC (Ret), MAJ James Rotramel USAF (Ret), Group Captain James N. Sawyer RAF (Ret), David W. Schill USN (Ret), Group Captain Michael J. F. Shaw, CBE, RAF (Ret), Frank Shelton USN (Ret), COL James R. Sherman USMC (Ret), Peter St.Cyr USMC (Ret), CAPT Federick G. Staudenmayer USN (Ret), MGEN Michael P. "Lancer" Sullivan, USMC (Ret), Steve Tack, Norm Taylor, Mrs Susan Teague, CDR Bruce Thorkelson USNR (Ret), Tony Thornborough, Richard Tipton USMC (Ret), John Trotti USMC (Ret), CAPT Roscoe L. Trout USN (Ret), Vance E. Vasquez, CAPT Fred Vogt USN (Ret), Ray Wagner, Simon Watson, LT Don J. Willis USN (Ret), RADM John R. "Smoke" Wilson, Jr., USN (Ret), Mike Wilson.

Book Design by Ian Robertson.

Copyright © 2000 by Peter E. Davies.
Library of Congress Catalog Number: 99-68305

All rights reserved. No part of this work may be reproduced or used in any forms or by any means – graphic, electronic or mechanical, including photocopying or information storage and retrieval systems – without written permission from the copyright holder.

Printed in China.
ISBN: 0-7643-1021-6

We are interested in hearing from authors with book ideas on related topics.

Published by Schiffer Publishing Ltd.
4880 Lower Valley Road
Atglen, PA 19310
Phone: (610) 593-1777
FAX: (610) 593-2002
E-mail: Schifferbk@aol.com.
Visit our web site at: www.schifferbooks.com
Please write for a free catalog.
This book may be purchased from the publisher.
Please include $3.95 postage.
Try your bookstore first.

In Europe, Schiffer books are distributed by:
Bushwood Books
6 Marksbury Avenue
Kew Gardens
Surrey TW9 4JF
England
Phone: 44 (0)181 392-8585
FAX: 44 (0)181 392-9876
E-mail: Bushwd@aol.com.

Try your bookstore first.

Contents

Foreword
(for the United States Marine Corps)

The McDonnell Phantom II (F-4) will certainly be recorded in aviation history as one of the greatest military aircraft ever produced. There are several facts that support this forecast. First, and probably the most amazing was that it took only six years from the first contract (No.55-272) until the first aircraft was delivered to the Fleet in June 1961. Also, the F-4 set more World Performance Records (15) than any other military aircraft, and it was used as a tactical fighter bomber by more domestic and foreign air forces than any other aircraft in the world. Over 5,000 were produced, and most all were delivered on time and within cost projections. A truly remarkable accomplishment, and an amazing aircraft.

The Navy and Marine Corps normally prefer to operate multi-mission tactical aircraft because of the space and support limitations when operating from ships at sea and expeditionary airfields ashore. This does present some unique design requirements in the early stages of aircraft development. In the early 1950s the birth of jet-powered military aircraft generated some additional, unique problems for operations aboard ships at sea and short expeditionary airfields ashore. Naval forces also faced a new threat from high- and fast-flying, jet-powered enemy bombers. The Navy's primary concern was for the protection of its ships at sea. The Marine Corps' primary concern was for the protection of its highly mobile forces ashore. Navy aircraft could provide protection for these Marine forces so long as they remained in the near vicinity. However, the threat of high and fast-flying enemy aircraft and the possible use of nuclear weapons made it highly probable that Navy aircraft carrier forces would not remain in fixed locations for any lengthy period. Consequently, the Marine Corps considered it essential to have land-based aircraft capable of defending against this enemy aircraft attack, as well as providing an attack capability for the support of its ground forces ashore.

During the same period that the jet engine was coming into use there were some significant new avionics advancements in radar, and fire-control computers were becoming available that would significantly improve all-weather operations. Both the Navy and Marine Corps were anxious to improve their all-weather offensive warfare capabilities. The Marine Corps was especially concerned about its all-weather interdiction and close-air-support for its ground

Lieutenant General Thomas H. Miller, Jr., USMC (Ret). (U.S. Marine Corps)

forces. Initially, the Navy considered this to be the most urgent requirement, and in October 1954 placed a letter contract with McDonnell for two AH-1 all-weather attack aircraft.

By 1955 the enemy's high-flying bomber had become the primary concern, as current Navy and Marine fighter aircraft lacked the capability to counter this threat. As a result, the Navy modified its contract with McDonnell and redesigned the aircraft as the F4H-1 (vice AH-1), with primary emphasis on its performance as a fighter. Along with this modification came a change to the more powerful General Electric J79 engines vice the J65 engines originally planned. In addition, because of the complexity of the all-weather fighter mission, a two-seat version with the most advanced long-range radar was called for.

Although the primary mission design requirement changed from attack to fighter, most of the early design criteria for the AH-1 were left intact, and the uprated engine power and larger landing gear all proved to enhance its attack capability. Marine Corps combat experience proved it to be an excellent fighter/bomber.

There are two truly amazing facts about this aircraft and this book. First, the decision process and procedures used in the design, development, testing, and production of the aircraft. Many who are acquainted with the procedures and decision-making process in use by the Navy during the F-4 development and early production phases will confirm that it would be impossible to achieve such overall program efficiency in today's decision-making process. The F-4 Program came to be known as the "Before McNamara Program." Secretary of Defense McNamara introduced new procedures of "System Analysis" and "Cost Effectiveness" into the decision-mak-

ing process when he came into office. Decision-making positions were not filled with the civilian aeronautical engineers in industry and the Navy, together with the personnel that would be operating that aircraft, but with scores of cost analyst and civilian politically-appointed officials within the Department of Defense. This system of decision-making (still in use today) adds untold costs and delays to most all military weapons-system programs, even though the basic problem was highlighted by the Chairman of the House Armed Services Committee of the Congress in the mid-1960s when he said the problem with the system was that the, "system and cost analyst knows the cost of everything and the value of nothing." As an example, the current V-22 Osprey Aircraft Program will have taken twenty years and untold numbers of excess funds by the time it reaches Fleet introduction. The F-4 Program was most fortunate to have come into being before the current system.

Secondly, the author has done a masterful job in bringing together the thoughts and experiences of those who have extensive experience in the multi-mission roles of the F-4 Phantom. As we all know, every pilot is different and has different views and opinions of the aircraft they fly and the tactics pilots use. The author is to be commended for his unmatched style, understanding, and insight in presenting and preserving the history of this truly great aircraft from its beginning. In my opinion, once readers begin this book they will have a difficult time stopping before the end.

I consider it to have been a great honor and good fortune to have had the opportunity to play a part in the early development and operational phase of this great aircraft weapon system, and to provide a Foreword for this book.

T. H. Miller,
Lieutenant General,
U.S. Marine Corps (Retired).

Foreword
(for the United States Navy)

Gray or grey? Which side of the Atlantic are you on? It doesn't matter; a rose is still a rose...

On a Friday night at the Miramar Officers' Club in San Diego, CA, on 29 March 1985, an extravaganza was taking place the likes of which will never be seen again. Hundreds of people, in and out of uniform, had shown up for the event and filled the club to capacity. The common thread bringing all these individuals (mostly naval aviators) together that night was the F-4 Phantom. This once-in-a-lifetime event was to commemorate and raise a final toast to the F-4's role in the United States Navy. It signified the last time the F-4 would see active or Reserve U.S. Navy service. The saga of the F-4 in naval aviation had ended, and that "Phantom Farewell" paid tribute to that aircraft in its Navy versions (the F-4B, F-4J, F-4N, and F-4S). I was there that night as a Reserve officer with VF-302, just starting to fly the F-14A Tomcat that had barely a month earlier replaced our gray F-4S Phantoms. I still remember the din,

the sound of music, the endless loop videos of these war machines launching from carriers, or flying tactics over the deserts of California and Arizona near El Centro and Yuma. For the U.S. Navy this gathering unofficially marked the status of the Navy F-4 as a "Ghost." From that point on, only memories and fragments of memories, museums, static displays, drones, and perhaps the knowledge that Phantoms still fly with foreign countries would keep the F-4 alive. Photos in books like this one will help to preserve the recollections of an era of Phantoms which were painted gray....Gray Ghosts.

The importance of the Navy F-4, whose element was salt air and blue water and whose home was the great, gray carriers, is incontestable. The role of the F-4 in war and peace was so important because, until the arrival of the F-14, it was the sole all-weather, fighter-interceptor-bomber in the Fleet. It was superior to anything we had in service at the time; even the venerable F-8 Crusader (which

CDR Bill Freckleton (holding cup) with RADM John Ed Kerr (former VF-302 CO, left), ADM Jay Johnson (CNO), and Garry Weigand at the retirement ceremony for RADM Pettigrew, January 1998. (Brenda Freckleton)

co-existed with the Phantom, one might imagine, like Neanderthal man may have co-existed with homo sapiens).

From hangar bays, where countless dedicated individuals worked on the plane, to flight decks, the F-4 flew for almost three decades from such renowned carriers as USS *Kitty Hawk, Coral Sea, Enterprise, Midway, Constellation, America, Saratoga, Independence, Ranger,* and many others. Carrier aviation was what the F-4 Phantom was all about—it was what naval aviation was and is about. In 1969 at Pensacola, FL, I remember an Admiral addressing us newly-commissioned officers. He began his speech with words I'll never forget: "Gentlemen, when you hear me referring to 'Naval aviation' you will understand and interpret me to be saying, 'carrier aviation,' because that *is* naval aviation." The point is that the F-4 was a warplane that exemplified carrier aviation. Author Peter E. Davies has brought the Navy F-4 to life again with his abundantly visualized "Gray Ghosts." He has performed monumental research and brought together material which has never before been in print.

Why "Gray?" The gray tactical paint schemes used on F-4s towards the end of their service lives not only gave the Phantom a sophisticated look, but more importantly gave it a decisively tactical advantage over the brightly colored paint schemes of earlier years, including Vietnam. The tactical colors unified all the squadrons that flew Phantoms. Even the Reserves of Carrier Air Wings Twenty and Thirty had F-4Ss painted in exactly the steel-gray colors of other Fleet Phantoms. The influence of the gray paint scheme sort of snuck up on the Fleet, but before long every F-4 flying in the Navy and Marine Corps had a complete makeover. This new, subtle blend of light grays fit over the deserts where we trained and out in the Offshore Warning areas. In Vietnam I flew F-4Bs that were painted in an incredibly fantastic scheme with shark's teeth and eyes on the radome and a full sunburst covering the entire vertical stabilizer and rudder—no half measures! What a work-load for maintenance personnel to keep this looking good.

After twenty-eight and a half years of active and reserve naval service (and retirement), I still find great pleasure in reading informative and well-presented books on the historical aspects of our naval air forces. In the pages that follow, Peter E. Davies has brought together a fresh and impressive history of the F-4 Phantom, often drawing on the memories of those who were associated with the aircraft. His extensive research makes this volume a collector's item for those who have flown "gray ghosts," or those who are interested in the historical and technical aspects of the Phantoms which were flown by Navy and Marine Corps aircrews. The detail is impressive; this is a work of commitment and dedication, and it will be a lasting testimony to those who flew and fought in the front or back cockpits of the F-4, and to its maintainers. Peter Davies gives many insights into the role of the Phantom in war and peace—Phantoms painted in that ghostly, light, sea-going gray.

Bill Freckleton, CDR. USNR (Ret).
VF-121, January-December, 1970 (F-4B, F-4J)
VF-111 December 1970-November 1972 (F-4B)
VF-302 January 1973-August 1993 (F-4B, F-4S, F-14A).
MiG kill on 6 March 1972 near Quan Lang airfield.

Introduction

Forty years after its U.S. Navy service entry, the feral howl of F-4 Phantoms still splits the air over the Naval Air Warfare Center, Point Mugu, where some of the earliest examples of the type continue to perform vital work. Although they constitute a relatively small proportion of the Phantoms still flying with the world's air forces, they are a reminder that the aircraft was originally designed and built for the U.S. Navy and Marines. In their service the F-4 revolutionized fighter operations. Of the many thousands of hours flown by light grey F-4s, a large number were spent on types of combat missions for which the aircraft was never originally intended. As a fighter, its crews devised tactics for it to win a healthy advantage over more nimble MiG opposition, though far more of its combat time was actually spent as a bomber. Other Phantoms fulfilled the aircraft's original function and protected the Navy's Fleets, while specialized variants provided the Marines with an unparalleled reconnaissance capability

Among its crews and maintainers the Phantom inspired a special respect and loyalty which lives on decades after their original contact with the "Spook." Some remained with it up to the end of its front-line service, flying it in Reserve units. A representative selection of their impressions and experiences, from the Phantom's earliest days, is included in these pages. Inevitably, many of them focus on the war period, reflecting the time in which the Phantom's potential was explored and developed. For many squadrons, almost all their years with the F-4 were spent on war cruises or land deployments.

It is hoped that these insights will be helpful in illuminating aspects of the Phantom's service life and technological development which have not been fully covered elsewhere in print. They should also help to re-establish some of the facts about this great aircraft whose fame has inevitably generated a degree of self-propagating mythology.

To attempt a complete technical and operational history of naval Phantoms, giving due emphasis to every squadron or event, would require many more volumes. However, in casting the net as widely as possible the author has sought and received the guidance and memories of a cross-section of those who knew the Phantom intimately and contributed to its undoubted status as the world's most illustrious jet fighter.

Peter E. Davies, 1999

1

Birth of a Bruiser

Ghostly Ancestors

Some great fighter designs are revolutionary, others evolve from a clear line of antecedents and a lengthy process of project definition between manufacturer and customer. In most respects, McDonnell's F-4 Phantom falls into the second category, though it included many radically new features. It was initially the product of a complex pattern of design and re-design, investigation of the customer's needs and responses to technological advances. The target customer, the U.S.Navy, had already bought jet fighters from McDonnell, including its first, the FH-1 Phantom. Its successors, the F2H Banshee and F3H Demon, were also ordered, but by 1953 it was less clear how the company could best supply the Navy with new fighters.

If the USN had accepted McDonnell's proposal at that date its "Phantom II" could well have been a single-place, single-engined, long-range attacker with a quartet of 20mm guns and wall-to-wall ordnance on eleven hard-points. By 1961 the Navy had received a very different aircraft which, at the peak of its production in the summer of 1967, was rolling out of the St. Louis factory at sev-

enty-one copies per month. Among them were many of the 1,262 Phantoms which were eventually built to Navy contracts. The F4H-1 Phantom II emerged as a Fleet defense interceptor with two crew, twin engines, all-missile armament and minimal attack capability.

The advent of jet bombers in the late 1940s made the aircraft carrier's self-defense task much harder. In WWII, a deck-launched, piston-engined fighter would have taken at least fifteen minutes to intercept an incoming attacker, even if that aircraft had been detected at around 100 miles from the ship. The first jets were unable to improve significantly on that performance, and the development of stand-off missiles soon after WWII, particularly those with nuclear warheads, meant that different tactics had to be employed. Since the missiles could not easily be intercepted their launch aircraft had to be met and destroyed or deterred at much greater distances from the Carrier Group. Combat air patrols (CAP) by fighters orbiting between the carrier and in the direction of potential threats had been flown at times during WWII, with CAP durations of up to two hours on station using fighters like the F6F Hellcat.

The Phantom's immediate McDonnell naval forebear, the F3H-2N Demon. VF-14 went on to fly Phantoms for ten years. (via C. Moggeridge)

Many Demon design features were carried over into the F3H-G mock-up in mid-1954. (Boeing/McDonnell Douglas)

The first F4H-1 (BuNo 142259) undertakes taxi trials with its main landing gear doors removed. (Boeing/McDonnell Douglas)

Early jets like the FH-1 had much shorter endurance, though the twin-engined Banshee could achieve respectable patrol times by shutting down one engine. In the next generation of fighters CAP times had to be increased with larger external fuel tanks and by the development of in-flight refueling techniques. Detection and destruction with long-range missiles (rather than guns) of targets up to 100 miles from the carrier required larger, more powerful radars, particularly at night. This was especially true if the interception

was to be carried out from a deck-alert launch rather than from a CAP position.

Some of these needs were met by the Douglas F3D Skyknight. Built in response to a 1945 Bureau of Aeronautics (BuAer) requirement for a radar-guided interceptor, capable of meeting its targets 125 miles from a Task Force at 40,000 ft and 500 mph, the F3D carried Westinghouse's weighty APQ-35 radar complete with 300 vacuum tubes. Twin Westinghouse J34 engines provided only 6,500

Sporting an early fiberglass helmet and some distinctive argyle socks, Bob Little boards the Number One Phantom ahead of its first flight. (Boeing/McDonnell Douglas)

Another shot of test pilot Bob Little with BuNo 142259, which had acquired some extra day-glo paint, possibly to match his flight suit. The walk-around pre-flight inspection of production F-4Js would require twenty-nine items to be checked before signing the yellow sheet and taking charge of the aircraft. (Boeing/McDonnell Douglas)

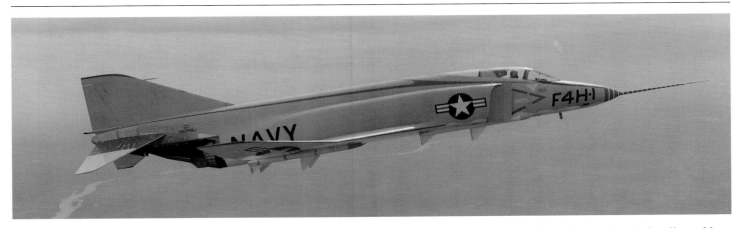

On 27 May 1958 Bob Little took the first aircraft on its initial test flight, curtailed after twenty-two minutes by a minor hydraulic problem. (Boeing/McDonnell Douglas)

lbs of thrust total. With drop-tanks they gave fuel economy sufficient to offer up to seven hours on station, but only snail-like acceleration. The F3D's weight and lack of power meant that it was unsuitable for the aircraft carriers of the day. However, it did provide the US Marine Corps with five "kills" in Korea during the first jet combat at night and experience with radar-directed all-weather interception, which was to prove invaluable in training two-man crews for the F4H Phantom. In 1955 the F3D-1M and F3D-2M with updated APQ-36 radar pioneered the first long-range, radar-guided missile, the Sperry Sparrow 1 using continuous-wave guidance. This too was of great importance in the genesis of the F4H Phantom. Meanwhile, McDonnell's radar-equipped version of the F2H Banshee, the F2H-2N, remained a single-place fighter. Its APS-46 radar was derived from the wartime APS-6 set. During the Korean War the F2H-2 was the only radar-equipped jet night-fighter operating from U.S. carrier decks.

The McDonnell Aircraft Company's role as a producer of Navy jets dated back to WWII. Its founder, James Smith McDonnell, Jr., had worked for the Glenn L. Martin Company for eight years before setting up his own company in 1938 at St. Louis. It expanded rapidly during the war, making sub-contracted parts for Douglas bombers and transports in new premises built with Government money. In October 1940 J.S. McDonnell, keen to promote his own designs too, received contracts from both the Army and Navy to research jet-powered aircraft. Other Navy contracts followed, and by January 1943 McDonnell's experience in studying applications for the early Westinghouse jet engines led to a Navy request for a carrier-based jet fighter design. Unlike the larger manufacturers, McDonnell's design capacity was not tied up with major combat types already in production for the war, so his firm had the capacity to investigate jet propulsion. The result, the FH-1 Phantom, was essentially a proof-of-concept aircraft. Although it was capable of

Test pilot Irv Burrows (orange suit) and other McDonnell personnel, including Bill Ross (third from left), proclaim the F4H-1's early record successes. BuNo 142260, the second F4H-1, won two of them, and had a busy career as a propulsion system development aircraft and chase plane until 1965. (Boeing/McDonnell Douglas)

Researcher David W. Schill with the inglorious remains of BuNo 142260 at Sellersville, PA, in January 1993. (David W. Schill)

F4H-1 Number 3 (BuNo 143338) receives attention from McDonnell technicians. Allocated the equipment evaluation section of the test program, it was also bailed to General Electric in 1960 and eventually displayed at NAS Quantico. (via A. Collishaw)

BuNo 145310, the eleventh aircraft, was used for primary and secondary armament system evaluation. The Bullpup missile, seen here on the inner pylons, was seldom used on Phantoms. (Author's collection)

480 mph at sea level, the straight-winged fighter's twin Westinghouse J30 engines developed only 1,365 lbs thrust apiece, insufficient for a realistic combat role. Sixty were built (from an original order for 100, curtailed by the end of WWII), and they gave the Navy and Marines extremely useful experience of jet operation. The first ejection from a Navy jet happened to be from an FH-1 when LT. J.L. Frewin punched out at high speed in August, 1948.

Encouraged by its first encounter with the turbojet the Navy ordered the F2H Banshee. Its design had been submitted for a September 1944 BuAer competition which also led to small production batches of two rival designs, the North American FJ Fury and Vought F6U-1 Pirate; both straight-winged and single-engined. Production of the Banshee ran to 895 units in four versions. Its maximum weight grew from 18,940 lbs to 25,214 lbs, and its fuselage length was progressively lengthened from 39 ft to 47.5 ft., though the Westinghouse J34's power was only increased by 600 lbs thrust

Donald D. Engen was one of the pilots involved in the fly-off competition between the F4H-1 and Vought's F8U-3 Crusader. (VADM D. D. Engen)

apiece to support the extra bulk. Banshees remained in service with the USN and USMC up to the mid-1960s and with the Canadian Navy until 1970. The fighter brought in considerable profits to the St Louis company and gave McDonnell designers a solid grounding in developing a design which could be readily adapted to a number of missions. F2H variants became Fleet defense day-fighters (the F2H-1/2 versions), radar-equipped night fighters (F2H-2N), and photo-reconnaissance vehicles (F2H-2P). The considerably redesigned F2H-3 used a different radar (APQ-41) with a larger 28 inch dish, an ejection seat and eight under-wing stores pylons. Inbuilt versatility of this kind was to reappear in the F4H Phantom II concept.

Declining maximum speed was the main casualty as the Banshee's weight increased, but greater speed was actually required for Fleet interceptors to meet the kind of threat which BuAer foresaw in 1947: a missile-armed bomber approaching at 550 mph and 40,000 ft. This required an interceptor to shoot down the intruder within five minutes of the warning of its approach, from a deck-launch. In May 1948 a Request for Proposals was issued for a transonic all-weather interceptor. Afterburning engines were prescribed to provide rapid acceleration to supersonic speed, either from the carrier deck or from a CAP station, giving the fighter a chance to meet the adversary before stand-off weapons could be launched. Designers were also asked to investigate rocket or missile armament as a means of extending the interception point still further away from the Task Force. Sperry's Sparrow missile promised a ten-mile range. A mere three years previously the FH-1 had been armed with four 0.5 inch guns and its performance offered no considerable improvement on contemporary propeller-driven fighters. The McDonnell Aircraft Company (MAC) now had to devise enormous performance improvements using new aerodynamic advances, including the swept-back wing and slab tail surfaces, powered controls, and both gun and missile armament.

The key to the whole concept was a far more powerful engine, the Navy-sponsored Westinghouse J40-WE-8. With 6,500lbs of thrust (9,200lbs in afterburner) promised from this new powerplant, McDonnell opted (for once) for a single engine to save weight and fuel consumption in their new design submission for the 1948 competition. Some spectacular ideas emerged from the other contenders. Douglas submitted the manta-ray J40-powered F4D Skyray, and Vought offered their radical twin-finned XF7U Cutlass, prototypes of which had already been ordered following an earlier competition in 1945. At a time when designs could be prototyped fairly speedily the Navy had already decided to back the F4D, which promised exceptional performance and seemed capable of meeting the BuAer rapid interception criteria. Two prototypes were ordered in December 1948, and it became the first carrier-borne supersonic fighter. Grumman, traditionally the Navy's principal supplier of fighters, had only their swing-wing XF10F Jaguar proposal on the drawing board, and this was not submitted. McDonnell's design offered the best hope of fulfilling the all-weather interceptor role, and the company's impressive record in producing well-made airplanes encouraged the Navy to order two prototypes of their submission, the Model 58 XF3H-1 Demon, also. In the early stages of planning, the Navy envisaged the F4D Skyray as its short-range "dash" interceptor, requiring the F3D Demon to take on a longer-range air-superiority role. This necessitated extra fuel and the heavy but comparatively reliable APQ-50 radar.

When Chief Test Pilot Bob Edholm gave the first XF3H-1 (BuNo 125444) its initial flight on 7 August 1951 the YJ40 engine had already begun to exhibit problems. Test flying had to be done with the lower-powered J40-WE-22 interim production version. Nevertheless, in March 1951 MAC received orders for 150 F3H-1Ns, plus another batch from a second source (originally Goodyear Aircraft, then Temco). This commitment, spurred by the Korean War, was something of an act of faith. The Navy had every confidence in the F3H airframe, but its success depended on Westinghouse producing an engine with the requisite power for both the Demon and F4D. In 1952 the F3H order increased to 528, with a planned total of 1,138 units from both production sources.

As J40 delays persisted MAC urged the Navy to specify an alternative engine, the Allison J71-A-2 (as used in the B-66 bomber) with afterburner. Douglas too realized that the J40 was a bad risk, flew their XF4D-1 with an Allison J35 in January 1951, and by March 1953 decided on the heavier but proven P&W J57 for both their bat-wing Skyray and the A3D Skywarrior bomber (another planned user of the fated J40). Fortunately, the F4D's designer, Ed Heinemann, had engineered its fuselage to take a slightly larger engine than the J40, so the J57 slipped in with little modification. MAC was not so fortunate. When the USN finally canceled further development of the J40 in September 1953, fifty-six F3H-1Ns had been built using the lower-powered J40-22 engine, and only twenty-eight were eventually retro-fitted with the J71 engine as F3H-2Ns. Although the aircraft completed carquals successfully in October 1953 and an advanced F3H-1M with Sperry AAAM-N-7 Sparrow 1 beam-riding missiles was included in the order, the troubles persisted. Six crashes cost the lives of two company test pilots, and there were several in-flight turbine failures. Production and development slipped badly. The last twenty-one F3H-1Ns were transferred by barge to NAS Memphis for airframe training purposes and, worst of all, the resolutely subsonic F3H lost a fly-off competition with Vought's XF8U-1 Crusader for the Navy's supersonic day fighter requirement.

The ninth aircraft, BuNo 145308, was used for aerodynamic and equipment evaluation under the Phase IV NPE during 1960. (Ostrowski via Norm Taylor)

Testing of the crucial centerline external fuel tank required a number of launches during the carrier suitability trials in November 1960. (Boeing/McDonnell Douglas)

With the J71 engine installed the Demon became the F3H-2, using the Hughes APG-51 radar which was better suited for directing guided missiles. Its Fleet Introduction Program at NATC was completed on 3 March 1956, and the Demon finally entered service with VF-14. In all, 519 F3H-2s were delivered by November 1959. Among them were ninety-five F3H-2Ms whose armament was the AIM-7C Sparrow III missile. This was a much-improved variant which had adequate range to allow one fighter to destroy several incoming targets. Much useful development work on the Sparrow and APG-51A system was done by VX-4 at NAS Point Mugu, preparing the way for the Demon's successor, the F4H Phantom II. Significantly, the most numerous Demon variant, the F3H-2N had its guns removed in favor of missile and air-to-ground ordnance.

MAC's experience with development of the F3H's airframe, radar, and armament was crucial in maintaining the company's lead in designing all-weather naval interceptors, and it made possible the next stage in that process; the F4H. So too was the continued support of the Navy throughout the J40 fiasco. In the end, the Demon proved to be a versatile and successful aircraft which would even have branched into photo-reconnaissance if the F3H-1P had not been chopped in the contract re-scheduling following the powerplant difficulties. MAC carried this multi-mission philosophy into its next phase of fighter development.

Fundamental to the sketches which began to appear on MAC drawingboards in 1953 was a return to twin engines for supersonic performance and greater safety over-water. Early proposals evolved from the successful F3H airframe with its high-mounted tail, broad 45 degree swept wing and flank-mounted air inlets. Knowing that they had lost to Vought in the supersonic day-fighter contest, the design team tried to keep their options open and their fighter proposals also included a substantial attack capability. Some of the answers came from their F-101 Voodoo for the Air Force, which used twin, afterburning J57s, and all-weather interceptor versions housed a second crew member to operate a sophisticated Hughes

MG-13 fire control system for Genie and Falcon air-to-air missiles (AAMs). It also had a secondary nuclear attack role, though this was not developed in service. Assured of continued Navy support, but without a clear BuAer requirement, MAC's advanced design team produced numerous outline proposals which J. S. McDonnell frequently submitted to George Spangenberg's Evaluation Division at the Bureau of Aeronautics in Washington in the hope of securing instructions to build a prototype. The earliest, in 1953 used a single Wright J67 (Bristol Olympus) engine and introduced the extending nosewheel leg to provide sufficient angle of attack (AOA) for catapult launch. A further development was presented on 19 September 1953 as the F3H-G/H "general purpose VF [fighter] aircraft." Other fighter manufacturers were also after follow-on business. Douglas began work in 1953 on the F5D Skylancer, a much-improved F4D with twice the range, Sparrow missiles and an advanced X-24A fire control system. Although it might well have become a superb long-range interceptor the Navy had to acknowledge Douglas' already-huge share of the naval aviation contracts and accept that it could not afford both the F5D and F8U Crusader. No procurement initiative was offered therefore to the F5D or the F3H-G/H. McDonnell persisted, and by 15 June 1954 BuAer was persuaded to run another competition, this time for an all-weather aircraft. The requirement remained somewhat vague, but implied two seats, radar, a secondary nuclear attack mission, and internal missile or bomb carriage. MAC's response was a redefined F3H-G/H with twin engines and a back-up F3H-E2 sketch using a single engine. The Chief of Naval Operations (CNO) preferred the twin engined "Double Demon" idea, favoring MAC over both Grumman and North American in this round of submissions with an order for two prototypes on 23 July. In fact, the BuAer requirement could have been written around the F3H-G/H (by then referred to as AH-1 under the McDonnell Model 32 Series), for which it issued a letter of intent rather than a contract in September and a request for much more precise design definition.

Throughout the remainder of 1954 and 1955 negotiations concerning the basic configuration and mission of the AH-1 continued. Basically, it was seen as an attack aircraft; MAC's first. Two engine options were offered; the F3G with twin Wright J65s, and the F3H propelled by a pair of General Electric J79s. Developed primarily for the Air Force's XB-58 Hustler supersonic bomber and XF-104 Starfighter, the J79 was bench-run in June 1954 and gave genuine Mach 2 performance. It was chosen by CNO as the AH-1's powerplant on 15 April 1955. It took until 7 June 1955 to decide whether the aircraft should have either one or two crew. MAC's original F3H-G/H mock-up had one seat but allowed for a second. In other respects the design still resembled the Demon and was intended to use many F3H-2 parts, though it had a much larger vertical stabilizer and air inlets. A four-gun armament package was retained, but ordnance was to be suspended on no less than eleven hardpoints rather than carried internally. Herman D. Barkey oversaw the evolving design as Senior Project Engineer, with Art Lambert as Chief Aerodynamicist and David S. Lewis in overall charge. In 1946 Barkey had designed the revolutionary XF-85 "parasite" fighter carried by the B-36 bomber, and in due course he would do much of the design ground-work on the company's F-15 Eagle.

23 June 1955 brought another shift of emphasis. The AH-1 designation was altered to F4H-1, reflecting a change in the Navy's perception of the project. USS *Forrestal*, America's first post-war aircraft carrier, was launched on 11 December 1954 as the first "super carrier," with a length of over 1,000ft and a displacement of 80,000 tons. It was designed to operate the Douglas A3D Skywarrior jet nuclear bomber which could gross 70,000 lbs, and thereafter a later generation of bombers with long range and large internal bomb capacity to fulfill the USN strategic nuclear mission, which the Air Force would in due course win as its monopoly. The success of the *Forrestal*-class carriers led to the even larger *Kitty Hawk* class and the nuclear-powered USS *Enterprise*. With greater deck space at their disposal BuAer philosophy on attack aircraft went off on a route which led to the subsonic Grumman A-6 Intruder with a 60,400 lbs take-off weight and the supersonic North American A-5 Vigilante, which was conceived as a high-altitude nuclear strike bomber. McDonnell's F4H-1 was shifted to the Fighter Section of the Navy's project offices, though it retained the attack and multi-mission versatility of earlier McDonnell designs. Various armament options, including guns, air-to-air rockets, plus a photo-recce version, were foreseen, possibly using interchangeable plug-in forward fuselage sections with either single or two-place cockpits. The latter issue was settled in June and July 1955 via CNO letters requiring a two-seat cockpit and also an emphasis on the Fleet defense mission.

As its orientation shifted once again the F4H-1 retained the AH-1 (Model 98B)'s basic large airframe, enabling it to carry a substantial air-to-air armament, sufficient fuel to loiter on CAP, and twin afterburning engines to enable it to move rapidly between incoming threats at the outer periphery of the Fleet's defenses. Perhaps following the USAF example it was decided that a second crew member would enable the aircraft's long-range radar to be used most effectively in detecting and tracking each of those threats. The armament was to be four semi-recessed Sparrow III missiles launched from "trapeze" extending rails to ensure separation, and on 26 August 1955 the Raytheon/Aero X-1A missile-only fire control system was specified. Guns and unguided rocket options were deleted, and external stores were limited to a single B28 nuclear weapon or a fuel tank on the center-line hardpoint. CNO's letter of 19 July 1955 defining the F4H-1 configuration also specified Mach

2 speed and a "single semi-automatic lightweight navigation device."

The long-awaited order for prototypes came on 25 July 1955 (Contract NOa (s) 55-272, Job Order 338) and secured the first seven F4H-1s. BuAer authorized procurement of the aircraft on 2 September, and detailed work on the F4H-1 cockpit mock-up began soon afterwards. Further design revisions came in May 1956 as MAC prepared engineering drawings, with the confirmation of the AIM-9 Sidewinder as the secondary armament. Until November 1957 it was hoped that a folding-fin Sidewinder variant could be carried, but the Naval Ordnance Test Station (NOTS) informed Navair that this would be unreliable, so a configuration using four external AIM-9s and two extra AIM-7 Sparrows on the same wing pylons was agreed.

Airframe Design

Wind-tunnel tests using two 13 per cent scale models in Summer 1956 yielded encouraging results, and the company was confident that the first aircraft could be ready to fly by the end of 1957, well within the contract time. Shifting the emphasis from attack to Mach 2 interception required Herman Barkey's team, and in particular Art Lambert, to make some changes to the flat-winged "Double Demon" start-point design. The wing itself had to be thinner, and the outer four feet on each side (from the wing-fold point) were given an increased chord via a saw-tooth leading edge to prevent wing-tip stall at high angles of attack. Directional stability was enhanced by giving these outer wing sections 12 degrees of dihedral. Although BuAer only required a single stores carriage point the eleven hardpoints designed into the AH-1 were kept to save a complete wing redesign. This was to prove fortuitous in later years when operational demands required the F4H to become an attack aircraft. No doubt McDonnell also had an eye to wider sales potential for its aircraft. The F4H's other distinctive aerodynamic feature, the 23 degree droop in its horizontal stabilizer, was also introduced to enhance stability at moderate angles of attack and to prevent the canted outer wing inducing a rolling effect in yaw conditions. The wing itself had been designed with the strength for high-g "snap-up" launches of unguided rockets at targets above 50,000 ft, as well as for the rigors of carrier landing. Its massive one-piece center forg-

Project *Top Flight* in October-December 1959 yielded a new World Absolute Altitude record. (Boeing/McDonnell Douglas)

ing was meant for manufacture by numerically-controlled machinery when this became available in 1958. Such large-scale forgings were possible because of the huge U.S. investment in machine tools, initially to make the Century Series fighter programs possible.

Each wing had an integral fuel tank holding 315 gallons with a further 1,357 gallons in flexible non-self-sealing fuel cells in the fuselage. All flying controls were hydraulically powered and included ailerons on the main panel of each wing with flaps outboard of them. Ahead of each aileron on the upper wing surface was a large spoiler (perforated on many F4H-1s). Ailerons moved downwards only while spoilers flipped upwards so that a rolling motion, for example to the right, would be imparted by lowering the left aileron and raising the right spoiler by moving the pilot's control stick to the right. Roll coupling problems were prevented by a stability augmentation system. Beneath the wing were two speed brakes, known as "boards." The wing leading edges could be drooped for low speed handling and, like the flaps, were "blown" by high-pressure air piped from the engine compressors. This kept the boundary layer air moving over the wing fast enough to give the aerodynamic effect of higher airspeed and kept the wing flying at a suitably low approach speed for deck-landing, though it only reduced the F4H's landing speed by seven knots.

Other structural changes which were incorporated were dual, steerable nosewheels, increased amounts of titanium structure, and provision for "buddy" refueling. The F4H-1 was also required to catapult-launch in zero-wind conditions without afterburner. In the April 1957 detail specification was a requirement for "hard" production tooling to be available for manufacturing the entire F4H run, rather than hand-building unique prototypes. It was another token of the Navy's confidence in McDonnell.

Although the F4H was foreseen as operating from the new "super-carriers," it also had to fit the deck-lifts on the smaller *Essex*-class vessels, though it never served on any of them in operational use. Fuselage length was therefore restricted to 49 ft. Wind tunnel tests and aerodynamic theory at the time dictated reflexed "area rule" contouring of the center fuselage to ease acceleration in the transonic range. This "pinched waist" was not considered a major design feature in the way that it had been for fighters like the Convair F-102. Twin engines and their massive air inlets required a broad fuselage which also allowed for much of the fuel needed to allow the aircraft to perform its specified three-hour CAP. At the aft end Lambert's team worked around the high-mounted tail with a keel area above the efflux. Used on the company's F-101 and F3H, this feature saved weight and prevented the rear fuselage from scraping the ground on take-off or landing. Together with the low-mounted wing it meant that undercarriage members could be kept short, but an extending nose-gear leg was included to "rotate" the aircraft to the correct attitude for catapult launch. Titanium was utilized to clad the lower keel area parts, which were to be subjected to the blast of the J79s' afterburners, with some hasty beefing-up of that area when the prototype cladding rapidly failed through sonic fatigue.

A great deal of original design effort went into the complex air inlet system for the J79s. With a military thrust increase of around 6,000 lbs, compared with the twin J65 installation of the first pro-

posals, a much larger duct was needed. The new Mach 2 speed requirement posed much greater airflow problems and resulted in the first variable-geometry inlets installed in any jet fighter. Following the F-101 example, a boundary layer plate was fixed four inches out from the fuselage to separate the slow-moving boundary air. Attached to the rear of this was a moveable ramp which extended outwards to constrict the airflow, dependent on Mach number and air temperature. Between the two sections was a slot to bleed off low-energy boundary air, directing it through louvers above the inlet. The process was also assisted by 12,000 perforations in each moveable ramp. Further ramps inside the inlet created shockwaves to "discipline" the airflow into the correct mass and speed for the engines to operate smoothly over a wide range of airspeeds. As it neared the engine a further quantity of air was redirected around each J79 to cool it, rejoining the main airstream in the afterburner area to provide extra thrust if required. Its quantity was controlled by a perforated bell-mouth diaphragm in front of each engine, with

With such a prodigious ordnance-carrying capacity, it was little wonder that naval Phantoms were to spend much of their careers as attack aircraft. This early selection contains M117 bombs (rear), Mk 81 and Mk 82 "slicks" (front), SUU-7 and LAU-10 rocket pods (center), Mk 77 napalm, AGM-12B Bullpups, and AIM-7E and AIM-9D missiles. (Boeing/McDonnell Douglas)

its area controlled by a computer which received basic data from sensors at various points in the inlet tunnel.

Considering that the inlet's designers were unable to replicate Mach 2 airflow in the wind tunnels of the day and had no access to sophisticated computers which tackle such problems today (though MAC did invest in the analog machines which were becoming available), the inlet design was remarkably "right first time." Throughout the flight testing program improvements were made quite easily by adjusting ramp angles and modifying the shape of the intake lip and forward ramp.

Power for the Spook

General Electric began work on America's first Mach 2 jet engine in March 1952. Their J47 had run to 36,500 units, powering the B-47 and F-86 among others, and its successor was designed to give an economical Mach 0.9 cruise with a major thrust increase for Mach 2 dash through modulated afterburning. GE engineered the required increases in pressure ratio by adopting a single-shaft design with variable-incidence stators. Tested beneath a B-45 bomber in May 1955 and then in an XF4D Skyray, the engine was developed rapidly. It replaced the Wright J65 in Lockheed's YF-104, making its first flight in February 1956, and became available as the X-24A for the Air Force's XB-58 bomber in August 1956. In its J79-GE-5A version for production B-58As it provided 10,300 lbs (mil) thrust and 15,600 lbs in maximum afterburner. World speed and height records soon began to fall before it, and the engine eventually set forty-six new ones, many of them in the F4H.

The J79-2 was specified for MAC's F4H-1 with similar thrust to the "Dash 5" version. Of its seventeen compressor stages the first six had variable-incidence steel vanes to ensure the incoming air hit them at the optimum angle of attack, depending on engine speed and inlet temperature. Further airflow shaping was given by variable guide vanes, connected directly to the variable stators so that they rotated in harmony to provide the 12:1 pressure ratio compressor stages with a smooth, stall-free airflow. Air was then burned in ten can-annular chambers, passing through a three-stage axial flow turbine and into an integral, close-coupled afterburner stage. A fully-variable petal-type nozzle was the externally visible section of each engine, operated by hydraulic rams driven by the engine lubricating system. Each J79 had two gearboxes to drive its controls, accessories, pumps, and 20 kV alternators. Versions installed in Navy F4Hs relied on high-pressure air from an external pneumatic starter to get the compressor turning for an engine start. USAF versions had a cartridge-start option. J79s weighed 3,500 lbs each and could theoretically be run in afterburner for thirty minutes below 30,000 ft and two hours above that altitude. Selection of the J79 on 14 December 1954 was BuAer's responsibility as a GFE (Government furnished equipment) item, and the decisions about the engine's configuration were made directly by the Navy. It proved to be an ideal, reliable power source for the F4H, and by 1958, when it was installed in the first aircraft, most of its development problems had been rectified in the F-104 and B-58 test programs. F4H pilots were unanimous in praising the engine's reliability. MiG-killer Guy Freeborn told the author that he felt the twin J79s were the aircraft's "biggest asset, particularly in situations where the aircraft's handling characteristics were pushed to the limits."

Weapons and Radar

Basic to the Sparrow III-oriented armament of the F4H was its search radar. Westinghouse's start point was the APQ-50A, which had been successfully employed in the F3H Demon and was already matched to the Sparrow missile, although the AIM-7C Sparrow fired by the F3H-2M had used the similar Hughes APG-51. Westinghouse's set had also earned a solid reputation in the Douglas F4D Skyray where, in the Aero 13F system, it had become popular as one of the first radars to combine all the fire-control elements in one nose-mounted package rather than scattered around the airframe in inaccessible places. The radars of the time were often far from reliable, and an integrated system of this kind simplified the "fixer's" job each time the system had to be "pulled" for maintenance. After brief consideration of an Autonetics system which had been selected for the AH-1, McDonnell and BuAer chose the APQ-50, linked to a 24 inch dish like the Autonetics unit, partly because of this integrated design. In the F3H-2M the dispersion of the fire-control system around the airframe had posed many maintenance horrors. Westinghouse simplified the task even more by hanging their equipment from an I-beam so that it could be slid forward from its nose compartment for all-round access.

Also borrowed from the F3H-2M was the Raytheon APA-128 continuous-wave (CW) system which linked together the missiles and APQ-50, using the latter's antenna to transmit a CW beam at a target which it had already detected in "search" mode. The reflected return of the beam from the target provided bearing and range information which the APA-128 passed to the missile's own miniature CW tracker, enabling it to lock onto the target and launch towards it. This data was also displayed on radar scopes in both cockpits, showing the position of the target in relation to the missile-launching aircraft as a "B-scope" display. With this, the crew were given a two-dimensional version of what the APQ-50 and APA-128 were "seeing," interpreted as a center dot on their screens, with range-to-target indicated on a vertical scale and bearing from the target along a vertical line at the base of the display. From this fairly crude data, presented as three luminescent dots, skilled radar operators were able to steer the aircraft to the optimum position and attitude in the sky for a Sparrow launch.

Overall control of the airborne missile control system (AMCS), engines, intake system, and fuel system was handled by a Garrett A/A 246 central air data computer (CADC) which drew its information from sensors reading temperature, airspeed, and aircraft attitude. It was also linked to the Lear AJB-3 bombing and navigation system. This twin-gyro system was optimized for the nuclear "toss" bombing mode, which was expected of the F4H when it was first ordered. It was also capable of conventional bombing using manually-entered co-ordinates and ballistics data. As a nuclear system it had a small display which indicated the aircraft's position in relation to a pre-determined route to the release point in the form of two dots. A pilot merely had to keep the dots centered on top of each other and wait for the bomb-release tone in his headset. Clearly, this tactic lacked the accuracy required for a standard dive-attack with "iron" bombs. However, it was also a valuable navigational aid.

Missiles: AIM-7 Sparrow

In January 1947 the Navy funded development of a missile project known as Hot Shot, by the Sperry Gyroscope Company. Under the new designation AAM-N-2 the missile became the Sparrow 1, using semi-active radar homing. It "rode" a beam of electromagnetic energy which was emitted by the launch aircraft at its target and then reflected back to the missile and detected by a passive antenna in the missile's outer shell. BuAer specified overall dimensions for the missile airframe: 12.5 ft in length and 8 inches in diameter. The Douglas Aircraft Company was contracted to design the vehicle for Sperry's guidance system, the signals from which were conveyed to four steerable fins and a miniature autopilot. Sparrow 1 reached production in 1951, and the USN funded a new plant to build it in quantity. It armed the F3H-2M Demon, F3D-2M Skyknight and F7U-3M Cutlass from 1952 onwards, allied to their APG-51 radars. Encouraged by this design experience Douglas proceeded with their own version, the AAM-N-3 (later AIM-7B, the Sperry model becoming AIM-7A) to equip their projected F5D Skylancer fighter. It employed a fully active system which gave the missile, in the AIM-7A airframe, its own K-band miniature radar to detect and track its targets independently of the aircraft's systems. Tests showed that this radar lacked the power to do this, particularly when it was not operating in clear-air, clutter-free conditions. In any case, it was only effective within a one-mile radius, as opposed to five miles with the AIM-7A, and production was canceled. The Navy turned instead to Raytheon, who advocated a continuous wave (CW) emitter for their AAM-N-6 Sparrow III (AIM-7C) proposal, which began flight tests in 1953. Reverting to semi-active guidance, the design used the aircraft's APG-51 search radar to find its target and a continuous tracking beam from the missile itself to lead it to the target, whereupon its sixty-five pounds continuous rod warhead exploded with lacerating effect. One major benefit from this more flexible combined search and track method was an increase in effective range to 24 miles. It would be twenty years before this figure was significantly increased in the AIM-7F variant.

The technology and manufacturing accuracy needed to make this missile work were formidably demanding. Its Aerojet General KS-7800 rocket motor accelerated it to 1,300 ft per second in its first two seconds of flight, during which time it had to maintain "lock" on the target and launch directly at it. By this the time the motor burned out it was traveling at Mach 3.7. Sparrow III was chosen for the F4H, and mass production began at the old Sperry/USN plant in 1957. Initially, the missiles were to be launched from the F4H using extending trapeze rails to take the missile three feet clear of the aircraft's boundary layer air for a clear shot, undisturbed by turbulence. (This became an urgent necessity when BuAer announced that a launch should be possible at speeds above Mach 2.) A similar method was used in the Convair F-102 and F-106, trialed in the F-111 (with AIM-9 Sidewinders), and revived for the Lockheed F-22. However, for the F4H the risk of a missile becoming stuck on the trapeze launcher with its motor firing was considered too great. Instead, the company employed ejection racks which used small explosive charges to propel a pair of actuator rods downwards, gently thumping the missile clear of the rack and fracturing the clips which held it in place. These "kickers" were located in four semi-recessed wells under the fuselage, housing the missiles with their upper fins fitting into slots in the belly. When carried, the missiles actually improved the aerodynamics of the lower fuselage. The difficulty of ensuring a smooth ejection which also lined the missile up perfectly for its trajectory required a small autopilot in each Sparrow III to correct its course during the first vital seconds of flight. The device also rotated the missile through 45 degrees so that its fins were able to provide better control in turns. In combat, the missile was limited to 3g maneuvers, but it was anticipated that its targets would mainly be high-flying bombers approaching in straight and level flight. It would be over ten years before the Sparrow was given real dogfighting agility. Though the missile grew in potency, its name stuck. As one Phantom veteran remarked, "Sparrow? Sounds kind of meek, doesn't it?"

An AIM-7 Sparrow leaves its launch aircraft, BuNo 149455, on a plume of flame. (via A. Thornborough)

Sidewinder

The U.S. Navy was behind the other principal air-to-air missile of the 1955-75 period and beyond. Designed in 1951 "out of hours" by a Naval Ordnance Test Station (NOTS) team led by William McLean, it relied on McLean's idea that an aircraft's exhaust would give off enough heat for a lead-sulfide infra-red seeker to find it and lock onto it, guiding a missile towards the heat source. A prototype was constructed from off-the-shelf items, but McLean's homemade missile attracted little official interest at first from the Navy which, at the time, was funding a large, long-range interception weapon, the Sparrow III. The NOTS (China Lake) team saw their idea more as a replacement for the gun in shorter-range tail-chase or maneuvering combat. However, early tests in September 1953 showed that the missile obviously worked, and the Navy was attracted by its simplicity and low-cost. It also appeared to be far more reliable than Sparrow. In the May 1956 re-configuration of the F4H-1, Sidewinder (as the new missile was named in recognition of its twisting flightpath as it homed on its target) was chosen as the aircraft's secondary armament. The first order for the Sidewinder, or XAAAM-N-7, came from the USN with 300 units purchased, soon to be followed by the production AAAM-N-7 Sidewinder 1A which, with additional orders from the USAF, reached a production figure for Philco/Ford-General Electric, its manufacturers, of 72,000 units. Raytheon fought back briefly with an infra-red seeking version of the AIM-7C in 1957, but it never reached production status.

The idea of sharing the guidance task between the aircraft and missile, as the Sparrow/AMCS did, was seen as a way of improving the "field of vision" of the AAAM-N-7 (AIM-9B after 1962), so that its target could be found within a wider search "cone" ahead of the aircraft than the Sidewinder's small IR detector could "see." BuAer therefore instructed Raytheon to modify the AMCS to take signals from an ACF Electronics AAA-4 with a six-inch diameter IR seeker, mounted below the nose radome of the F4H. Information from the AAA-4 was transmitted to the Sidewinder's own seeker via a computer linked to the AMCS so that the missile knew where to "look" for an IR lock-on. In practice, Sidewinder development advanced so quickly that it soon made the AAM-4 redundant, and although it was installed on the fifth and subsequent F4H-1F and some F-4Bs, it had little use and was removed soon after the F4H entered service. Colonel James R. Sherman, who flew the F-4B from the earliest days of its U.S. Marines service, found the AAA-4 complex but effective. As he explained: "The idea of the AAA-4 was to enable the pilot to make a passive intercept without the use of the radar [which would warn the enemy that he was 'locked up']. The seeker head was cooled by liquid nitrogen, and if the coolant failed the seeker head burned out in a matter of seconds and was very costly to replace. I only flew two or three sorties using the AAA-4 and could acquire an F-4B in a head-on aspect at 50 miles plus. It was a great concept, but technically thirty years too soon."

Phirst Phantom Phlight

By the time the first of two YF4H-1s (BuNo 142259) was ready for its first flight at Lambert Field, St Louis, on 27 May 1958 over 6,800,000 man-hours had been expended in its design and construction. GFE (government-furnished equipment) and CFE (contractor-furnished equipment) items had come from fifteen hundred subcontractors, and many aspects of the design specification were ready to apply to the production aircraft. However, some modifications were already under way. Most obvious visually was a design change to the nose area. In February 1958 MAC accepted Westinghouse's contention that the 24 inch radar dish for the modified APQ-50 (now re-designated APQ-72) was too small to provide the required search range. This was confirmed by early Naval tests. A 32 inch dish was specified instead, requiring a much larger Brunswick fiberglass radome to house it. However, the first YF4H-1 and the sixteen development aircraft which followed it all flew with the smaller nose cone. In the case of BuNo 142259 this also lacked the AAA-4 housing, radar, and missile control systems. It also had a single ejection seat, as the rear cockpit was used for test gear. Seats in early F4Hs were supplied under a 1955, five-year contract with the Stanley Aviation Company, but they lacked the zero-zero capability which BuAer specified in 1957. An alternative Martin-Baker Mk 5 seat was chosen in May 1958 and retro-fitted to most early production aircraft.

McDonnell test pilot Robert C. Little, a WWII P-51 Mustang veteran, began taxi trials with the new fighter on 17 May 1958 and was ready for the first flight on 27 May. Lift off was at 184 mph, and a steep climb-out commenced, interrupted only by a rpm fluctuation on the right engine (later found to be FOD-damaged). On 29 September 1954 Little had taken the first F-101 Voodoo (53-2418) to Mach 1.07, the first time an aircraft had done this on its initial flight. He had hoped to take the YF4H-1 to an even higher speed on the first hop, but this was not to be. A ruptured hydraulic line in the Number 2 reservoir took pressure of the #2 power control system to zero, and the nose gear door failed to close fully. Little confined his 22 minute flight to exploring basic handling characteristics, using airbrakes and flaps. He brought the prototype in for a safe landing. Bad luck dogged the second flight too, as an undercarriage safety pin was left in place, preventing gear retraction. On the third flight the aircraft went to Mach 1.68, but cockpit pressurization failed. It also became clear that the NACA-type flush-mounted cabin air conditioning intakes on the nose became "blocked" by boundary-layer air at supersonic speeds, causing overheating. New intakes, offset from the fuselage, were designed.

The first photo of a "bombed up" Phantom (Bu No 145310). Test pilot William S. Ross flew this load of twenty-two water-filled dummies a couple of times to determine flight stability. (McDonnell Aircraft Corporation via LGEN Tom Miller)

Eleven test flights were performed at St Louis before the YF4H-1 was moved to Edwards AFB for the start of the five-phase Navy Preliminary Evaluation (NPE) from 15 September to 10 October, 1958. Naval Air Test Center pilots from Patuxent River flew forty-three flights during Phase 1, which included a fly-off against the aircraft's main rival, the F8U-3 Crusader. Chance-Vought/LTV had been asked by BuAer to develop their F8U-2 Crusader design further, partly as an insurance against problems with the F4H. Built around the 25,000 lbs thrust P&W J75, the engine of the mighty F-105 Thunderchief, Vought's fighter was four feet longer than earlier Crusaders. It had a larger wing with boundary-layer control, Sparrow missiles in place of guns, and an enlarged vertical tail plus ventral fins. Apart from some internal systems it was a new and superior aircraft, though not a gunfighter in the F8U tradition.

BuAer's Evaluation Division was enthused by the potential of both the F4H and F8U-3 projects, ordered prototypes of both for extensive comparative testing, and wanted to purchase both in quantity when the tests proved to be so promising. The NPE fly-off was the Navy's considered answer to this proposal: one fighter had to be chosen. McDonnell's aircraft was the first to be ready, but Vought managed to get their Super Crusader ready for the beginning of NPE on 15 September 1958. The fighters' interception capability, their handling and maintenance characteristics, and the comparison of the two seats/two engines versus one of each were all carefully assessed. The Crusader proved to be the faster and more maneuverable. Admiral Don Engen, who flew both aircraft for NATC's

Flight Test Division, reckoned the F8U-3 would have made Mach 3 if its acrylic canopy could have stood the heat. Much of the two manufacturers' teams' time was taken up in fine-tuning their aircrafts' air inlet systems to prevent intake stalls. Water injection systems were installed in both prototypes to increase top speed. Don Engen experienced some dramatic supersonic inlet stalls which he compared to, "the sound and fury of a 40mm cannon going off right between your legs" in the F8U-3. Although the majority of the personnel involved in the NPE still advocated acquiring both aircraft, the Navy's decision in favor of the F4H was to some extent predetermined by its 1955 requirement for a two-seat aircraft. The F4H's far greater potential in ground attack was not an important consideration, though its longer-ranging radar and twin engines were. Vought's design was canceled on 17 December 1958 (shortly after the inlet and windshield had been re-designed), though the company did receive a production contract for their F8U-2N (F-8D) in September 1961.

RADM Robert Pirie's announcement, also on 17 December 1958, that the F4H had won the fly-off led to a contract which took the total order of F4H-1s to twenty-three, including the two YF4H-1s. They included five Production Block 1 aircraft (BuNo 143388 to 143392) which resembled the YF4H-1, but some were modified to include the revised production-style inlet and splitter plate. Two received "blown" (BLC) flaps and leading edge modifications. Five hardpoints could be fitted with pylons for fuel tanks and other stores. These aircraft joined the busy test program with McDonnell,

Tom Miller made the first drop with these bombs east of Edwards AFB and then flew to St Louis, where engineers checked the aircraft before he went on to make the first live drop at Camp Lejeune on 24/25 April 1961, with three more days of demonstrations in June. "Shortly after this," he recalled, "the USAF decided they wanted the F-4." (McDonnell Aircraft Corporation via LGEN Tom Miller)

Raytheon, General Electric, and NATC (Point Mugu). BuNo 143391 was equipped for carrier suitability trials, while Block 2(b) airframe Number 7 (BuNo 143392) had a large parachute fairing on its rear end for spin trials and the enlarged Brunswick radome to take the production-standard APQ-72 thirty-two inch radar dish, along with hydraulic drive for the dish. One of the busiest F4H-1s was Number 11 (BuNo 145310) from Block 2, which conducted primary and secondary weapon tests at Edwards AFB, including Special (i.e. nuclear) Weapons carriage right through to the end of 1963.

The twelfth aircraft off the line (BuNo 145311) was officially devoted to Aircraft and Engine Performance. In practice this meant the World Speed Record program. Spurred by the results of the second stage of NPE from July 1959, the Navy naturally wanted to put the F4H-1's spectacular performance in the record books. Tragically, the first attempt at a world altitude record, Project *Top Flight*, cost the life of a company test pilot, Gerald Huelsbeck, when on its 295th flight the first prototype shed an engine access door in a high altitude zoom climb above Mach 2, and the subsequent fire destroyed the aircraft. "Zeke" Huelsbeck was imprisoned in the cockpit by unsurvivable g-forces. The second prototype (BuNo 142260) was equipped with water injection and flown by CDR Lawrence E. Flint in a series of twelve ballistic zoom climbs from 50,000ft at maximum power between October and December 1959. The aircraft was stripped of radar and other non-essentials, including the rear seat. Engine rpm, fuel flow rates, and afterburner nozzle settings were all tweaked for higher thrust. CDR Flint reached Mach speeds up to Mach 2.41 just before reaching Edwards AFB airspace and pulling up for the 3.5g, fifty degree zoom climb to be measured. Towards the peak of each vertical climb fuel flow was cut to prevent overheating as the afterburners "blew out" from lack of oxygen above 65,000 ft. In the pure, unpolluted atmosphere, darkness was so intense that when Flint's shadow fell on the instrument panel he could not see the dials at all. At the top of the zoom the engines were shut off, speed dropped back to around 45 mph, and the aircraft entered a dive to 45,000 ft for an engine re-light and pull-out to controlled flight. When the aircraft landed much of its orange paint had been stripped off by the heat generated on Flint's low altitude run-in. The outer windscreen panels were also distorted by the heat. On one flight CDR Flint was unable to shut off the right engine as he neared the apex of his climb until it had exceeded 120 per cent rpm and the TOT had gone off the scale. Luckily it held together, and on post-flight inspection it was found that up to half an inch had melted off the outer end of each turbine blade. The highest altitude reached during the Project was 98,557 ft on 6 December 1959, though this new record was broken by a USAF F-104C before the end of the year. To a potential enemy these spectacular flights were a dramatic proof of the new aircraft's potential as a high-altitude interceptor. Larry Flint, who had commanded VF-11 *Red Rippers* in the Korean War, had been a test pilot since 1944 and he went on to become Chief Test Conductor for the Apollo space program and manager of the F-14 Tomcat Aircrew Training Program for the USN and Iranian Air Force.

1959 also brought the first order for "production standard" aircraft (twenty-four were requested in February (BuNo 148252 to 148275)) and for the "definitive production version" in September. Contract NOas60-0134 called for seventy-two F4H-1 aircraft using the more powerful J79-GE-8 (17,000 lbs afterburning thrust) in place of the "Dash 2" or -2A of pre-production aircraft. In addition to the enlarged radome and straight-lipped inlets of the revised F4H-1, the fuselage was modified to accommodate a raised and enlarged cockpit canopy, giving the back-seater better visibility but also improving vision over the nose, since the front cockpit was also raised by more than a foot. Aircraft 19 (BuNo 146817) introduced these changes from Block 3 onwards, and they caused a minor redesignation to distinguish the design differences in the early aircraft. Build Numbers from Number 3 to 47 became F4H-1F (later F-4A in the 1962 DoD type redesignation for all the U.S. service aircraft), while aircraft 48 (BuNo 148363, Block 6) was the first true F4H-1 (later F-4B). By then the aircraft also had a name. McDonnell celebrated their 20th Anniversary on 3 July 1959. Employees were allowed to vote on suggested names with "traditional" McDonnell supernatural connections, such as Goblin, Sprite, Ghost, and Satan. J. S. McDonnell allowed himself the final vote, and the new fighter was christened "Phantom II."

Phantom Afterlife

Larry Flint's BuNo 142260 still existed in 1999. Historian David W. Schill was driving past a scrapyard in Sellersville, Pennsylvania, in 1993 on the lookout for a reported TBM Avenger when he spotted an F-4 fuselage resting in the yard. On closer investigation it proved to be the *Top Flight/Skyburner* aircraft, in poor shape but recognizable. After its record-breaking flights it had become a "chase and target" aircraft at St Louis before being struck off flying status in February 1965 to perform arrester gear barrier trials at Lakehurst, New Jersey. It had come from NADC, Warminster, on a truck around 1978, the wings having been disposed of elsewhere. David attempted to initiate a restoration project, using parts from the No 11 aircraft (BuNo 145310) which had been used for weapons trials and was frequently depicted carrying twenty-two 500lb bombs. In 1999 the Number 2 Phantom, the oldest in existence, was still the subject of a dispute over payment and languishing with an uncertain fate. BuNo 145310 remained with a Californian aircraft restoration facility who was attempting to sell it. David recorded other development airframes still extant. The Number 3 aircraft (BuNo 143388) was stored externally at the USMC Museum, Quantico, Virginia, though it never served in Marine colors. BuNo 145307, the *Sageburner* aircraft [see Chapter 2], remained with the Paul Garber Restoration Facility at the National Air and Space Museum, Washington, DC, and the Number 9 machine (BuNo 145308) belonged to the Florida Military Aviation Museum at Clearwater. BuNo 145315 was put on display in the USS *Lexington* Museum, while 148267 was scrapped at NADC, Warminster, in 1993. At least two F-4As survived in museums; 148273 at Lamberton, New Jersey, and 148275 at the Naval Academy, Annapolis.

2

Early Days

F-4B Deployed

The final phase of F-4 testing before it joined Fleet squadrons was the Navy Board of Inspection and Survey (BIS) assessment. These trials began at Patuxent River in July 1960 using eight F4H-1F aircraft, Numbers 14 to 21 (though aircraft 19 spent most of 1960 with MAC on missile, carrier suitability, and structural work). They allowed the Navy to explore the full range of the Phantom's equipment, performance, and use in Fleet service at sea. In fact, the first carrier launch and landing had been made on 15 February 1960 by LCDR Paul Spencer aboard USS *Independence*. Improvements were continually built into the aircraft during the 1958-61 period. Main landing gear brakes were enlarged after overheating problems, the already-massive tailhook forging was beefed up to prevent skipping and bouncing over arresting cables, and the undercarriage was strengthened to allow a sink rate of 24 ft per second at 34,000 lbs landing weight. The last of these was in recognition of the F4H-1's weight increases as extra equipment was added and the structure was reinforced with a wing "strap" to allow fatigue life to increase from the original 5,000 hours limit.

When the BIS trials were completed five months later it was clear that the F4H-1 had exceeded the Navy's requirements by a handsome margin. It was time to make the aircraft's spectacular performance more public. The Deputy CNO (Air), VADM R. B. Pirie, and RADM P. D. Stroop, Chief of the Bureau of Naval Weapons, set out to smash as many records as possible using the Phantom II. It was also time to consolidate Marine interest in the aircraft, which had been strong from the outset. Like the Navy, the Marines had flown both the F3D Skyknight and F4D Skyray (with its efficient radar but limited weapons system) and the F8U dayfighter. They were attracted to the F4H-1's all-weather capability, but also recognized its considerable potential in the attack and reconnaissance roles. USMC influence helped to retain the aircraft's in-built ordnance-lifting ability and versatility, both vital for Marine operations. Therefore, there were two Desk Officers in charge of service acceptance of the Phantom in 1960; CDR Jeff Davis for the Navy, and LTCOL Tom Miller for the USMC.

BuNo 142260, with LT COL R. B. Robinson, USMC, exceeded 1,606 mph and gained a new Absolute Speed record in Project *Skyburner*, November 1961. (Boeing/McDonnell Douglas)

In Project *High Jump*, F-4B BuNo 149449 set eight new time-to-climb records in 1962. (Boeing/McDonnell Douglas)

Project *Sageburner* in August 1961 was flown by BuNo 145307 at over 900 mph and a mere 125 ft altitude. This, the first *Sageburner* F4H-1 (BuNo 145316), crashed during a speed run on 18 May 1961. (Boeing/McDonnell Douglas)

Speed and Height Flights

Following the success of Top Flight a new series of record contest flights took place in September 1960. First to be attacked was the 500 km Closed Circuit Record. Tom Miller had flown fighters in WWII and Korea, and first flew the F4H on 2 December 1959; one of the first two Marines to do so. (LTCOL Bob Barbour had been the very first on 6 October, 1959). CDR Davis and LTCOL Miller (Class Desk Officers, BuWep) applied for their "sporting licenses" to attempt the 500km and 100km courses, commencing in August 1960. For the former event a triangular course totaling 335 nm, marked by pylons, had to be flown. Since the flight had to be made at 50,000 ft to allow for high fuel consumption there were obvious problems for the "umpire" in deciding whether the contestants had

flown around the outside of those pylons. Eight test flights were made, and they soon determined that the Phantom would have to take off carrying maximum fuel, including all three external tanks. The largest of these would have to be dropped at supersonic speed, a procedure which had not secured McDonnell or USN clearance. Tom Miller received a waiver on that requirement and began test flights, with Captain Jim McDivit, USAF (later an astronaut) in an F-104 chase plane. Although the Starfighter was unable to keep up with the F4H-1 at its maximum speed, McDivit was able to keep a crucial watch on the Phantom's tail section during turns when the afterburner exhaust impinged on the stabilator. LTCOL Miller reached a release speed of Mach 1.6 for the 600 gallon centerline tank. Observer Jim McDivit reported that the tank disappeared so

Another view of the *Skyburner* aircraft with its early windshield and canopy but production-style air inlets. (Boeing/McDonnell Douglas)

One of the LANA F4H-1Fs, with yellow bands on its wings and fuselage, which set new trans-continental records in May 1961. (via Peter B. Mersky)

Project LANA depended on air-to-air refueling by A3D-2 tankers. Pre-LANA practice is provided here by a VAH-9 "Whale." (Boeing/McDonnell Douglas)

fast that if he had blinked his eyes he would have missed the ejection. It was known that the aircraft could accelerate to Mach 2 with the tank aboard if required. In any case, it was necessary for the Phantom to enter the course with full internal fuel in order to complete it. Although many areas of Edwards AFB would have allowed a dead-stick landing if the gas had run out, LTCOL Miller preferred to avoid this, and very careful calculations were made to allow for a full circuit of the course at Mach 2 or above. The previous record had been set by another McDonnell product, an RF-101 Voodoo at 816 mph. Gradually the F4H-1's performance improved over a series of practices in varying temperatures. Finally, on 5 September 1960 (Labor Day) the conditions were right for the official attempt.

F4H-1 Number 12 (BuNo 145311), a Block 2 airframe with the original J79-2 engines, blasted off and climbed to 33,000 ft over Salton Sea to drop its wing tanks. Tom Miller pushed the speed up to Mach 1.7, released his 600 gallon tank at 48,000ft near George AFB and entered the "start gate" at Mach 1.76 with 11,000 lbs of internal fuel. The Phantom soon reached Mach 2.0 at 50,000 ft, made a 3g turn around the second pylon and a 4g swing around the second turn. Entering the final leg, still at Mach 2, Miller had 5,200 lbs of fuel remaining, enough for 7.5 minutes. He passed the final gate at Mach 2.15 with a cockpit temperature of 125 degrees F and only 900 lbs of fuel remaining. He decided that, "the only way to get on the ground with the engines running was a split-S maneuver with a near-vertical dive, speed brakes out, and engine at idle power. This maneuver provided positioning for a straight-in approach to the runway at Edwards. Flaps and wheels were lowered at the last minute when I knew that I had the runway made even if the engines quit. Fortunately, they did not flame out until I touched down on the runway." His circuit of the 335 nm course had taken 15 minutes 19.2 seconds at a speed of 1,216.78 mph, beating the previous record by 400 mph. The aircraft continued to perform useful service. ADM "Smoke" Wilson came across it years later in the Carrier Suitability Branch where, still without radar or back seat and with considerable structural lightening, it performed arrestment trials.

CDR Davis took the 100 km record at Edwards on 25 September flying a sustained 3g turn over the circular course at a constant 70 degree bank. He entered the course at Mach 2.31 and 45,000ft, leaving it 2 minutes 40 seconds later at Mach 2.21, 2,000 ft higher. His speed over the 104.9 km course actually flown was 1,459 mph (calculated to 1390.24 mph over 100 km).

The Navy next used the entire continental USA as the stage for Project LANA, a public celebration of the "50th (hence "L") Anniversary of Naval Aviation." Others read the acronym to mean "Los

By May 1963 when this formation was flown it was clear that the F-4 would be the leading fighter for all three U.S. air arms. The nearest F-4B (BuNo 151001) went on to fly with VMFA-542 and was destroyed in a mid-air collision in May 1969. BuNo 151005 (center) survived until 1963 with VF-121. (Boeing/McDonnell Douglas)

In April 1965 Sanford McDonnell (nephew of J.S. McDonnell) handed over the 500th F-4B from a production run of 649. Like many of its kind this aircraft (BuNo 151459) was eventually re-lifed as an F-4N, flying with VF-302 until 1978. (Boeing/McDonnell Douglas)

This *Diamondback* F4H-1 was assigned to LCDR E. T. Wooldridge, who took part in early tests of the Phantom as a Fleet interceptor at Patuxent River. Retiring from the Navy as a Captain, he became Assistant Director for Museum Operations at the National Air and Space Museum. (U.S. Navy via Norman Taylor)

Angeles to New Amsterdam," the extent of the record flight. On 24 May 1961 five F4H-1Fs from Block 4d and 5e took part. The Phantoms chosen as LANA 1-5 flew the 2445.9 miles from Ontario, California, to Floyd Bennett NAS, New York, non-stop. LANA 1, 2 and 4 experienced problems on their three flight refuelings. Their five A3D-2 tankers had arrived early, spending too long at 35,000 ft; above their optimum refueling altitude. When the Phantoms descended from their 50,000 ft supersonic cruise altitude they found the tankers had frozen refueling pipes. LANA 5, with LCDR Keith Stecker and CWO J.H. Glace, landed in Kansas, lacking a serviceable tanker. LCDR Dick F. Gordon and LTJG Bobbie R. Young were more fortunate and managed to complete the course in 170 minutes at an average speed of 869.73 mph, winning the Bendix Trophy. The race won enormous publicity for the Navy and the F4H-1, and it demonstrated the rapidity with which Phantoms could be deployed over long distances. This point was not lost upon the

Air Force, which was already impressed by the fighter and ordered it in September of that year.

The LANA aircraft were decorated with distinctive colored bands, but they also bore the designator letters of the first two Navy F4H-1 units, VF-101 *Grim Reapers* (coded AD) and VF-121 *Pacemakers* (NJ). On 29 December 1960 aircraft Number 28 was the first squadron Phantom to be delivered when it arrived at NAS Miramar for VF-121. The following day their boss, CDR Eugene R. Hanks, officially took delivery and made the acceptance flight. Training of the first cadre of F4H-1 crews began in January 1961, and Miramar received the first Weapons Systems Trainer in June 1961. Meanwhile, development work continued throughout the summer of 1961. In May and June two live "iron bomb" demonstrations at MCAS Camp Lejeune, NC, included aircraft Number 11, flown by Tom Miller, dropping twenty-two Mk 82 (500 lb) bombs. This was part of a project involving an Engineering Change

F4H-1 BuNo 148390, assigned to CDR Julian S. Lake, commander of the first deployable Phantom squadron. (Boeing/McDonnell Douglas)

A *Be-devilers* F4H-1 gets some attention from USS *Forrestal's* deck crew. (U.S. Navy via CDR James Carlton)

F-4B BuNo 148381, the 66th Phantom, flying with VF-74 in November 1961, but written off in July of the following year. (Boeing/McDonnell Douglas)

Proposal (ECP) enabling the Phantom to have multiple weapons capability and take advantage of its prodigious lifting power. Phase V of the NPE in February 1960 had already demonstrated the F4H's basic attack "muscle" (though no bombs had been dropped).

Combat capability demonstrations were the parallel agenda for the next round of record attempts, also. LT Hunt Hardisty and LT Earl D. DeEsch showed that an F4H-1 could fly at Mach 1.2 over a 3 km course at a maximum altitude of 125 ft. Project *Sageburner* (28 August 1961) convincingly showed that, with a crew who could

VF-41 *Black Aces* were the third F4H-1 unit, standing air defense duty at NAS Key West during the Cuban Missile Crisis before going to war aboard USS *Independence* in 1965. (via Norman Taylor)

stand up to such a rough ride, it would be hard to stop a nuclear weapon-carrying Phantom in that flight regime. Their aircraft (BuNo 145307), McDonnell's Structural Demonstration vehicle, never exceeded 500 ft altitude, even on the turns. In an earlier attempt, on 18 May 1961, BuNo 145316 had crashed in New Mexico due to a pilot-induced oscillation (PIO), killing the pilot, CDR J. L. Felsman.

The year ended with a new World Absolute Speed Record. LTCOL Robert B. Robinson, USMC, flew the faithful second prototype (BuNo 142260) at speeds up to 1,700 mph in Project *Skyburner* on a series of sprints across a 15/25 km course at Edwards AFB. Robinson had to keep his altitude to within 500 m of a set figure while at the same time accelerating and shedding his drop tanks. His speed averaged 1,6063 mph, another new record. Also in 1961 came the parallel sustained altitude record over the same course, to within 100m of a set altitude. CDR George Ellis, USN, hit the entry gate at Mach 2.2 and held 66,443.6 ft for a new record, which was a clear 11,000 ft above the previous one and also a salutary warning to potential high-altitude intruders.

In the first three months of 1962 a team of Navy and Marine pilots flew Project *High Jump*, using the Phantom almost as a pyrotechnic device to smash eight time-to-climb records for altitudes from 3,000 up to 30,000 ft at Brunswick, Maine, and Point Mugu, California. On each no-flaps take-off the aircraft was restrained against full afterburner power until the pilot released a steel bridle using an explosive bolt. On the 30,000 ft attempt the Phantom's acceleration took it into a zoom climb above 100,000 ft,, breaking its own previous record. Many of the aviators who took part in these record events went on to become senior figures in the Phantom community, and the aircraft itself was firmly established as a world-beater.

Fleet Phantoms

The two RCVG (Replacement Carrier Air Group), or "RAG" squadrons, worked hard to build up a core of Fleet-ready crews as the definitive F-4B variant, which first flew on 25 March 1961, began to leave the line at St Louis. Aircraft Number 50 (BuNo 148365, later a QF-4B drone) was the Fleet's first front-line Phantom, de-

An *Aardvarks* F-4B tugs the arresting wire aboard USS *Kitty Hawk* in 1966. When the aircraft was converted to F-4N configuration in the 1970s its inboard leading edge flaps were secured in the "closed" position. (via Norman Taylor)

In clean configuration a *Diamondbacks* F4H-1 is poised for launch while VF-74 aircraft queue behind. The Phantom's complex system of catapult launch and hold-back cables was greatly simplified in later "nose-tow" launch methods for the F-14 and F/A-18. (Boeing/McDonnell Douglas)

livered in June 1961 for use by VF-74, the first carrier-deployed squadron. VF-121 under CDR Pat Casey took charge of F-4A BuNo 148256 on 29 December 1960 to begin flying training for the West Coast squadrons, while LCDR Gerald O'Rourke formed VF-101 at Key West, Florida. He soon moved the training program to Oceana with VF-101 Det A in June 1960, leaving Key West to complete its training cycle on the F4D Skyray and F3H Demon. Phantoms later returned to Key West, and in May 1966 the Oceana Det was re-established for Air Combat Maneuvering (ACM) and weapons syllabuses and to teach Carrier Qualification (carqual), inflight refuel-

ing, and radar training. Eventually, in 1977 these roles passed to VF-171 *Aces* who had received the first USN FH-1 Phantoms back in July 1947.

As F4H-1 production built up to ten per month LCDR O'Rourke's *Grim Reapers* pushed hard to get the first class of forty-six Phantom aviators ready, assisted by his Executive Officer (XO) "Scotty" Lamoreaux, one of the LANA pilots. Tom Bennington, a VF-101 Det A Crew Chief on one of the LANA aircraft, remembered Jerry O'Rourke as a, "very subtle, persuasive character, and a 'Judge Advocate.' He's one of the few people I met who could

In 1964 VF-84 were among the first F-8 Crusader squadrons to transition to the F-4B, and they remained with the Phantom until 1976. (via Norman Taylor)

The Phantom's tough landing gear usually gave very few problems, but this VF- 102 F-4B (BuNo 150427) suffered a collapsed nose gear after deck arrestment in March 1963. Modelers should note the extensive maintenance stencils, gradually deleted as the aircraft matured in service. (via Peter B. Mersky)

VF-74 F4H-1s, with two still awaiting their full squadron colors, in November 1961. (Boeing/McDonnell Douglas)

make a suggestion that something should be done and next day it was done." One of his problems was acquiring spare parts for the new aircraft; Tom recalled that they were in very short supply. Also, most pilots came from Skyray or Demon squadrons, but providing occupants for the vital back seat position was a more complex task, since the Navy had no real tradition of two-seat interceptor operations. O'Rourke had to come up with a completely new curriculum for them and the equipment to teach it. The only precedent was in the role of Naval Air Observer in the Marines' F3D Skyknight, which had achieved six air-to-air victories at night during the Korean War and pioneered the Westinghouse radar and CW devices which led to the Phantom's equipment. In all, fifty-five F3D-2T2 (TF-10B) were acquired, and some were equipped with the APG-51 radar and AD or NJ tail-codes of the two RAGs. VF-101 Det Alpha had at least ten. Several F-10Bs (including BuNo 124610) were later fitted with complete F4H-1 radomes and APQ-72 sets.

The Navy decided to call its Phantom radar operators Naval Flight Officers (NFO), giving them equal rank with pilots. McDonnell preferred to call them Radar Intercept Officers (RIO), and this term became widely used. Their training began with basic theory work at Connelly AFB, using USAF night-fighter manuals, but by mid-1961 LCDR O'Rourke had a full USN syllabus in place for RIOs after their basic pre-flight training at Pensacola. The Naval Air Technical Training Unit at Glynco, Georgia (coded 4B, later becoming VT-86), put the new RIOs through navigation and radar training in slow-but-safe TF-10Bs between 1961 and 1965. Some of the instructors were USMC radar operators with Korean War experience. Training then moved to Oceana, where the second Naval air Mobile Trainer was delivered in December 1960, providing a "cockpit" with very basic simulation of interception procedures on the RIO's radar screen.

The change to the two-man cockpit concept came quite logically to some of the new batch of flyers. CAPT Orville "Tex" Elliott, who had been in VF-101 in the days of the F2H-2 Banshee and F4D Skyray, recalled that the F4D "had a primitive weapon system. We generally flew a 120 degree lead-collision intercept. At the proper moment in time and space we would fire 2.75 inch rockets and they would collide with the target. The radar scope gave you

minimal flight information (an altitude line from the radar side-lobe and the wing position). On a 1957 cruise on USS *Saratoga*, while making a low-altitude, in-the-rain night attack against a NATO 'enemy' (I think it was a Lancaster bomber), it occurred to me that I was doing two full-time jobs and doing neither one as well as I wanted to. Therefore, I became a proponent of the two-seat interceptor. When the F8U-3 was in competition with the F4H-1 I was concerned that the Navy might opt to save money and buy the F8U-3. I was very pleased when they selected the F4H-1, and I immediately started putting in every six months for assignment to the first F4H-1 squadron to be assigned to the Atlantic Fleet. In 1960 I was delighted to receive orders to report to VF-74."

The Commander of that squadron, CDR Julian S. Lake, had watched the F8U-3 and F4H-1 contest with interest, and his exchange tour with the USAF flying two-seat radar interceptors such as the F-89 Scorpion and F-94C Starfire had helped to convince him that the "back-seater" was a valuable asset. During a stint at the USN Test Pilot School, Patuxent River, he had a chance to fly in the Number 5 F4H-1F (BuNo 143390), which Raytheon was using to develop the Sparrow missile. He then took over VF-74 *Be-Devilers* in 1960 towards the end of their time on the Skyray and began work-up on the F4H-1. Tex Elliott had first met CDR Lake when he was XO of an F4D squadron in 1956 and was impressed with his leadership and his ability to innovate. His first aircraft were from a batch of twelve which, in April 1961 were divided between VF-114 *Aardvarks*, the first West Coast squadron, at Miramar and VF-74 at Oceana.

"Pilots who were going to be the cadre of VF-74 were all second-tour, or first-tour with at least one deployment with the F4D or F3H, so the group was hand-picked and had already reported to VF-74 when the Phantoms arrived at Oceana. Needless to say, the afternoon that the first Phantoms landed at Oceana it didn't take long before CDR Lake was in the air for the first flight. The decision had been made that the initial flight for pilots would be in the back seat with an instructor in the front. Several days earlier, on 22 April, I had to go to Patuxent River and, in a pressure suit, ride on a 'burner blow-out' flight with a Marine captain in the front seat. At 76,000 ft the afterburners blew out and he had to shut down the engines. The airplane glided over the top and back down. I had my first indoctrination flight on the way and was ready for the front seat at Oceana.

VF-74 completed training in September 1961 and detached from VF-101. In October we did day and night carquals aboard USS Saratoga for the first time as a squadron and became involved in November in a TAC evaluation, under General Sweeny, of the F-106 against the Phantom. Identical missions were flown by F-106s from Langley AFB and VF-74 from Oceana. A joint team of USN/USAF umpires decided the F4H-1 could do anything that the F-106 could do and so much more that General Sweeny and his team decided to buy the Phantom. I thought it ironic that we had bought the plane as an interceptor, but the USAF would use it as a TAC bomber. [The Phantom's superiority over the F-106 was hotly contested by Convair, who pointed out that F-106 pilots out-maneuvered the F4H-1 on many of the 153 sorties flown and that the Delta Dart's AIM-4 missiles were better in the all-aspect mode than the AIM-7 and more resistant to ECM] VF-74 deployed to the Mediterranean in 1962-63, and CDR Paul Spencer relieved Julian Lake as CO. CDR Spencer had done carrier suitability work for the F4H at Pax River."

Tom Miller pointed out that the APQ-72 radar was:

"...far better than the F-106's radar. The USAF could not believe that F4H-1s could lock on to a target measuring three square meters at fifty miles when the F-106's radar acquisition range was only 25 miles. The F-106 had serious instrument flying limitations, and many of their missions had to be canceled. It was Air Defense Command (ADC) that first became interested in the F-4. At the time, a Colonel Graham, USAF, was permitted to fly the F-4 very early in the program [a flight authorized at Edwards AFB by Tom Miller as Desk Officer], and he pushed it for ADC vice the F-106. His effort culminated in the so-called fly-off between the two aircraft. The word got around the USAF as to what a great aircraft the F-4 was, and the Navy advised the USAF that it was also a great attack aircraft because it could carry such a large tactical bomb load. This was what prompted the bombing demonstrations that I flew at Camp Lejeune. At that time the Tactical Air Command (TAC) of the USAF had greater pull within the Air Force, and they were so impressed with the F-4's fighter/attack capability that they stole it away from ADC. The F-106 remained in ADC, and the F-4 was purchased for TAC."

Green-jacketed deck personnel hook up a VF-121 aircraft (BuNo 153915) to the cat shuttle. Before tensioning the catapult the aircraft's nose-gear would be extended, requiring at least 2,750 psi of hydraulic pressure. (Boeing/McDonnell Douglas)

With nose gear still extended, a practice bomb-laden F-4B leaves USS *Ranger's* **catapult. The acceleration could force fuel out of the external tank and into the fuselage fuel cells at a rate beyond the tank's venting capability. The subsequent vacuum could collapse the drop tank unless the external tank transfer switch was set to "off." (via Norman Taylor)**

CDR Joseph Konzen oversaw the work-up of the second East Coast F4H-1 unit, VF-102 *Diamondbacks* in 1961, and he handed the squadron over to Jerry O'Rourke the following year. In the third unit to transition, VF-41 *Black Aces* under CDR Whit Freeman, was RIO Fred Staudenmayer. He had the distinction of being the first Phantom crew member to achieve 1,000, 2,000, and 3,000 hours on the type. CAPT Staudenmayer later commanded VF-33. In his RIO training he came up against a familiar phenomenon at that time, the reluctance of pilots to accept the need for a back-seater: "Many a pilot has wished the space was filled with petrol instead of a RIO!" In truth, the second cockpit did cost fuel space and made the 600 gallon centerline tank virtually a fixture on Navy Phantoms. As single-seat jockeys, often with their own rather primitive radar scopes to manage, the F4D, F3H, and F8U pilots who climbed into Phantoms often reckoned they could handle intercepts without the need for a radar "nanny" behind them. Unkind tags such as "talking ballast" were applied to hapless RIO candidates. CDR Bill Knutson, also a VF-33 Skipper later in his career, came through to Phantoms via the single-seat route.

"I started flying in the F6F-5N Hellcat with APS-6 radar and went through the F2H-3 Banshee, F3H-2 Demon, F4D-1 Skyray, and F8U Crusader series. I had flown the aircraft and operated the radar all my flying days, so I 'tolerated' the RIOs and gave them lip-service, but at crunch time in an intercept or dog-fight I did what I knew had to be done without listening or taking any directions from the RIO. On one intercept mission the RIO had locked onto the target and I simply said, 'I've got it,' and proceeded to maneuver into position for a Sparrow shot. The RIO became furious. He turned off the radar and said, 'See how good you are without me operating the radar.' Naturally, the radar could not be operated from the front cockpit. These problems smoothed out over the years, and essentially went away when it came time for combat."

CAPT Roger Carlquist was LSO on VF-74's work-up and first deployment to the Mediterranean, having flown as initial Project Test Pilot for the 1960 BIS trials at Patuxent River. On his first cruise as a CO with "nugget" (inexperienced) RIOs in his own squadron he recalled:

"I spent a lot of time telling the pilots, who had mostly transitioned from the F8U-2N Crusader, that they needed to work with their RIOs if we were going to get anywhere, and telling the RIOs that their talents were the critical ingredient in the success of the squadron, so every opportunity should be seized to improve those talents. We also had frequent 'All Air Crew' meetings devoted to squadron progress. I made it a point to fly with each one of the RIOs once every ten weeks to evaluate strengths and weaknesses."

In Tom Miller's recollection, "Many experienced Marine NFOs from USMC F3D squadrons served as instructor NFOs at both East and West Coast RAGs as the Navy had very few current, experienced all-weather NFOs."

Fred Staudenmayer well remembered his RIO training on F3D Skyknights at VF-101 Det Alpha, Oceana, while awaiting the arrival of the first Phantoms, and other sources of expertise:

"Experienced flight officers (NFOs) came to us from AEW squadrons who were flying in Super Constellations, some ECM crewmen from AD-5Q 'Queer Spads,' a few from P2V Neptunes, and even flying-boat ASW squadrons. The most savvy of this bunch were the Super Connie airborne air controllers (AACs) who had some idea of what was going on. These gents set up most of the intercept syllabuses, along with pilots who had flown the Demon and F4D-1 'Ford' with the Fleet. After training with the AACs most of us could control intercepts in the air if the carrier's radar wasn't working, or (more likely) couldn't see us—a big problem in the 1960s when, without IFF we wasted a lot of time. The former AACs who had the greatest influence were Duke de Esch (who flew in the

VF-31's lineage dates back to July 1935 and the Boeing F4B-4. From 1948 onwards the squadron flew F8F Bearcats, F9F Panthers, and F3H Demons before graduating to the F-4B in 1963. Based on USS *Saratoga* from 1964 to 1981 with Phantoms, it then converted to the F-14A. This F-4B is depicted in July 1967. (Steven Peltz via Norman Taylor)

back seat on Hunt Hardisty's Sageburner flight), Tom Johnson, Jerry Ryan, and Bill Alhorn, and each of these men went to initial F-4 squadrons to lend their experience as they transitioned to Phantoms. I myself came to VF-101 Det A without benefit of basic or pre-flight training, grabbing what I could along the way. Flying in the F3D served as advanced training, I guess. The F3D 'Drut' (spell backwards) would in fact go high with its straight wing—over 42,000 ft—but it took an awful long time to get there. We usually flew at altitude because in that era that's where the jets flew, and also for endurance; getting more intercepts per training mission. We had, however, a substantial low-level intercept syllabus and often ended the training hops with self-contained radar approaches to the duty runway with the tower's permission. We didn't have to be under positive radar control, which I believe enhanced our creativity. We got fairly proficient in navigating towards the land for direct approaches from out at sea, using our APQ-51 radars and some 'feet-dry' check points at the coast plus tall apartment buildings. If the duty runway had to be approached from over land, forget it! Almost all our hops were over water, and we had plenty of training room at sea over the Virginia Capes. We quickly learned how difficult it was to pick out a low-flying aircraft on a breezy day with a sea state above 2. With side-by-side cockpit configuration in the F3D pilot instructors who had operated the same APG-51 in the Demon, while also flying a demanding airplane, were not reluctant to offer suggestions on tuning, adjusting the radar antenna position, etc. We also flew some low-level over-land missions, but the emphasis was always a blue-water one. Later, at Key West when the Phantoms were there, we had the best training areas possible over the Gulf of Mexico and in the Straits of Florida. There was no overland flying at all, except when we borrowed some Air Force air-to-mud targets at Avon."

In Bill Knutson's view:

"One of the biggest problems with NFOs at the time was that they were getting senior in rank. They might out-rank the pilot. They were made Department Heads (Operations, Maintenance, etc) to give them experience and qualify them for command [of a squadron], but would they get command? The CO was the leader of the squadron on the carrier and especially in the air. The RIO had no flying controls in the cockpit and could not lead a section, flight, or

One of the innovations with VF-74 was the introduction of the KD2B airborne target (beneath this F-4B) for missile practice. (USN via J. T. Thompson)

The last F-4B in production Block 12, BuNo 150435, first flew in September 1962 and served with VF-151, later re-emerging as an F-4N. (Author's collection)

strike group the way a pilot did. They could only direct traffic. It was a major effort to get them to think like pilots so that they could assume command responsibilities."

There were all manner of basic F-4 operational techniques to be established on those early cruises including, for example, the choice of external fuel tank combinations, using two wing tanks and the centerline "bag" for carrier operations. CAPT "Hap" Chandler:

"We worried about the external tank configuration; three tanks, two tanks, or one. Three tanks were dismissed right away. Two tanks restricted Mach, and one tank restricted engine access but had the least effect on speed. We ended up with one centerline tank, as the J79 engine was so gripe-free we didn't have to get to the engine hardly at all." Access to the mighty turbines was via two large under-fuselage doors, hinged downwards at the aircraft centerline.

Back at St Louis production continued to gather pace. Phantoms were built on a car factory-type production line using a jig suspended from an overhead track, which placed the aircraft at a particular work station for about a day before it was moved on to the next stage by the night-shift team. The 22 foot long center and aft fuselage assemblies were mated to the massive 27 foot wide wing center section, and also to the forward fuselage, which comprised two clam-shell halves, split vertically. Most of the wiring in these forward areas was pre-installed. Radar and other electronic equipment was tested over a three-month period after roll-out of the basic airframe. Deliveries of the F-4B began in mid-1961 and continued through to the end of 1966, by which time 698 had emerged from St. Louis, built alongside the Air Force's order for 583 basically similar F-4Cs.

On the West coast VF-114 *Aardvarks* became the first Pacific Fleet squadron to deploy with Phantoms in September 1962. Their XO when they transitioned in mid-1961 was Roscoe Trout, who explained the origins of the nickname adopted by the squadron soon after it switched from the F3H Demon:

"I was the author of the letter that originally requested and received permission from Johnny Hart for VF-114 to use the aardvark logo. This aardvark character was featured in the newspaper comic strip 'B.C.,' authored by Hart. The idea sprang from some peoples' view that the F-4B, parked on the line with the canopies up (suggesting ears), its high, droopy tail and long nose resembled that character. And the missiles could zap prey at long range just like the aardvark's tongue. In those days almost all squadrons had mascot symbols of one sort or another, and we chose that one." A few years later Air Force pilot Al Mateczun observed similar characteristics in the General Dynamics F-111, which finally adopted the Aardvark nickname as its official title in 1997.

RADM Peter Booth, who was also in VF-114 at the time, recalled that, "the squadron went from the 'executioner' logo when we were in the F3H Demon to a complicated Greek one that few understood and which never appeared on F-4Bs, and then to the aardvark." Their CO CAPT "Hap" Chandler and his RIO were actually responsible for choosing the logo:

"While I was serving with VX-4 I received orders to become CO of VF-143 Pukin' Dogs, who were to be the first West coast F4H-1 squadron. For unknown reasons the aircraft were shifted to VF-114, and my orders happily changed to VF-114, also. Before I could join the squadron they received Phantoms and entered the RAG. Technically, CDR 'Tiny' Granding, who was then their skipper, was their first F4H-1 CO. I joined them early in the RAG and relieved Tiny, so I was their first skipper during the F4H-1's initial Westpac deployment. Joe Konzen was my XO. My RIO, Kirk Sheehan, suggested the aardvark, as its tongue could reach out a long way and zap the ants, as the F4H could do to its enemies." This process of choice underlines the importance to squadrons of their trademark logos, and similar tales could be told about many other squadron identity plates.

Another *Aardvark* member at the time, Fritz Klumpp, recalled the work-up under the command of CDR Konzen, who succeeded "Hap" Chandler:

"I joined VF-114 just before a visit from President Kennedy. For the firepower demonstration we had four F-4Bs in formation fire AIM-7s at a QF9F target drone directly across the bow of the USS *Kitty Hawk*. There were three direct hits. Our mission was solely air defense, and we spent most of our time working on low

Early *Gray Ghosts* jets prepare for a mission. (via Peter B. Mersky)

altitude intercepts. We would fly at approximately fifty feet and then climb to 100 ft to fire missiles."

Former F8U Crusader pilots like Ken Baldry, flying with VF-96, had to get used to the Phantom's very different handling characteristics:

"The biggest problem with the Phantom was that it was very nose-heavy. For instance, it was impossible to hold the nosewheel off the ground after a normal landing, whereas the F-8 Crusader nose could be held off for several thousand feet while you held full back stick and got pretty effective aerodynamic braking. We used the drag chute on the field with the F-4, and turn-offs after 6,000 ft on the runway were very easy. The nose-heavy problem was compounded in a 'carqual-weight' (around 29,000 lbs) catapult shot, as the lower end-speed on the catapult required full back stick to get the nose coming up after leaving the flight deck. As a matter of fact, during carquals the catapult officer would give a 'pulling the stick into your gut' motion with his right arm and would not fire the catapult unless the horizontal stabilizer was in the full aft position. The nicest cat shot the airplane took was the full 54,800 lbs max gross weight shot, as the cg was far aft and the end-speed allowed you to go off the front end with a completely neutral stick, which is the way all us F-8 jocks liked it anyway. The F-8 took the world's most perfect cat shot! Neutral trim, neutral stick, and it just went off like an arrow. The reason the early Phantoms came aboard so fast was not because the airplane wouldn't fly slower, it was because you just ran out of back stick to hold the nose up!"

For the next Navy Phantom model, the F-4J, McDonnell devised a slotted leading edge to the stabilator, which had the reverse effect of a slatted wing and provided an increased down-force at lower speeds. It was retro-fitted to some F-4Bs (reducing landing speed by about 20 kts to 125 kts), including those of VF-96. Ken Baldry:

"It gave all the nose authority you needed and really improved the low-speed handling characteristics. Even with the slot, the 'carqual weight' shot took all the back stick you could muster, and still it felt like the nose was never going to come up! At night the nicest and most popular cat shot was off the waist cats, as you still had several hundred feet of lighted ship on your right side to give you a nice horizon. Off the bow cats, you just flew into an ink-well, and it was a lot easier to cage your own gyros, which immediately

VMF(AW)314's F4H-1 BuNo 149453 provides a useful modeler's guide to maintenance placard placement. Note the side-number 2 repeated on the fin leading edge. The squadron dropped the "AW" designator in favor of "VMFA" in August 1963. (USMC via Norman Taylor)

Armorers load AIM-7 and AIM-9B missiles on a *Gray Ghosts* F-4B. (via Peter B. Mersky)

told you (falsely) that you were going down. You could usually see who had gotten the old 'falling out the bottom' sensation on the cat shot because the afterburner would still be going for a lot longer than normal."

Ghosts and Knights
While the Navy F-4B squadrons began to proliferate, the Marines also commenced their Phantom induction process. VMF(AW)-531 *Gray Ghosts* moved to F-4Bs at MCAS Beaufort (via VF-101 at Oceana), while VMF(AW)-314 *Black Knights* also received the new fighter at MCAS El Toro, following conversion at Miramar. Although the issue is hotly debated still, it is likely that the *Knights'* receipt of F-4Bs on 29 June 1962 put them slightly ahead of the Beaufort unit. Colonel Jim R. Sherman recalled the early stages of the Phantom transition:

"In the Fall of 1962 VMF(AW)-531 received the first East Coast F-4Bs to replace their F4D Skyrays. The RIOs were enlisted personnel, many of whom flew the Grumman F7F Tigercat and later the F3D Skyknight in Korea. When the F4D replaced the Skyknight most of these RIOs went into the 'air control' field, working as GCI/GCA controllers and tower operators. With the receipt of the F-4B most of these old professionals volunteered for flight duty and became Warrant Officers. So it didn't take more than 60 to 90 days to bring the squadron to 'combat-ready' status, and they deployed from MCAS Cherry Point to Key West, Florida, where from 1 February 1963 they operated for several months under the control of NORAD (USAF)."

The Key West deployment, commanded by LTCOL Robert F. Foxworth, was a reaction to the Cuban Crisis in 1963 and to Washington's desire to have the most advanced interceptors in place to counter the threat of Russian and Cuban aircraft. VF-41 had previously occupied this slot, and VF-102 on USS *Enterprise* helped to maintain the blockade of Cuba. VF-41's move to the base, according to Fred Staudenmayer, "had been initiated by one of SecDef MacNamara's whiz kids without going through the usual Navy chain of command, a harbinger of the micro-management that was to fol-

low. With the most sophisticated and effective airborne weapons systems available we assumed Alert status under the operational control of the Air Force to intercept anything coming up from the South. The MiGs never did, except to defect, but we did intercept an interesting assortment of Piper Cubs, transports, balloons, and boats." However, the Navy Phantoms' superiority over its USAF counterparts was not something the Air Force seemed to want to publicize at that time: "During a 'keep up the good work' visit just after the maximum crisis time President Kennedy toured the squadron lines of airplanes that had not been flying and bypassed us!"

By the time the *Gray Ghosts* arrived the threat had diminished, but there was still some action against Cuban MiG-17s which provided Phantom crews with some valuable early experience of a future adversary. MGen Mike "Lancer" Sullivan, who became the high-time Naval F-4 aviator with 5,000 hours on Marine Phantoms in a 23-year career on the type, experienced one such encounter:

"C.C. Taylor, my RIO, and I launched on an airborne scramble, vectored 120 degrees for a bogie at 47 miles. We caught them before they went feet dry over the Cuban Keys. During the intercept we lost contact with GCI, so our wingman climbed to an altitude where he could relay our instructions. We were at 800 ft doing Mach 1.1 in full afterburner at three miles with an overtake speed of 470 kts and we got a 'Break X' just when GCI said to maintain a five-mile trail on the bogies. I didn't have a visual on them, so I chopped power to idle with speed brakes out and did a high-g barrel roll to keep from flying out in front of the bogies. Luckily, we stabilized on their tail at about half a mile, and I saw them slightly high at 12 o'clock.

C.C. Taylor was screaming that GCI said to maintain a five-mile trail and seemed very nervous about being so close, with Cuba only a few miles away. I told him that we should drop back, but about then the two MiG-17s went into a 60 degree port turn. I immediately thought they saw us and I slid to the outside of their turn, re-confirmed I was 'in Sidewinder' and periodically, when at the bogies' six o'clock, I would hear that beautiful 'Grrrrrr' Sidewinder tone. I thought the fight was on but then realized that the MiGs weren't pulling g in their turn. Watching them for a few seconds more it appeared the wingman on the right was learning how to fly

F-4As on the St Louis production line in October 1960 with BuNo 148266 in the foreground. (Boeing/McDonnell Douglas)

On the point of launching, a VF-121 F-4B. (Boeing/McDonnell Douglas)

formation. We were now only seconds from being over Cuba, so I started a hard turn back towards Key West with C.C. feeling cocky and wanting me to get closer to the MiGs so that he could get a good look at them. But it was too late, and he griped all the way back to base. The flight lasted only about 25 minutes, and we landed back at base with only 800 lbs of fuel. Both Sidewinders came back with their seeker-head glass blurred [by low altitude, high Mach heat friction]."

VMFA(AW)-531 pioneered the USMC's "Hot Pad" alert of four aircraft (which would become a familiar arrangement in Vietnam), aiming to get jets airborne within 150 seconds of the alarm sounding. On one occasion the two Alert Pad aircraft picked up some MiGs which had been firing rockets at boats full of refugees trying to escape from Cuba. The Hot Pad deployment brought some excitement for the Phantom's first cadre of Marine maintainers, too. Peter A. St Cyr had enlisted in 1959 and trained at Jacksonville on the 32-weeks Aircraft Electrician course. Assigned to Marine Aircraft Repair Squadron 27 (MARS-27), he worked on a variety of types until the F-4B entered service and then retrained at El Toro, at Long Island (the Sperry Company), and at Oceana before joining the *Gray Ghosts*.

"It was soon after we became operational that we were deployed to NAS Boca Chica, Florida, as part of the blockade of Cuba. Our mission was to man the Hot Pad with two F-4Bs ready at all times. Though only a few scrambles were commanded, we worked 'round the clock. Night time sorties were awesome; the Phantoms roared down the runway in afterburner, and once airborne they looked like stars. It was a definite 'high' each time Phantoms were launched."

With the squadron at the time was the first RAF exchange officer to fly the Phantom. Flt. Lt. Jim Sawyer (later an RAF Group Captain) and his RIO, Sqn. Ldr. Ian Hamilton, were excluded from action during the Cuban crisis "unless in extremis" by the British Government. However, their experience of the aircraft came at a time when the British were already showing an interest in the F-4, leading to a purchase in the mid-1960s. Compared with their usual

steed, the English Electric Lightning interceptor, Jim Sawyer found the Phantom far less forgiving.

"Where it really scored over its contemporaries was in its weapons system, which was probably, in the APQ-72, a pulse radar at its ultimate in development. Further improvements in airborne interception radar had to await the advent of Doppler techniques. It was a significant advance on what we had [as a radar interceptor] at the time, the Gloster Javelin. The USMC had been sending crews to us and the Royal Navy (on the DH Sea Vixen) for some time before this. Despite its several shortcomings the Phantom is the first love of all the aircraft I have flown. She was an ugly and mean-looking thing, and you had to keep on top of her all the time. If you did that she would never let you down. One of the problems that the newer pilots had to face was the tendency of the aircraft to yaw and thus roll from the direction of the 'up' aileron/spoiler. Unless no other action was taken, the aircraft at high g (and therefore high angle of attack) could rapidly fall out of control. Hence the development of the term 'rudder with judder.' As angle of attack increased the rudder became more important for roll control. It is the only jet aircraft I know which, when under symmetrical power, depended so much on the use of rudder. I have witnessed some very impressive inadvertent maneuvers by pilots who forgot this feature. Happily, all escaped, but some had to go as far as having to pop the drag 'chute to recover. Indeed, 'in extremis' in a spin this was the recommended technique.

The pitch sensitivity of the Phantom was legendary in its middle and later years, but early 'convertees,' despite their general flying experience, could find it bit them hard. I recall the first day that Ian Hamilton and I turned up on the squadron and were told that we would be seeing the skipper as soon as possible after he landed. He

The *Gray Ghosts* patch, 1963 vintage. (Peter St Cyr)

NAS Key West patch, c. 1963. (Peter St Cyr)

and his RIO did so in due course—but on the ends of parachutes. They had flown their way into pilot-induced oscillation (PIO). All previous experience suggested one should just let go of the stick and the aircraft's natural stability would dampen PIO down. Not with the F-4! In this situation, with the stick free it could move and thus exacerbate the event, with the consequences I relate. It did not take long for the rest of us to learn a new technique, which was to retain a firm hold on the stick slightly aft of the neutral position."

The USMC F-4 training program involved "carquals" for crews on a regular six-monthly basis and a worrying moment for the "Brits" while completing their first carrier sortie off USS *Forrestal*:

"We had rolled onto the forward catapult and all checks were complete, at which point I asked Ian Hamilton if he was ready to launch; he confirmed it. I signaled to the Deck Officer that we were prepared, and within seconds we were belting down the 160 feet of the catapult's stroke, at the end of which all acceleration appears to stop. I selected full burner, and at the same time I asked Ian how things were with him. There was utter silence. As I discovered afterwards Ian had rested his metal kneeboard on the cockpit coaming, and in his excited state he had forgotten to strap it on. The ensuing launch pushed it into his face, and for several long seconds he was dazed, to say the least. At the end of the sortie we went below decks, and the Flight Surgeon treated us both to a much-needed medicinal brandy!"

From the outset the Marines sought to use the Phantom's attack power and trained for bombing techniques, which were to be vitally important in the succeeding years. Jim Sawyer remembered that:

"Dive bombing was an uncomfortable exercise in an aircraft as heavy and as fast as the F-4, and 45 degree angle bombing was more so. It was on the latter sortie, as I initiated the pull-out that the aircraft's stability augmentation system tripped out. In those days the switch was held on electrically and a power 'trip' could deselect it. The nose reared up to show me blue sky ahead, followed almost immediately by a view of the terrain below as the nose plunged rapidly down. An experience to remember." Jim was able to apply the recovery technique involving holding the stick aft of neutral to stabilize the F-4B.

Jim Sherman joined the *Gray Ghosts* in June 1963 after they returned to Cherry Point to prepare for the first of the foreign deployments which are a major aspect of Marine activity. "All the RIOs remained to go overseas, but eight pilots joined them from Skyray or Skyhawk squadrons and had a year to learn the air intercept business from twenty-one of the best RIOs there were. In August 1963 the squadron was redesignated a Marine Fighter Attack Squadron (VMFA-531), though we didn't have any bombracks until September 1964. (The first ordnance I dropped from an F-4 was on my first combat mission)."

Their first deployment to the Far East came in June 1964 to Atsugi, Japan, following VMFA-513. Jim Sawyer also made the "Transpac" flight:

"We staged through El Toro, Kaneohe Bay, Hawaii, and Wake Island to Atsugi, which was to be the squadron base. The whole trip went fairly smoothly, although on the leg to Kaneohe Bay I discovered that I had a faulty refueling probe, which meant I could not lock into the tanker's 'basket.' Still, we made it with only a little less fuel than the others."

On arrival Jim and Ian were informed that they should not be there, as the U.S./Japan defense agreement did not allow the participation of third parties. They returned to the USA to help work up a new squadron, VMFA-323, and were replaced by another RAF crew, Mike Shaw and Keith Shipman. Carquals for the deployment to Japan had their problems for Jim Sherman, too. Prior to the "traps" on USS *Forrestal* (20-23 March 1964) in the Virginia Capes area the squadron did field carrier landing practice (FCLP) at Cherry Point.

"We screamed over Cherry Point from sunset to 1a.m. for two nights and became extremely unpopular. When the populace learned we were going to do this for six weeks the Commanding General banished us to the USMC Auxiliary Landing Field at Bogue, North Carolina, twelve miles to the south west and carved out of the pine trees to retain that expeditionary look. My first couple of nights at Bogue were very memorable. It was pitch black, and you could not fly visually but had to rely on instruments. It took me several passes to find the field, much less land on it. When you were in the downwind leg you could not see the runway, as the tall pine trees blocked your view of the landing lights. Upon landing at Cherry Point I told my RIO what was already apparent to him—'I need your help!' He responded, 'I thought you would never ask.' We sat in the Ready Room that night and came up with a solution. I would fly instruments around the pattern, never looking outside the aircraft. Ray, my RIO, did the looking, and as an experienced GCA controller he could get me lined up with the runway and established on glidepath. From a near-perfect set-up a quarter mile from touchdown I tried my best to hold what I had set up until the touchdown. When

VF-84 F-4Bs in October 1964 with AIM-7s on their inboard pylons. (USN via CDR James Carlton)

VMFA-513 F-4Bs prior to the adoption of their *Flying Nightmares* tail design. BuNo 149456 was lost after a AAA hit in November 1968 while flying with VMFA-115 from Chu Lai. The other F-4B, BuNo 150466, was lost in South Vietnam after a refueling accident. (via Peter B. Mersky)

we hit the runway I went to full power and back in the air and I was back on instruments."

Colonel Sherman and his RIO were among many Phantom crews who were developing the crew co-ordination techniques which would maximize the role of the back-seater and make him far more than just a "scope watcher."

"Our system worked extremely well, and we flew out to the USS Forrestal on 20 February full of confidence that this 'day stuff' would be a piece of cake. Mother Nature had a different idea and planted a huge area of low pressure over the operating area. On the first day the ceiling was 600 ft with one mile visibility. The fantail of the ship was going up and down 10 to 15 feet. The skipper brought us aboard, and that was all we did that day. Next day the weather was even worse, but by 1300 hrs the ceiling had lifted to 1,500 ft, though the deck was pitching quite badly. A RA-5C flown by a test pilot from Patuxent River joined the pattern. I had completed my second trap and had 2,500 lbs of fuel. The Air Boss said to catapult me and they would refuel me after my next trap. On my next pass I boltered due to the pitching deck. As I was turning through the 90 degrees position the RA-5C hit the deck on its upswing, and it became several pieces as if it had been stepped on by a giant.

Without waiting to be told I cleaned up the landing gear and turned West for Oceana, supposedly 80 miles away. Our TACAN was locked on the carrier and our VHF radio was OK. My RIO was busy firing up the radar. Passing 18,000 ft the low fuel light came on. I leveled at 32,000 ft, still on IFR as the radar blossomed, and Ray held Oceana on the nose at 120 miles. The carrier had erred in establishing their own position by at least fifty miles. As we continued at 32,000 ft at the best endurance speed our TACAN was spinning [not locking onto the station], and Oceana approach control would not respond to my calls on the radio. At sixty miles out I commenced a descent, and by then I was concerned that I was penetrating the Air Defense Identification Zone (ADIZ) without notification, so I switched my radio to Guard channel (international distress frequency) and my transponder to Code 77, which identified me on radar as 'in distress.'

I thought our luck had changed when our TACAN locked on to Oceana twenty-four miles out and Oceana came up on Guard channel. But it had not. Oceana advised us that the field was closed because the weather was below operating minima, and so were four other nearby airports. I informed them that I was on emergency fuel and must land at Oceana. They then directed me to turn left for

Tensioned and ready to launch from USS *America*, a VF-102 F-4B with a cluster of practice bombs on its centerline rack. (USN via CDR James Carlton)

radar identification. I told them 'Negative, we are eight miles out on a self-contained approach to Runway 10. If we are unable to land at the first attempt we are going to proceed straight ahead and upon reaching the beach we will eject, so please have the Search and Rescue (SAR) helo cranked up!' RIO Ray did his usual unbelievable job. He positioned us 1.5 miles abeam the runway with 3,000 ft altitude. I started a standard-rate turn from there and dropped the landing gear at 90 degrees out, passing 1,800 ft. about 40 degrees from the runway heading. Ray had the runway on radar and called for half flaps. He rolled me out on the runway heading as the GCA controller made his one and only announcement; 'You are on the centerline passing the end of the runway. If you do not have the runway in sight execute a wave-off and climb straight ahead.' At that instant we hit the runway rather smartly, and I was surprised to see the runway centerline beneath my nose. As we rolled out, hailstones the size of golfballs were hitting the runway and bouncing back up to eye level in the cockpit. We turned off at the end of the runway. I added throttle to taxy in, and both engines flamed out for lack of fuel."

Jim and Ray were then treated to a series of martinis rather than the medicinal brandy, but they still managed to return to the ship next day. The poor weather meant that only nine of the squadron's twenty-one pilots had managed to achieve day carrier qualification when they returned to Cherry Point and had to await the arrival of USS *Saratoga* off Jacksonville a couple of weeks later to complete their allocation of "traps."

Gas for Ghosts
In making the first Transpac for the F-4 the Marines pioneered many of the long-distance deployment and refueling techniques which were to enable regular exercises of this kind. Self-supporting deployments to the Far East were made possible by the acquisition of the GV-1 (KC-130F) tankers, which the USMC bought in the late 1950s. Although this tanker will continue in use well into the 21st Century, it had some disadvantages compared with the Boeing KC-135. Jim Sherman recalled its role in that first long haul to Atsugi:

"The main drawback with the KC-130 was that it cruised much slower than the receivers. Therefore the tankers departed El Toro en route to Hawaii 1.5 hours before the receiver F-4Bs, doing 200 kts while the Phantoms flew at 480 kts. Ideally, the receivers caught up with the tankers at the intended refueling areas."

Jim Sawyer: "With three external tanks to fill the Phantom at the latter end of the refueling 'bracket' was very heavy and did not like flying around at 200 kts. Half flap was used, and the worry there was that in the cold air at 20,000 ft the bleed valves of the boundary layer air recirculation system would not shut off when the flaps were selected 'up,' thus pumping very hot air into the wing. Our worst fears were not realized, and we all arrived in Japan on schedule." Group Captain Mike Shaw added that full flap could not be used for inflight refueling despite the awkwardly low speeds, "Because if the F-4B over-ran the 'basket' the hose would be caught between the leading edge flaps. When heavy, as it would be when taking on fuel with three external tanks, the pre-stall buffet was present, and the lateral controls had to be treated with great caution. Over-use of the aileron/spoiler at high angle of attack could

cause severe adverse yaw and possible loss of control. A partial remedy was for the KC-130 to 'toboggan' downhill to build up speed to 230 kts IAS. This eased the F-4B's difficulties but was not always an option, for example, if cloud tops were at 20,000ft."

Towards the end of the refueling process handling became tougher still, requiring minimum afterburner on one engine while maintaining position with the throttle of the other engine. In this state, as Colonel Sherman recalled:

"You were burning an enormous amount of fuel to top off all the tanks. As soon as your tanks were full you wanted to get off and accelerate to the best climb speed and then climb back to cruise altitude. If one or more receivers had problems getting plugged in the rest of the flight had to unplug and then climb without the last F-4, which could take an hour or more to catch up. The best way to prevent this was for each receiver to report when he was plugged into his tanker, and when all were connected the tanker commander announced, 'Commence refueling,' and tankers would turn on their transfer pumps. Ideally, all F-4s would top off within thirty seconds of each other. The refueling evolution, from the time you plugged in until you were back at altitude consumed at least 6,000 lbs of fuel, and normally you would take on around 13,500 lbs."

Just finding the tankers in the first place could be a problem: "With the F-4 we had VHF radios with a range of at best 80 to 100 miles. Without INS the F-4s didn't know where they were, only that they were making good speed. They had to rely on their radars to pick up the tankers below 20,000 ft. Tankers used their UHF/ADF once you had established communications with them and could tell you if you were on course, behind them or to one side. In VFR conditions this normally doesn't pose a problem, but if either or both aircraft are IFR the rendezvous could get interesting, with six tankers and six receivers converging with an overtake of possibly 100 kts."

VMGR-352 provided the KC-130Fs for the Transpac, which was made as three flights of five F-4Bs with six tankers. Departing Cherry Point on 16 June at 0800 and 0838 the first two flights refueled seventy-five miles west of St Louis. Jim Sherman led the third wave out of Cherry Point at noon and hit problems with the tank-

VF-101 Det A provide a neat trio of F-4Bs in March 1963. As the Atlantic Fleet F-4 RAG, the *Grim Reapers* trained F-4 crews for nearly seventeen years. (Boeing/McDonnell Douglas)

Variations on VMFA-513's tail logo included this on BuNo 150420. (via Steven Albright)

ers, who had gone further west to seek clear weather VFR for the rendezvous. By the time the *Gray Ghost* flight had them on radar they were 250 miles west of the designated refueling position and still in IFR weather. As the Phantoms were getting critical on fuel they "plugged" in anyway, regardless of the weather, and completed the first 2,300 mile leg to El Toro. The rest of the transit (2,500 miles to Kaneohe Bay, 2,400 miles to Wake Island, another 1,500 to Guam, and a final 1,500 mile leg to Atsugi) required the tankers to pre-position at each of those locations. Tanker crews were flying up to fifteen hours per day, and the whole process took eleven days, including stand-down days to get fifteen Phantoms to Japan for a twelve-month stay.

Their main task was air defense. As Jim Sherman put it, "The *Gray Ghosts* still had the all-weather fighter mentality, as we did not have bomb-racks at the time. Accordingly we conducted about half of our normal training at night and complained when a night sortie was canceled for lack of aircraft or weather." Air defense was still the main item on the syllabus back at the Phantom training squadrons, too, though as Fred Staudenmayer judged:

"With the advent of the Phantom giving the Navy much more of a blue-water capability, tactics were developed on a number of fronts in the early 1960s. In the early transitions, training sylla-buses would be a mixture of old F3D and F3H tactics, intercepts developed by VF-101 and pass-downs from squadrons who had deployed with Phantoms. When shore-based, squadrons would add low-level navigation hops to a number of low-level, high level, or high-and-descending intercepts with targets usually simulated by wingmen. Once embarked for 'work-ups' flight missions were much more controlled by the carrier's air operations, although each squadron's operations officer had a large input to what missions were flown. The emphasis was always on carrier landing proficiency, with plenty of instrument flying, night flights most nights, tanking practice, and coordinated operations with the Early Warning (EW) and ECM types."

The focus on missile-armed Fleet Defense tactics for the Phan-tom flyers had its compensations, though. Tex Elliott was given a crowd-pleasing task by his VF-74 CO, CDR Julian Lake, as they prepared for their first deployment:

"I developed at his direction a maneuver for use in our airshows around the carrier, an 'Immelman to an intercept.' We'd have an RF-8 Crusader or RA-5C Vigilante 'target' at 12,000 ft coming bow to stern down the port side of the carrier at some distance out so that people from the flight deck could view it. I would come in from about ten miles astern. The target would be doing about 350 kt, and I was at 500 kt as I came along at flight deck level or below. We would lock onto the target, pull up into an Immelman as we passed the ship, heading up at 30 degrees and in range of the 'tar-get' at six miles. We'd simulate firing an AIM-7 and continue on into the Immelman, rolling out right behind the 'target' for a simu-lated Sidewinder launch. When RA-5Cs got involved as targets they would dump some fuel at the moment of a 'Fox 2' (Sidewinder launch call) and torch the fuel with their afterburners. It looked spectacular from the ship. One feature of the Immelman was that if the cloud-deck developed you could just rotate the target out and lower the altitude so that it became more of a chandelle. You could save about 4,000ft and still have a good maneuver to watch. This demonstration was adapted by other Air Wings, too."

"Show the Flag" airshows and firepower demonstrations were very much in vogue in the early 1960s. When Tex detached from VF-74 in Summer 1963 to join USS *Enterprise* as Flight Deck Of-ficer (FDO) for its round-the-world cruise he found that, despite the airshows involving forty or fifty airplanes which expended quite a lot of ordnance, they still had plenty left on return to Norfolk, Virginia.

For the Marines, air interception training followed much the same pattern as the Navy's. Their Phantoms had been bought ini-tially to provide air defense when the Marines were embarked, and they could provide helicopter escort and "beach preparation" with ordnance for a seaborne landing. Once ashore and based on a for-ward air strip, Phantoms could continue to provide air superiority over the battlefield and join other fixed-wing types in close air sup-port (CAS) and battlefield air interdiction (BAI) of enemy assets.

However, it was the Navy's air interception syllabus which dominated training in the 1960s. Jim Sherman:

"Most of our intercept training was conducted east of MCAS Cherry Point, out over water and well clear of airway traffic overland. In the winter months we were required to wear either cold water exposure suits (poopy suits) or our full-pressure suits. Many of us preferred the latter, as they weren't hot like the exposure suits. The full-pressure suit plugged into the aircraft's ventilation system, and you could control the volume and temperature of the air circulating through the suit. Like the astronauts [and U-2 or SR-71 crews] we had a special van that took us to our aircraft, as well as a hand-held suit ventilator which we carried. The only negative thing was the 100 per cent oxygen in the entire headpiece, which dried your eyes and made them itch, but you couldn't do anything about it."

Part of the F-4's interception envelope was at very high altitude. Jim Sherman and a few colleagues decided to take advantage of the availability of the pressure suits and the Christmas break to make a personal attempt on the existing Phantom altitude records.

"My RIO and I developed our own profile based on the results of other flights which never got higher than about 83,000ft. Our profile had us climb to a point near Kittyhawk (birthplace of aviation), and at 38,000ft we turned south and accelerated in full burner to Mach 1.25. As the F-4 accelerated supersonic the computer closed the ramps in the engine air intakes, reducing the amount of air entering the engines. It did so in stages, and you could feel the change in acceleration as the ramps re-positioned. It seemed that the best performance was at Mach 1.25 or at Mach 1.7. We climbed at Mach 1.25 to 50,000 ft and then accelerated to Mach 1.7. Now, here comes the secret of success. We knew the engines began programming back at about 62,000 ft and completely shut down at 70,000 ft, so we wanted to be established in our zoom climb before the engines reduced power. Our climb angle had to be at least 45 degrees, and 60 degrees if possible. The problem was that the air got so thin that full-back stick required four or five times longer to bring up the nose of the aircraft than it would at 30,000 ft. Others had begun their zoom at around 60,000 ft and never got the nose more than 25 degrees up before the engines died, and they were lucky to reach 80,000 ft.

Once you established your trajectory you could push the control stick anyplace in the cockpit and it had no effect on the aircraft. Cockpit lights had to be full on, as there was no air to reflect light. The sun made one side of the cockpit so bright that it hurt the eyes, while the other side was so dark that you needed light to see by. The rate of climb was well in excess of 6,000 ft per minute, and Mach 1.2 was still indicated when, without warning, the nose fell through and pointed straight down, though the rate of climb was still pegged at 6,000 ft upwards. A further check of the instruments showed the engine rpm at 100 per cent, zero tailpipe temperature, and a fuel flow of up to 6,000 lbs. As I had forgotten to bring the throttles back to idle the fuel system was jettisoning raw JP5 out of the tailpipes, which was pretty spectacular. Soon the rate of climb went to the opposite extreme and we were descending at Mach 2 with no power and no control. At about 40,000 ft you could sense something on the stick, and the nose started to come up ever so slowly.

Speed slowed to around Mach 1.4 at 40,000 ft, though the engines were still windmilling. Finally, we were straight and level at 20,000 ft, and when we slowed to sub-sonic flight we hit the igniters and the engines came to life."

Colonel Sherman's best altitude was 93,000 ft in an F-4B with two wing tanks and 96,000 ft in a "clean" aircraft, verified by the local GCI squadron. The exercise didn't quite match the performance of the lighter, modified "High Jump" F4H-1, but it was a reminder that a service Phantom could make the upper atmosphere uninhabitable by high-altitude intruders.

Interceptor

Crucial to the Phantom's success as an interceptor at any altitude was the APQ-72 radar. Although it was generally seen as being vastly superior to previous sets, the electronic technology of the time limited its in-service performance. In Group Captain Mike Shaw's opinion, the Phantom's biggest weakness during its early days with the USMC was its APQ-72 radar:

"It was, when working, a very good airborne interception (AI) radar, but its reliability was not too good. Of the fifteen F-4Bs on the squadron (VMFA-323) in 'A' status (there were three more in 'B' status, or deep servicing), we could count on eight on the line each morning. Only four would have fully serviceable radar. This was not too much of a problem, as those with doubtful radars would be used as 'targets' for interception training, or sometimes in the ground attack role. Reliability of the Aero-1A system was about 50 per cent, and half of those which were serviceable on take-off needed some form of rectification before the next radar sortie. We did have a Westinghouse or a Raytheon 'rep' on each squadron, and very clued-up men they were."

Airborne interception techniques were taught quite early in the U.S. Navy Training and Operations (NATOPS) syllabus, which was also the basis of USMC Phantom instruction. It began with eight sorties in the weapons systems trainer (a fairly basic mobile simulator), overseen by a specialist NCO. The last session was supervised by the pilot who would be providing the familiarization (FAM) sorties. Mike Shaw described the rest of the course from the first FAM flight onwards:

An alternative *Flying Nightmares* design on BuNo 152278. (via Steven Albright)

"The new pilot would occupy the rear seat with no flying or engine controls and precious few instruments. Forward view was negligible, too. A demonstration profile would then be flown showing how the aircraft handled at various heights and speeds. 'Wall to wall buffet' is a phrase that springs to mind. The next sortie, FAM 2, was a moment of truth. The new pilot, who probably hadn't flown anything for six months since leaving flight school at Pensacola, was put in the front seat with the hapless check pilot in the back. A FAM 1 profile was then replicated. This was a tense trip for the pilot in the back, as the F-4B was very different from the TF9J Cougars and F11F Tigers the trainees had flown previously. After the excitement of FAM 2, provided that one was flown satisfactorily, new pilots would be launched with an experienced RIO who tended to have well-developed powers of self-preservation and could be of immense help. This system seemed to work, but the total absence of 'two-stick' F-4s [with rear-cockpit flight controls, like all USAF Phantoms] was not what I would personally have recommended.

After about ten trips 'exploring the flight envelope' the trainee progressed to airborne interception sorties. These were nearly all at medium level (30,000 ft), and the object was to train the crews to conduct Fleet Air Defense. The best ranges for air-to-air weapons, for the Cherry Point squadrons were in the Caribbean, so they would deploy for ten-week stints to NAS Roosevelt Roads in Puerto Rico. There they could use the excellent Atlantic Fleet Weapons Range, firing Sparrows against AQM-34 target drones, which our own F-4Bs could launch, and Sidewinders against flares, also self-dropped. Then the emphasis would switch to ground attack, using 30 degree dive bombing and rocketing techniques on nearby ranges on the island of Vieques and Culebra. It was at this stage that crews were introduced to air-to-air refueling from the KC-130.

The remainder of the approximately one hundred sorties in the NATOPS syllabus related to the handling of high-flying or very high speed targets, or to cross-country (airliner type) flying. The last sorties were flown on weekends, and were useful to build up cockpit time without involving the use of the complex weapons system."

Naval Phantoms carried a variety of air-to-air gunnery and missile targets, including the rocket-powered KD2B (AQM-37), launched from a LAU-24 trapeze ejector on the centerline pylon and providing both radar and IR signals to equate to a larger target. External configurations other than the preferred 600 gallon tank could cause the aircraft to "over-rotate" with a nose-high attitude on catapult launch. CDR Curt Dosé found this a difficulty with the AQM-37:

"Over-rotation off the cat was not a problem, usually. The F-4 was more sensitive when flying with two wingtanks instead of our normal centerline tank, and it was unstable when launching with the small AQM-37 drone on the center station and wing tanks. I was one of two VF-92 pilots certified for the AQM-37 mission, and 'vultures row' [the spectator area on the carrier's bridge tower] was always full. We would always over-rotate badly on these flights, wing-walk, kick up some ocean spray with the burners, then recover and be on our way. It was like a slow-speed 'scissors' with the ship's bow."

One of the earliest target methods was devised by VF-74's CO, CDR Lake, who mounted a Delmar target reel on the F-4's centerline hardpoint. As CAPT Tex Elliott recalled, Julian Lake was, "very much the innovator. He came up with a combining lens for use with the radar scope camera, and the dart target on a line for Sidewinder practice." The Delmar tow line could be winched out up to three miles behind the F-4B, enabling regular, cheap live-firing practice compared with the costly free-flying drone targets.

Interception practice tended to be one-on-one with the emphasis on attacks in the forward hemisphere. In a real Sparrow launch this gave the missile its best chance of a BVR interception, as its twelve-mile range could be used. In a rear-hemisphere attack the range reduced to about three miles, as the target "ran away" from it, increasing the danger of damage to the launch aircraft by debris from the stricken target or premature missile detonation. Officially the missiles' warheads could not arm within 3,000 ft of launch, but exceptions occurred from time to time.

The majority of the sorties flown from Cherry Point and other F-4 training stations in 1963-64 were "conversion," or air defense (AD) radar sorties. Mike Shaw:

"We planned the program one day in advance, hoping for eight aircraft on the line. Briefing would be about 1.5 hours before take-off time; there would be about 45 minutes on the details of the sortie, then 45 minutes to sign up for the aircraft, walk to it, strap in, start up and taxi. Starting up was a slow process, because the F-4B had no self-contained starter. The method was to use high-pressure air from a small jet engine on wheels (air transportable) which impinged on the turbine blades to provide a slow build-up of rpm. During this, the pilot would open up the high-pressure cock with the igniters firing, build up a few rpm, then shut down the fuel when the EGT (exhaust gas temperature) reached the limit. Eventually the engine would become self-sustaining and would idle at 65 per cent. Because the exhaust nozzle was wide open there was little thrust even at this engine speed.

Take off was usually in a five second stream, with climb-out according to a cleared standard instrument departure (SID). Once in the 'play area' we would either set up our own attacks, with 'target' and 'fighter' splitting about 50 miles apart before taking up

Four VF-96 crews carried out cross-deck operations on the British carrier HMS *Hermes*. "Roller" landings were performed, as the ship could not launch F-4s. (USN via CDR James Carlton)

pre-determined heights and headings. We would alternate as 'fighter' and 'target,' if we both had good radars, and recover as a pair to base when we had about 6,000 lbs of fuel left; enough for a diversion to Seymour Johnson AFB if necessary. Once in the pattern we would fly touch-and-goes, often using the deck landing mirror, until we were down to 2,000 lbs for the final landing. after landing, aircraft would taxi straight to the fuel pits where, with engines running, we would put the refueling probe out to de-pressurize the tanks, set all the fuel valves to 'receive' and top up. We would then return the aircraft to the flight line, shut down, sign up the Form 1 after noting any gripes and go to debrief. This could take another thirty minutes, making the elapsed time for the sortie about 3.75 hours, including 1.5 hours in the air."

The emphasis on Fleet Air Defense and BVR missile tactics in Navy training tended to rule out any thought of dogfighting as a role for the F-4 in the early 1960s. The Phantom had been designed as a continuation of the line of interceptors typified by the F-102 and F-106 (but with an inherent ordnance-carrying capability), which used their long-range missile overkill to render traditional "visual" close air combat an archaic practice. In Mike Shaw's opinion, pilots, "tended to be trained to fly like airline captains rather than combat pilots. We did very little ACM, and I suspect that many pilots went to Vietnam without ever having turned their Phantoms upside down. This observation probably applies more to the USAF than to the USMC, but all their pilots showed some reluctance to explore the performance limits of their aircraft. Too many of their flying hours were spent straight and level on cross-country flights. As most probably became airline captains this experience may not have been wasted!" It should be noted that Mike Shaw was more used to flying the nimble BAC Lightning with the RAF's 74 *Tiger* Squadron.

Atsugi Alert

Operating from their Atsugi base in Japan, the Marines' first deployed Phantom squadrons, VMFA-314 and VMFA-531, began to put their interceptor training into operational use. Flying was somewhat limited by local rules that there could be no touch-and-go or "carrier" landings, and by the lack of TACAN stations, though the five-hour "mean time between failure" rates on the F-4B's radio and TACAN sets meant that self-contained radar approaches to Atsugi were not uncommon. Colonel Sherman clearly recalled the procedure: "Initial approach over Mount Fuji at 20,000 ft, heading south. Power back to 80 per cent, speed brakes out at 250 kts. 30 degree standard rate turn; 6,000 ft per minute rate of descent; cross the island south of Atsugi at 1,500 ft and 200 kts heading 360 degrees. Transition to landing configuration at 140 kts. While passing over a large car-park (which the RIO would have used as a radar lock-on), start a descent of 500 ft per minute. The RIO broke radar lock on the car-park, and the radar antenna, in search mode, started looking for the metal aircraft hangars on each side of the runway. We had also sited at the side of the runway four radar reflectors that the metal shop had constructed. Each was a three-foot triangle of sheet metal, and they presented an excellent return on the radar scope."

Flying from the Japanese base offered some challenging experiences. On one night intercept training flight Colonel Sherman and his wingman launched into a 250 ft cloud ceiling with a quarter-mile visibility on the assurance that the base would be above minimums on their return. They flew several practice intercepts before his wingman's radar went down, making him the "bogie." "About the time we headed for the initial approach fix over Mount Fuji to meet the expected approach time he lost his TACAN, and when we switched to Approach Control frequency he lost his radio, also. As was our standard operating procedure (SOP) he would fly my wing, and I would line him up on the runway to land first and go around to land behind him. My wingman was concerned because he couldn't hear what approach control told me. The weather had not improved." Colonel Sherman then had to make an approach in one-eighth of a mile visibility with his wingman about twenty feet away. "The critical point was where I saw the runway at one-eighth mile and 150 kts. I had to be to the left of the runway so that my wingman could hit it without having to make any major corrections that close in. All went well, and he landed and rolled in."

On another occasion (10 November 1964) Jim Sherman was sitting on Alert during a heavy snowstorm which had already deposited up to ten inches of snow on his Phantom. He called in to declare "Mandatory," meaning he could only launch in a dire emergency with minimal chances of returning safely to base. The snow depth had reached up to eighteen inches when the Klaxon sounded. Sherman was assured that the launch must go ahead, and he took

"1,000 hours" patch. (Marty Lachow)

off alone, his wingman having aborted with an electrical failure. Following a vector of 130 degrees to 30,000 ft he entered a communications snarl up which eventually delivered the information that a JAL Boeing 707 bound from Tokyo to Honolulu had suffered an onboard electrical fire. The *Gray Ghost* crew finally located the airliner on their radar and led it, devoid of all electrical power and communications, back to a safe landing at Tokyo airport.

For many years the U.S. Seventh Fleet had the annual task of sending an ASW Task Force through the Sea of Japan to demonstrate to the USSR the right to transit international waters. As the anti-submarine carrier (CV) could not operate F-4Bs the CAP task fell to VMFA-531 from Atsugi. Flying with two wing tanks and a full missile load the F-4Bs also had 50mm hand-held cameras with a motor-drive plugged into the rear cockpit electrical outlet. Jim Sherman and his wingman took the 1400-1600 hours CAP on Day 1 and had been on station for about fifteen minutes, flying "figure of eights" when he heard on his head-set, "Here they come!"

"As we continued our turn they appeared on the edge of our radar scope. Ray (the RIO) responded, 'Contact. Multiple bogies at our altitude'. Bringing the targets 'round onto our nose I said, 'My God! How many are there?' Ray replied, 'More than I can count. I don't believe what I'm seeing—there must be at least fifty aircraft.' Control directed, 'Chieftain 8. Intercept immediately to determine type of aircraft and armament. We have counted eighty, repeat eighty bogies.' I must admit that I had some trepidation about going into the midst of eighty enemy aircraft when we didn't know what their intentions were.

I joined on a formation of the aircraft, and it was readily apparent that they had Soviet markings and were Tu-16 Badger bombers. They had no external stores, and twin 23 mm guns in their tails were in the stored position, pointing aft and elevated 30 degrees." The Phantom crews had intercepted the Russian formation at about 170 miles from the Task Force and were concerned that the Tu-16s would be carrying beam-riding anti-shipping missiles which would have been launched at around 30 miles in a real engagement. On this occasion the *Gray Ghosts'* job was to move from bomber to bomber taking photos of them and their occupants. "The bodies that appeared in the gun blisters were dressed in fleece-lined flight gear and soft head-gear with large ear-flaps. Apparently the Badger was neither heated nor pressurized. We had been shooting pictures (1,200 in all) and had worked our way to the rear of the in-bounds in about twenty minutes when we realized that what had been a gaggle of eighty aircraft had now sorted itself out into waves of fifteen to twenty, and these were executing a simulated attack on the Task Force. At ten to fifteen miles out and 20-25,000 ft each wave would execute a hard turn to starboard and reverse course for Vladivostock."

Jim and Ray were on the 1200 hrs flight schedule for the following day, 15 June 1964, feeling less than enthusiastic after the previous day's encounter.

"The weather had turned bad throughout the Sea of Japan. We never stopped to think where the Task Force would be in relation to Russian airspace. The briefing didn't know where the Task Force was, either." After inflight refueling they were directed to a CAP station 100 miles from the Task Force and heard an E-2A Hawkeye report saying, "Here they come, low." He estimated thirty incoming Russian bombers below 2,000 ft. "When we leveled off and acquired the incoming aircraft there was a lone bomber leading the flight as on the previous day, and they had already formed into two waves. I told my wingman to go after the first wave, and I joined the lone Badger. We looked him over for any different configurations, antennas and transmitters, found nothing, but took close-up pictures anyway. As we expected the 'loner' went around the Task Force while the others simulated their missile delivery from 1,500 ft, 'launching' at ten to twelve miles and breaking away between four to six miles. The 'loner' fell in behind the two waves with me on his wing. After a short while he climbed to 2,000 ft IFR. I stayed with him, and within three or four minutes the entire crew of the Badger gave me the signal that they were dropping their landing gear. So did I, then the flaps, and we began to descend. Not until about 300 ft did we break out of the overcast to see we were on a very short final to landing in Vladivostock. I put on the throttle, cleaned up the aircraft and got the hell out of there!"

A quick-launch procedure for the Alert aircrews at Atsugi had evolved to suit the local air traffic conditions:

"In the Alert trailer we had a landline telephone direct to our Marine Air Control squadron, which was tied into the Japanese Air defense system. We were under the tactical control of the Japanese, and they all spoke English until something out of the ordinary happened, at which point they would revert to Japanese. If the hotline rang like a telephone you answered it. If it went off like a Klaxon you ran for the aircraft. Cockpits and switches were set up for personal preferences so you didn't have to plug in your helmet or oxygen mask, or turn anything on, and the trim was set for takeoff. The crew chief always beat us there and got the starter unit going. The RIO was first up the ladder, and when he was in the rear seat his first thing was to close his canopy, then strap in. A crew chief wouldn't pull the chocks away unless both canopies were closed. By the time I sat down the air from the external generator had the right engine turning at 35 per cent power, and I would hit the igniter, bringing the throttle 'around the horn' and noting from the tailpipe temperature that the engine was lit off. I would then strap in and switch the air to the port engine.

Meanwhile, the plane captain had taken away the boarding ladder and backed away. Shoulder harness was fastened, by which time the left engine was up to 'start' rpm, so I would hit the igniter and bring the left throttle up. While the engine reached idle I would put on the helmet and fasten the oxygen mask. As the port engine reached idle the ground crew turned off the air at the starting unit, disconnected the air hose and secured its access door. The plane captain would have pulled chocks, and the crewmen under the aircraft rolled clear as the pilot saw the crew chief's thumbs up signal and added full power, letting the aircraft accelerate at 20-25 mph. Throttles would then be snatched back as we approached the runway so that we could make the turn onto the runway heading without rolling a tire right off the rim. At the same time you glanced towards the tower, where they would be shining a green spotlight in your direction, indicating that you were cleared for take-off. Throttles then went to military power. At 80 kts the afterburners were pushed to full, the stick was pulled full aft at 135 kts and the nose came up. Total elapsed time from Klaxon to lift off was fifteen seconds either side of three minutes."

In its first four years of service the F-4B proved to be a very robust aircraft. Its muscular landing gear and 4.5 g-rated tailhook stood up to the sudden violence of deck landings, while the J79 engines were universally praised for their power and reliability. Pilots who were used to the more sluggish performance of early jets were impressed by the J79's rapid throttle response (achieved mainly because the first seven stages of stator blades were variable-angle) and very fast (1.5 seconds) afterburner light-up. However, there were inevitable mechanical problems. Engine shut-downs due to generator failure occurred. Mike Shaw analyzed the problem:

"The generators were driven by constant-speed drive units (CSDU) which used engine oil. A generator failure warning had to be assumed to have been caused by a lack of oil in a CSDU and therefore a similar lack in the engine. The real problem was that a CSDU, if run dry, might catch fire. So, if a generator would not reset, the associated engine had to be shut down. This, in my view, was a design weakness."

Jim Sherman noticed that CSDU failure tended to occur, "on take-off when you disengaged afterburner and you had an immediate generator failure. The manufacturers questioned our troubleshooting techniques, and it turned out that when they cut open one of the oil tanks the stand-pipe welded to the bottom of the tank had broken off. The system would still indicate an 'oil full' situation when there was, in fact, no oil in the tank."

In Mike Shaw's opinion the problem was symptomatic of a rush to get the much-needed Phantom into action:

"As a flying machine the F-4B showed all the signs of being put into service in a hurry. Only two years of development had been done between the first flight of the F4H-1 (1958) and the service entry of the F-4B (1960). The teething troubles were still with us in 1964. One of the worst was the auto-stabilization system. This was provided for all three axes, but was selected by one magnetically-held switch. If there was even a momentary power supply interruption the switch would trip and the aircraft would wallow about like a drunken pig—an uncomfortable feeling, particularly when the cg was still aft, just after take-off. Later, each axis was to be controlled by a separate up-and-over switch, a much better arrangement."

The effects of auto-stabilization failure at low altitude could be so severe that in one case an F-4B snap-rolled, pulling +13g and -8g. Both engines were dislocated from their mountings and the wing tanks were ripped away, though the tough Phantom managed to recover safely. "The hydraulics tended to leak a bit (It's OK sir, it's only a one-rag leak), and arrested landings after utility hydraulics failure were common. As this caused a loss of nosewheel steering, brakes and rudder inter alia, it was just as well that the hook was mechanically released, and the landing gear and flaps, which could be blown down with an emergency pneumatic system, were so good!" Despite his minor misgivings Mike Shaw stayed with the F-4, writing the Pilots' Notes at Boscombe Down for the F-4K and F-4M and eventually commanding the RAF's 228 OCU (training squadron) for over three years.

Other F-4B problem areas included the main gear tires. Marine squadrons got less than five landings out of a set and had to pre-position stocks of wheels and tires at several potential diversion bases. Utility hydraulics failures seemed to occur capriciously. Jim Sherman's wingman had five during a cross-country flight over a single weekend, while Jim himself never had one during 780 hours on the F-4B. The lack of electromagnetic shielding on the flight controls meant that F-4B "Yellow Sheets" in the early 1960s contained lists of UHF radio frequencies which could not be used by a wingman in a flight, as they could cause the flight controls of a nearby Phantom to actuate fully "up" or "down." As Colonel Sherman put it, "it was sheer terror to make a section take-off into IFR conditions if Lead switched the flight to a certain frequency.

A *Be-devilers* F-4B at NAS Corpus Christi in September 1963 with another variation on the unit markings. (Don J. Willis)

When you as the wingman acknowledged, out popped the leader's starboard spoiler to the full up position."

The warm-up times for vacuum tube radios in early F-4s constrained Alert duty take-off schedules. Jim Sherman: "Once you started the F-4B and turned on internal power it took about three minutes for the radio receiver to come on line, but it took five minutes for the transmitter to warm up." In order to hit the three-minute Alert take-off time, all this equipment was up and running in advance. Carrier-borne F-4Bs also encountered salt-water corrosion problems and galvanic corrosion where different types of metal were in close proximity, for example, in the tail section. Deployed squadrons in the far East occasionally had to have replacement aluminum panels made by Nippi Aircraft at Atsugi. F-4 maintainer Rod Preston had to manage this problem aboard ship: "It was especially common when we were at sea because of the constant low-level flying and the salt water that is always being deposited on the aircraft. Some of the worst areas were along the wingroot on top of the wing. Many times I would walk up to these areas and stick my fingers through the panels along there. This problem, in the old days before the Navy started a corrosion control program, probably killed more aircraft than the enemy." Hap Chandler's answer was simply to order all his pilots, if at all possible, to fly through rain clouds with the flaps, gear, and even the tailhook down. "I don't know how effective it was, but it didn't hurt."

F-110A

A small number of the first F-4Bs to enter Navy service also eased the Phantom's introduction to the USAF. Faced with the incontrovertible fact that it was better than any of the Air Force's own fighters, Tactical Air Command borrowed two F4H-1s (BuNos 149405 and 149406) in 1962 for a seventeen-week evaluation. They toured USAF bases, including Bentwaters in the UK (BuNo 149406) during 1962 to show the troops what their new fighter looked like. The F-4B had already made an impressive debut at the Paris Salon the previous year. Both were formally transferred to the Air Force and given new serials (149406 became JF-4B 62-12169). A further 27 F-4Bs were "bought" from the Navy order for $147.8 m on the understanding that the Navy would have them back once the USAF's own Phantom variant, the F-4C (originally F-110A under the pre-McNamara designations) began to roll off the line. There were in-

evitable changes to the design for the production F-4C, but they were comparatively minor. Structurally, the wing root of the F-4B was thickened to accept wider (11.5 inch) wheels with anti-skid brakes on the main gear in place of the 7.7 "skinny" F-4B tires. Anti-skid brakes didn't appear on Navy Phantoms until F-4J BuNo 157242 and up. Ground attack capability was enhanced by the AJB-7 bombing system, and cartridge-starting J79-15 engines were used. A control column appeared in the back cockpit, as both crewmen were regarded as pilots. The in-flight refueling system was converted to the standard Air Force flying boom system.

Under the command of record-breaking test pilot Colonel Pete "Speedy" Everest, the "borrowed" F4H-1s equipped the 4453rd Combat Crew Training Wing at McDill AFB, Florida, from 4 February 1962, training crews for the 12th TFW until "real" F-4Cs began to arrive in November. RIOs, known initially as Pilot Systems Operators (PSOs) in the Air Force and later as Weapons Systems Operators (WSOs) found the removable control column in the rear cockpit an uncertain advantage. Unlike the F-4C's stick it had to be unplugged and stowed before the radar controls could be slid out for use. Interestingly, "front seaters" were required to do time in the back to study the radar interception task. This twin-stick approach enabled the backseater to take control of the aircraft, but with no access to controls for the landing gear or brakes. The F-4C also had rudimentary throttle controls, but the backseater still could not land the aircraft alone.

The Air Force had some early adaptation problems with the F-4B's Martin-Baker H5 ejection seats, which were more complex than anything they were used to. Two fatalities and a serious injury to McDill personnel resulted from failure to appreciate that the "banana-link" mechanism located on top of the seat could initiate ejection if moved or compressed with the seat armed. In one case an F-4B's seat which had been inadequately secured slid up the rail during a negative g maneuver, fired the canopy jettison device, departed the cockpit, and then slid back along the fuselage. Fortunately, the pilot's parachute also deployed and he survived.

Like their Navy and Marines counterparts, the McDill F-4 crews were on Alert during the Cuban crisis, though their effectiveness in action would have been limited by the fact that they had not begun weapons training at the time. However, Phantom crews from all three services were soon to find plenty of action, much further away in South East Asia.

3

While Thunder Rolled

As the Summer of 1964 and the darkening situation in S. E. Asia wore on, the U.S. Navy's Phantom squadrons approached their first real trial. There were around 320 Phantoms in service with Navy and Marine units. The earliest deployed squadrons, VF-21, VF-41, VF-96, VF-102, and VF-114 had been joined by VF-14 *Tophatters*, VF-31 *Tomcatters*, VF-92 *Silverkings*, VF-142 *Ghostriders*, and VF-143 *Pukin' Dogs* in 1963, mostly ex-F3H and F4D units. They were followed in 1964 by two former F-8 squadrons, VF-33 *Tarsiers* and VF-84 *Jolly Rogers*, and several of the remaining Demon-flying units including VF-213 *Black Lions*, VF-151 *Vigilantes*, and VF-161 *Chargers*. More Crusader squadrons switched to the F-4B in 1965 (VF-154 *Black Knights* and VF-32 *Swordsmen*), a process which continued until VF-191 and VF-194 converted from F-8J Crusaders in 1976. USMC F-4B squadrons in place by the end of 1964 included VMFA-115 *Silver Eagles*, VMFA-251 *Thunderbolts*, VMFA-323 *Death Rattlers*, in addition to VMFA-314 *Black Knights*, VMFA-513 *Flying Nightmares*, and VMFA-531 *Gray Ghosts*, with more to follow.

In all, USN F-4 squadrons were to make eighty-four war cruises with Task Force 77 off the Vietnam coast. In 1964 VF-142 and VF-143 were making their second cruise on USS *Constellation*, while VF-92 and VF-96 had teamed on USS *Ranger*. In fact, the original VF-142 had been re-designated VF-96 on 1 June 1962 and deployed alongside VF-92 in Carrier Air Wing (CVW-9). An old Naval tradition kept initial digits of squadrons the same as those of their Air Group: a new VF-142 re-formed as part of CVW-14 with VF-143. VF-96 adopted black and gold tiger-stripe tail-cap and wing-tip markings, which supposedly aided formation flying, and headed for Yokosuka, where their CO, CDR William Mulholland, and RIO LT Robert Kelley had the distinction of making the first in-port cat launch for an F-4B. With the Laotian crisis of May 1962 brewing, VF-96 had joined Crusader-flying VF-91 in conducting air operations off Vietnam, but with no hostile outcome.

VF-96's first war cruise on USS *Ranger* lasted from 5 August 1964 to 6 May 1965, commencing operations on the night of 2 December 1964 in support of Operation *Yankee Team* over Laos. (USN via Peter B. Mersky)

The second cruise for VF-96 (from April 1966 on USS *Enterprise* under CDR Robert D. Norman) gave the *Tiger Stripes* plenty of opportunity to develop ground attack techniques. Rocket pods proved particularly successful. This F-4B was passed to VMFA-542 and lost to AAA over Laos in 1970. (US Navy)

VF-21 flew F-4Bs from USS *Midway* in 1965 and USS *Coral Sea* in 1966-67 before making five further war cruises on USS *Ranger*. This aircraft was lost over North Vietnam to unknown causes in September 1966. Its crew, LCDR J. R. Bauder and LTJG J. B. Mills, were reported MIA. (via J. T. Thompson)

Devoid of Modex numbers or crew names and showing evidence of some hard flying and erosion control, this F-4B nears the end of a long war cruise in October 1965. (Don J. Willis)

Crusading Zeal

Many of the Phantom's early cruises were as paired squadrons with F-8 units, which were regarded as "day fighter" (or in the F-8 pilots' estimation, "fighter") units, compared to the F-4's "all-weather interceptor" role. The rivalry between the two communities was intensified as the F-4 began to outnumber the Crusader and F-8 drivers started to move over to the Phantom. Jerry B. "Devil" Houston, who flew Crusaders with VF-11 before making the transition, summed up the situation:

"The F-8 stole your heart from the get-go with its beauty and, for its time, power (God, an afterburner!), but it quickly earned an ensign-killer reputation, rightly or wrongly. In the long run that reputation contributed greatly to fighter pilot development in the U.S. Navy: only the top ten per cent of pilot graduates were even considered for the F-8 pipeline—the créme de la créme. And as

luck would have it a damn solid base of mid-level stick-and-throttle talent groomed the hungry youngsters into frothing-at-the-mouth tacticians. Without experiencing it, anyone would be hard-pressed to understand the aura that surrounded that early F-8 Crusader community. An ensign in F-8s took no crap from a Lieutenant Commander who flew anything else. Period. Among that small, but growing privileged group, anyone in the first few squadrons knew damn near exactly where they stood in the overall tactics ladder. Killing capability meant everything: King of the Mountain, in spades. Gunnery, tactics, gunnery, tactics, just enough intercept training to get you into the sky with another victim. F-8s didn't bomb then—hell, the plane didn't even have hardpoints on the wings. In other words, everything funneled the best pilots and the best airplanes through a narrow training spout, and out popped the world's best fighter pilots. They all ate, breathed, thought, and dreamed about fighting airplanes. All the time.

F-4B BuNo 150485 of VF-101 runs some tests on deck. (Boeing/McDonnell Douglas)

The F-4, on the other hand, was an ugly, two-seated, gas-hog monster that hit the Fleet with a bunch of F3H Demon interceptor pilots. Consequently, the F-4s got off to a horrid start when running into F-8s. Their reputation was dog-s**t, and the ex-Demon drivers didn't have a clue about changing it. So they practiced intercepts, avoided tactics, and groomed follow-on clods in their own image. All the while that lovely top ten per cent advantage kept infusing the Crusader community with superior talent, which rapidly molded into long-clawed killers. The rich got richer—and even richer. Better people, better training, frightful competition. Crusader pilots didn't have to brag; they owned the sky. So it remained, until finally an F-8 jock got into an F-4, screaming and dragging his heels all the way. The rest is history."

The influx of F-8 pilots and tactics as their squadrons transitioned to the Phantom certainly accelerated the development of F-4 ACM training. As late as 1971, when Jerry Houston's VF-51 moved from the F-8J to the F-4B, he thought of the unit as a "double-barreled Crusader squadron with an extra set of eyeballs in each plane." However, the awareness of the need for ACM began much earlier than that, and certainly pre-dated the initiatives of the later 1960s which gave rise to the Top Gun ACM training project. Fred Staudenmayer:

"As the F-4A and F-4B replaced the older Skyrays and Demons at VF-101 Key West it was acknowledged that air-to-air ACM should comprise a larger portion of tactical training. Although I was at Oceana in 1962 when the first Phantoms arrived I had to wait until the third East Coast Phantom squadron, VF-41, for my Fleet assignment. Newly transitioned squadrons adopted VF-101 Det A's training syllabus at the outset, and soon added low-level navigation (with or without radar mapping), as well as integrated Air Wing flying when embarked. Permanent aircrew assignments were the goal in all the squadrons, with the idea of putting the inexperienced RIOs with the experienced pilots and vice versa. Learning the skills to get Phantoms aboard ship smartly and safely came first, of course, but RIOs controlled the interception of aircraft as well as navigation and, as experience was gained, picked up a large share of the communications. During those early days any ACM was strictly on an 'ad hoc' basis: one saw (or detected on radar) a bogie and 'jumped' it, with no pre-briefing other than an occasional remark to another aircrew as you were manning up on the carrier deck.

When I arrived at Key West in June 1965 assigned as an Instructor for replacement aircrews we were just beginning to devote a considerable part of the syllabus to ACM. At about the same time, air-to-ground tactics were also being refined as limited S. E. Asia experience began to filter back. Before too long the air-to-ground stuff was delegated to VF-101 Det Alpha at NAS Oceana where the original F-4 transition had taken place. Replacement pilots from the Advanced Training Command and new RIOs were led through a syllabus consisting of familiarization, instruments, formation, landing practice, then air-to-air intercept work (often at night), and finally six to eight ACM flights, with the final two-versus-one, or two-versus-two ACM exercise using the squadron TF9F Cougars as the simulated MiGs.

After the graduation 'hop' aircrews went to Oceana for air-to-ground and field carrier landing practice (FCLP). Development of ACM started with a section of Phantoms climbing out to around 35,000 ft, separating by 50 to 60 miles, turning in and using radar to make an intercept, then closing to visual, at which point the 'fight was on' and each crew attempted to maneuver to gain nose-to-tail advantage and work for the six o'clock position. The supersonic Phantom maneuvered poorly at altitude with not much instantaneous g available, and as speed bled off below 300 kts planes were forced to dive and 'unload' to regain sufficient energy for high-g turns.

As tactics became refined altitudes were lowered, the vertical plane was brought into play, weapons tactics were refined, and all

Deck crew scrutinize the arrival of a VF-92 F-4B on USS *Enterprise* at the end of their 1966 combat cruise, which cost six of the Air Wing's F-4Bs. (Boeing/ McDonnell Douglas)

Silver Kite **201 blasts the USS** *Enterprise's* **deck with hot tongues of afterburner as its cat shot is initiated. (Boeing/McDonnell Douglas)**

manner of deception was employed. The heart of the program was the creativity and ability of the fleet-experienced Lieutenants who instructed, briefed, and built a syllabus for inexperienced aircrew. The best among these were LTs Ken Mase, Ken Toby, Don Atkins, and W. M. 'Killer' McGuigan (our first MiG killer: destroyed a MiG-17 on 13 July 1966). These guys were all superb stick-and-throttle pilots and tacticians who would brief and debrief new and transitioning aircrews, lead the flights, instruct the new RIOs in the back seat, and get aircrews ready for flying ACM in the Fleet after only fifty or so hours on the type. In 1965-66 we were still experimenting with how to fight the 'Spook' and use its great energy potential to counter the much greater turn rate and radius of the MiGs. We soon learned to keep our energy level very high, trade speed for altitude, and fight in the vertical. These tactics led to some hairy maneuvers. I was once with a student who pitched the F-4 vertical until we reached zero airspeed. A bit later I observed white vapor passing by our wing. We were backing down through 30,000 ft and passing the contrails we had left going up, still at zero airspeed!

Unfortunately, this was the period of time when the Navy lost huge numbers of exceptional aviators to civilian life at the very peak of their effectiveness and leadership abilities. All four of these gents left active duty. We staff back-seaters would teach intercept techniques to new and transitioning pilots and fly with them on the training hops. 'Mac' McGuire, Ed Jackson, and I were among those who tried to maximize ACM time. Early in the introduction of ACM at VF-101 in 1965 we would try hard turns while sub-sonic and at altitude. Even with full burner, speed would bleed quickly, the wing would get into buffet as lift deteriorated, and we would get into wing rock. During one of these introductory flights the aircraft I was in stalled, departed controlled flight, and entered a spin from which we could not recover. We both ejected and were picked up by a shrimp boat heading out into the Gulf of Mexico.

With 'nugget' pilots these hops could get very exciting, as they were learning to fly the Phantom to its limits (and sometimes beyond). As tactics improved and we learned the lessons of energy

maneuverability, using the vertical and keeping our speed up, the Phantom was able to give a good account of itself in visual encounters. Once it got supersonic it would really turn. A typical maneuver was to unload the wings in an abrupt pushover while keeping the bogie in sight; quickly go supersonic and pick up the kinetic energy to turn back into the fray. As the aircraft nose got within 60 degrees of the adversary, RIOs would turn from visual to the radar scope to attempt a lock-on for a Fox 1 simulated Sparrow kill. This was extremely difficult to accomplish in the few seconds available, but a few of us got pretty good at it."

Some of the tactics for the East Coast ACM syllabus were passed informally from VF-121, the Miramar (West Coast) RAG. On the East Coast there was a greater emphasis on Fleet Air Defense, as VF-101-trained aircrews were usually destined for "blue water" cruises in the Mediterranean where air interception tactics

Switchbox **104 (BuNo 152226) on USS** *Coral Sea* **in August 1965 configured for CAP. The carrier had recently completed five months on station and received the Admiral Flatley Award for Aviation Safety. There were no F-4 losses among the twenty-one combat casualties. This F-4B ended its days at Chu Lai with VMFA- 314. (Don J. Willis)**

VF-96 F4H-1s still in *Tiger Stripe* markings during their 1963 cruise on USS *Ranger*. The nearer aircraft survived long enough to be converted into a QF-4B in the early 1970s. (USN via Norman Taylor)

could regularly be used against Bears and Badgers. The emphasis on radar-controlled intercepts carried over into ACM, also:

"We used radar extensively in ACM, and the RIO's eyes were in and out of the cockpit continuously until the battle closed to a vertical scissors, which it usually did. The maintainers were under great pressure to get us fully radar system-capable aircraft and, unlike the West Coast, we always attempted to use only 'up' radar aircraft for ACM hops."

Awareness of the need to use the F-4 as a fighter in the traditional sense as well as an interceptor emerged slightly earlier in the West Coast RAG, as Miramar-based squadrons were the ones which were scheduled for Westpac cruises in the volatile South East Asian scenario. CAPT Ken "Bullet" Baldry recalled the situation at VF-121:

"In late 1963 as the fight in Vietnam heated up, people in VF-121 became concerned over the fact that it was not going to be a war where you launched missiles at targets over the horizon on radar. You were going to have to go in and see who you were shooting at! This was called VID (or visual identification), and it meant that the people you were 'ID-ing' would also identify you and could engage at close range. This was a concept that the Phantom community (and the F3H Demon community before them) found very hard to swallow, as they had always had the Sparrow; a fine head-on weapon against a non-maneuvering target. The F-8 Crusader community had never had a weapon that could be fired head-on and had always preached the idea of a close-in fight with guns or Sidewinders. VF-121 had at this time a few forward-looking pilots, among them a British exchange pilot [who was evaluating the F-4 ahead of the British order for the F-4K/M]. With our CO's concurrence, Pete Carroll and I were assigned to a project to see what the F-4B was really capable of and how best to fight the aircraft. The British exchange instructor was called Geoff, and had a fine black beard and a 'Let's get it on' attitude. We launched in two flights:

one F-4B and one F-8D Crusader and went at it from every direction. We tried flaps down (both half and full flap for the F-4B) and variable incidence wing both up and down for the F-8D. I noticed that Geoff was in afterburner a lot, and at about forty minutes into the fight he called 'Bingo' [low fuel] and we returned to Miramar. Geoff led into the break, and as he landed with me close behind on the other side of the runway I saw a dark object fall out of the tailcone of his aircraft onto the runway and he called 'No chute.' In the debrief we found that Geoff had been in afterburner so long, with so much g on the airplane that he had literally cooked the drag chute. We also learned that the leading and trailing edge flaps on his F-4 had been severely scorched by the hot BLC air. Scratch the idea of fighting the airplane with the flaps down, except in extremis! They caused more drag than anything, and caused a lot of havoc with the BLC operation because of the high power settings.

We came to the conclusion that the F-4 really did best at low altitude and needed a lot of speed to be at its best. As I later learned, it had an amazing ability to accelerate when unloaded to zero g and turned its best in the 450-500 kts range. At lower speeds being close aboard an adversary in the F-4 was no good deal because there was no gun, and throwing rocks was not really an option. The F-4 was also an extreme gas hog, particularly at low altitude in afterburner. We guessed the fuel flow at sea level in afterburner at about 80,000 lbs per hour. The low fuel light came on at about 2,500 lbs, and it

Still secured to the deck padeyes, a pair of *Aardvarks* F-4Bs awaits a strike mission in the Summer of 1966 aboard USS *Kitty Hawk*. (Boeing/ McDonnell Douglas)

really meant it! Geoff had called 'Bingo' about 50 miles from Miramar and landed with approximately 500 lbs of fuel, enough for one pass or pull-up and eject. We had set 'bingo' for the F-4 at 4,000 lbs, a very realistic figure, and I later learned that Geoff had actually called 'bingo' at 2,500 lbs when the low-fuel light came on. The F-8, on the other hand, was happy at a 1,500 lbs bingo and could make an instrument approach to Miramar, miss the approach, fly to El Centro (about 80 nm) and land VFR with 500 lbs of fuel. There were several other engagements with other instructors, and that really was the 'germ genesis' of Top Gun. As a result of flights like these the Phantom community began to devise tactics to gain separation from an opponent, and also conceded that the AIM-7 Sparrow was of very limited use in a close-in fight, as it took too much concentration to get the thing in the interlocks and the RIO had a real problem trying to lock up an opponent in a vertical maneuvering situation.

Most of the conclusions we learned came to fruition some years later, but it took a big infusion of F-8 Crusader talent into the Phantom community before they really got serious about getting the airplane out to the outer edges of the envelope." VF-96 was one of the first squadrons to give priority to ACM, and its commander, from July 1966, CDR 'Lefty' Schwarz, had a big part in this change of philosophy. Ken Baldry flew with him. "Lefty changed the face of Phantom flying to a very large degree. When he took command he and I were the only two F-8 experienced pilots in VF-96. He immediately said, 'We are taking off the tanks and MERs and TERs and we are going to find out what the bird will do.' Since I had been a two- and four-plane instructor in VF-124 I had considerable input into the tactics syllabus with VF-96, even riding in the back seat on numerous occasions to help out folks that were being introduced to a whole new world of fighting airplanes. Most of the pilots took to the idea like a duck to water, and I cannot remember ever having more fun in an airplane than taking a clean F-4 all the way out to the edge of the envelope and sometimes a bit beyond. Fortunately, the F-4 is a totally honest machine and will literally beat you to

USS *Coral Sea's* flight deck as she neared Hawaii en route to join USS *Ranger* and USS *Hancock* off Vietnam for *Flaming Dart 1*, the opening strikes of the war, on 7 February 1965. The pristine rows of VF-151 F-4Bs, VA-153 and VA-155 Skyhawks, and VA-165 A-1s, plus a VAW-11 "Willy Fudd," would soon look worn by combat. (Don J. Willis)

A quartet of *Switchbox* F-4Bs aboard USS *Coral Sea* in April 1965. Aircraft 107 (as 62-12186) was one of twenty-nine F-4Bs loaned to the USAF for F-4C crew training at McDill AFB. (Don J. Willis)

death in heavy buffet before it stalls, and when it does stall it 'falls through' fairly predictably, so we had a lot of departures and recoveries, but no stall-spin events.

The F-8, on the other hand, was a very unpredictable critter and very sensitive to large inputs of aileron at slow speeds. When it did depart you had better be ready for a heck of a ride! The F-4 fought best at low altitudes (2,500 ft and below) and needed to have energy in the 400-500 kts range, but it did have amazing acceleration, and you could command 30 degrees of rudder if you had a big enough foot, so it was possible to stay in a very tight rolling scissors. We had to devise tactics for separation since we had no gun, and to be at six o'clock at 600 ft to an opponent didn't do much for you. Towards the end of our tactics phase we had many engagements with the local F-8 talent at Miramar, and I am glad to say we got a lot of bragging rights at 'Happy Hour,' much to the chagrin of the F-8 drivers. This was all in 1966, prior to Top Gun, but I like to think we were certainly in on the beginning of Top Gun.

It was also discovered that the best way to turn the airplane hard was to keep deep into the buffet, and since the airplane was very honest about stalling it wasn't hard to learn. At that time the F-4B still had a feel system in pitch [elevation] called 'downsprings,' which were heavy-duty springs, meant to keep the pilot from overstressing the airframe, particularly in the transonic area where the g available literally doubles as the speed drops from Mach 1.0 to Mach 0.9. Lefty Schwarz, after his tour with the *Blue Angels,* was convinced that this was a nefarious plot by McDonnell-Douglas to keep fighter pilots from using the airplane properly. VF-96 got permission to remove the downsprings to conduct tactics evaluation and later got permission to remove them altogether. This merely meant that the F-4 control forces in pitch were quite heavy as compared to horrendously heavy! When you had finished a high-g engagement in an F-4 you were physically tired. To this day my right forearm is still visibly larger than my left, even though I haven't flown the F-4 in over twenty years!"

Chapter 3: While Thunder Rolled

Laden with "slicks" a VF-21 F-4B heads off for its target. (USN via Angelo Romano)

There was still much to learn about the F-4's idiosyncrasies. Flying with VF-96 on Thanksgiving Day, 1964 William Greer and his wingman ended a short pre-dinner "hop" with some ACM to use up fuel. "Upon return to the ship I found I was unable to extend my tailhook or retract it once I had tried to extend it. We were sent to Cubi Point to inspect the problem. As we stood near the aircraft after landing the tailhook extended autonomously and worked normally thereafter. We returned to USS *Ranger* to find that dinner was completed, and the four of us had a sumptuous meal of bologna sandwiches. Further investigation revealed a tendency of the tailhook to bind when heated by the afterburner."

The best West Coast ACM tactics in the mid-1960s, in Fred Staudenmayer's opinion, originated from VF-21, "the first squadron to plan and execute a MiG attack early in the war." Among the VF-21 personnel preparing for the conflict in Vietnam was LT David Batson, pilot of one of the F-4Bs that took part in that "planned attack." His background is an indication of the brief time which elapsed between training and combat for the earliest crews of the Navy's complex new fighter. Having completed the normal jet training program between June 1961 and September 1962, he was assigned to the West Coast RAG, making his first F-4B flight at Miramar on 21 December, the day before his twenty-fourth birthday. On his sixth flight he was teamed with RIO LCDR Rob Doremus and continued his training alongside other "nugget" and ex-Demon pilots. There were many night intercept flights under ground radar control as part of the essentially "all-weather interceptor" program, and dog-fighting was "rare, unstructured, and unauthorized." The two men were assigned to VF-21 *Freelancers* in March 1963 soon after it became the fourth West Coast F-4B unit, and prepared for the squadron's first deployment, on USS *Midway* in November. Their training regime still consisted of "mostly intercepts and carrier landing practice," and this continued after the first deployment. "We did some visual bombing practice—both 30 degree dive and low-level, but no ACM."

Re-embarked on the *Midway* in March 1965, VF-21 was paired with an F-8D squadron, VF-111 *Sundowners*, as the carrier prepared for its first "on line" period off Vietnam from 10 April 1965. The situation had already turned hot after North Vietnamese patrol boats appeared to threaten the USS *Maddox* on 2 August 1964. U.S. Navy Phantoms flew their first missions over the Laotian mainland as escorts for *Yankee Team* RF-8A reconnaissance flights. USS *Ranger's* two squadrons, VF-92 and VF-96, began similar escort flights for RVAH-5's newly-acquired RA-5C Vigilantes on 2 December 1964, with permission to fire on any AAA sites which might attempt to interfere. VF-96 flew escort for the A-1H Skyraiders of VA-95 as the Navy began Operation *Barrel Roll*, limited armed reconnaissance flights over the already-established Ho Chi Minh Trail in Eastern Laos, on 17 December. Further Viet Cong attacks on U.S. personnel in South Vietnam triggered Operation *Flaming Dart*. A weather abort caused the first mission to be canceled on 7 February, but four days later VF-96 dropped its first bombs of the war when the CO, CDR Bill Frazer, and LTJG Chris Billingsley flew a flak suppression sortie against the Chanh Hoa barracks near

VF-84's CAG Phantom awaits its turn on the cat. (Boeing/McDonnell Douglas)

53

Dong Hoi. On Frazer's wing was LTJG Terence Murphy, with LT Giles "Red" Phillips "in back."

Meeting MiGs

As CVW-9's attacks continued the USS *Midway* was still heading for the area. As David Batson recalled:

"En route in March 1965 we had many situation briefings and an introduction to the Rules of Engagement. Basically, we could not attack a MiG unless he showed hostile intentions. The CAG, CDR Bob Moore, told us that any MiGs we saw were hostile—shoot them down. My first mission was on 10 April 1965 to look for Terry Murphy and Ron Fegan following their engagement with MiGs from Hainan Island."

The loss of Murphy and Fegan's F-4B (BuNo 151403) on 9 April remains something of a mystery, and aspects of the engagement remain classified at the time of writing. Three VF-96 crews were assigned Barrier Combat Air Patrol (BARCAP) duty, involving a racetrack pattern over the sea about 25 miles off the coast near Haiphong. The purpose was to prevent hostile aircraft from attempting to attack elements of the 7th Fleet, and these missions took a great deal of the fighter squadrons' flying time throughout the war. Pairs of aircraft cruised the "racetrack" at about 300 kts and 20,000 ft, flying the legs of the circuit in around three minutes, but often slowing down and nudging a little towards the coast on the inshore side of the circuit in the hope of drawing out a MiG or two. The monotony was somewhat alleviated by a tanker refueling after about an hour, and the crew would then hope that the relief pair of Phantoms showed up quickly. Sometimes they didn't, and another refueling was required. On the return journey to the carrier it was customary to burn off some of the 2,000 lbs combat fuel reserve in a little one-on-one ACM; probably the only opportunity for this tactic to be rehearsed.

Leading the 9 April BARCAP was CDR Frazer, with LT Don Watkins as wingman. The third F-4B sustained an engine failure on launch, and LCDR Greer and his RIO LTJG Bruning had to eject. William Greer explained the problem:

"Standard USN practice was to use full flaps for normal F-4 carrier operations. Curiously, when launching a heavy aircraft in near-minimum wind conditions half-flaps were to be used. The resident fly in the ointment was the inability of a single engine to sustain level flight when full flaps were selected. The air required by both leading and trailing edge blown flaps, when drawn from a single compressor, so reduced airflow through that engine as to reduce thrust below level flight requirements, even with afterburner. On the launch in question no unusual noise was heard either in the cockpit or on the flight deck, though later inspection of the TV tape showed a large fireball from the right tailpipe about two-thirds of the way through the catapult stroke. Engine failure at that point was the assessment. As I left the deck I rotated to a flying attitude, retracted the landing gear, and only then felt that something was amiss. By the time I realized I was decelerating I was too slow to safely reduce flap to half and shortly reached the classic conundrum of running out of airspeed, altitude, and ideas simultaneously. Lacking command ejection in that model of the F-4 [F-4B BuNo 151425] I called for my RIO to eject and followed as soon as he had responded. The television tape shows his ejection at some ten feet of altitude; mine seems to have been on the far side of the rooster tail of the engine exhaust and/or splash of impact. I recall looking face down at the water, followed by a hefty jolt as the parachute opened and snapped my feet forward. I hit the water on my back before I could enter a vertical descent. All worked as advertised, and my sole blemish was a cut on the chin where my pistol apparently struck me."

Terry Murphy and his RIO Ensign Fegan launched in *Showtime 611* as replacement section leader, with standby crew LT Watkins and LTJG Mueller on their wing. Heading north for Hainan Island Watkins tried to catch up with Murphy, but heard him transmit "Three in contrails" as Chinese Communist MiG-17s were sighted. Watkins' aircraft then came under attack from astern by a fourth MiG-17, and Murphy reported that he was being fired upon. Bill Frazer's section was called back from their BARCAP and engaged the MiGs. Two AIM-7s were fired; one went ballistic, and the other

Switchbox 101, loaded with Mk 82s, is tensioned for a cat shot. Don Willis, a bombardier/navigator with VAH-2 aboard USS *Coral Sea*, recalled that the launch was, "on or about 24 August 1965 when we followed four Phantoms in a small valley by a river and dropped bombs on smoke markers from a L-19 FAC." (Don J. Willis)

With Mk 82 LDGP bombs uploaded, F-4B BuNo 152263 awaits a September 1966 strike mission with a battle-worn VF-213 example. During this 122 day cruise at the peak of *Rolling Thunder*, USS *Kitty Hawk's* CVW-11 sustained twenty-five aircraft losses to all causes. (R. Harrison via Norman Taylor)

Another *Black Lions* F-4B prior to the 1966 deployment in which VF-213 flew both F-4B and F-4G data link aircraft. In-flight refueling probes were extended to de-pressurize the fuel system. (Clyde Gerdes via Norman Taylor)

one's motor failed to ignite. Two AIM-9 Sidewinders were launched at closer quarters, but one was evaded and the second failed to make contact. Low on fuel and keen to find a tanker the three Phantom crews realized that Murphy and Fegan were missing. Their last transmission had been, "Out of missiles. Returning to base." However, Watkins and Mueller had seen a MiG-17 explode and crash shortly after shaking off their assailant. As *Showtime 611* turned away from Hainan a second radar blip was seen behind it on the Fleet's CIC radar, and the F-4 then vanished off the plot.

Although the Chinese claimed the destruction of the Phantom, it was not considered politically prudent to counterclaim a MiG-17 and Murphy's kill remains unconfirmed. Other Communist Chinese reports stated that the F-4B was hit by a missile from one of the other Phantoms. China's "no fly" zone extended thirty miles south of their border, and other USN aircraft fell to their fighters,

including an A-6A on 21 August 1967. Though some of the aspects of this first engagement with the MiG-17, and the first F-4 combat loss, remain unclear it immediately high-lighted problems which were to persist throughout the Vietnam War. Primarily, missile reliability was suspect. Of the four which are known to have been fired, none found its target. Also, if (as the Chinese claimed) Murphy and Fegan did attempt to turn with the MiGs at fairly low speed and lost when their F-4 ran out of energy in the turn, they were perhaps the first of many to realize that this maneuver was potentially fatal in an F-4. Communications were also a perennial problem. The two elements involved in the clash, Frazer's and Murphy's, were operating on different frequencies and under different radar controllers. Frazer's section was authorized to engage the MiGs only after they had shown hostile intent and dropped their wing-tanks. With only one main radio per F-4 the mutual effectiveness of the two sections

VF-41 F4H-1s in October 1962 demonstrating the F-4B's short-range load carrying capacity, with twenty-four 500 lbs bombs each. (USN via CDR James Carlton)

A "clean" VF-84 F-4B awaits the signal to launch. The pilot was instructed to inform his RIO of the control stick position prior to his "go" salute to the cat officer. The manual stated, "holding the stick in any position that will not place the stabilator in a leading-edge down angle during a carqual weight launch will impart a nose-down pitch to the aircraft off the bow, from which it may be impossible to recover." (USN via CDR James Carlton)

may well have been compromised. The only consolation was that it was VF-96's only wartime loss to enemy aircraft, if that was indeed the cause of *Showtime 611's* demise. The U.S. Government's unwillingness to make an open military commitment also tended to obscure issues such as these opening moves in the air war. CDR William Greer:

"That 1964-65 deployment was indeed an experience. Being on the scene in the early stages of the war was a good deal safer than later with the massive air defense build-up. Flights over Laos were required to avoid North Vietnam, and all strikes were required to be recorded as 'miscellaneous, non-combat, non-training, not otherwise classified.' Mr. McNamara did not approve of flights that could be called 'combat,' even when aircraft were being shot down."

David Batson's attempt to find any trace of *Showtime 611* proved fruitless, and its occupants were both declared killed in action (KIA) shortly afterwards. His own MiG encounter several weeks later was a very different affair. The mission was a *Rolling Thunder* strike force of aircraft from USS *Midway* and USS *Bon Homme Richard*. As David explained:

"About half the missions were *Rolling Thunder* strikes. Briefing was in the Air Intelligence Center as usual about two hours before launch. It was a large strike on a military target south of Than Hoa. The U.S. Secretary of the Navy, Paul Nietze, was present. He came to our ready room for our individual squadron brief. Lou Page told him how we would shoot down MiGs."

CDR Lou Page's confidence was to be well-placed. As VF-21 XO he led the flight with LT John C. Smith as his RIO. Batson and Doremus were on his wing with two other sections of F-4Bs. The *"Bonnie Dick's"* F-8E Crusaders were given the flak suppression role, while the six Phantoms shepherded a gaggle of A-4C Skyhawks which were the main strikers. As TARCAP aircraft (keeping a watch for MiGs over the A-4s' target area), the F-4Bs carried a pair of AIM-9s on the left wing pylon, two AIM-7s in the fuselage wells, and a third on the right wing. A 600 gallon tank, which was virtually essential on all operations, was slung on the centerline. The fuel load had to be pegged at about 4,000 lbs less than full to keep

Foot races on "Ropeyard Sunday," a traditional relaxation day designated by the *Coral Sea's* CO, Captain George L. Cassell, as the ship steamed for home in October 1965. F-4B BuNo 152243 (later a QF-4N) served as a grandstand. (Don J. Willis)

An *Aardvark* F-4B BuNo 152241 traps on USS *Kitty Hawk*. (USN via Angelo Romano)

gross weight below the 47,000 lbs for launch. Usually the centerline tank was emptied before the force went "feet dry" (over land) in case it had to be dropped. David Batson described the rest of the mission:

"We launched, made a running rendezvous with the KA-3 tankers about 75 miles from the ship, and proceeded to our search area in the vicinity of Ninh Dinh, northwest of Than Hoa. We began patrolling in a north-south pattern in basic search formation; one mile abeam each other so that we could provide each other with protection from someone sneaking up on our tail. Using our powerful radar to look northward for targets we listened to the strike group arrive on target, carry out their mission, and depart. As the strike group called 'feet wet' (over sea), Lou called for, 'One more sweep north.' Up to then the flight was completely normal, except there was a significant amount of AAA, mostly from Ninh Dinh. As we rolled out of the turn, J. C. Smith spotted two radar targets about 45 miles north. Rob Doremus spotted them almost immediately, also. We had observed a slight pattern of MiGs appearing late in our missions when our fuel was getting low. I think we were all suspicious at this point. Lou called for me to move from the search position to the attack position; a three-mile trail and slightly below. We accelerated to approximately 500 kts for better maneuverability. The first plane (Lou and J. C.) was to set up a head-on attack, having made a positive visual ID. Rob and I would maneuver for a head-on Sparrow shot. 'J. C.' and Rob talked to each other regarding which radar targets to lock on to, and 'J. C.' took the farther target, creating a slight offset to the head-on attack. This caused the MiGs to make a turn into the lead plane. When they banked, the very distinctive wing plan of the MiG-17 was visible."

J. C. Smith was from the first cadre of Phantom RIOs, having spent a relatively unsatisfying period as a bomber pilot with VA-126. Trained in the vacuum tube secrets of the Aero-1A system at James Connelly AFB alongside eighteen others, he absorbed the knowledge of Air Force F-89 and F-94 back-seaters and joined the other trainees on a converted WWII B-25 bomber equipped with an intercept radar system. He also spent time at St. Louis with MAC technicians studying the system, which helped him to devise parts

VF-84 spent a hundred days on the line in 1965, and their CAG F-4B already shows an impressive mission scoreboard only eight days into the first line period on 7 July 1965. (USN via CDR James Carlton)

of the NFO syllabus for VF-121 in time for the arrival of their first F4H-1s. He rapidly established a reputation as an exceptional "scope" and tactician. On the 17 June mission he and Rob Doremus had acquired the MiGs at around 32 miles, head-on and 2,000 ft lower. Clearly Dave Batson's suspicions were confirmed. The MiGs had been hoping to tail-chase the departing strike force and shoot the last few in the back.

Within Sparrow range and sure of their target (which turned out to be four MiGs in V-formation rather than two), Lou Page led the attack:

"Lou fired at close to minimum range while shouting, 'It's MiGs!' I saw his missile fire, guide towards the formation, and the warhead detonate. At first I thought it had missed, but then the outer half of the right wing came completely off the MiG and it started rolling out of control. I then put my full attention to the steering information on my radar scope, and I fired at minimum range. Just before firing, the steering dot moved up the screen, causing me to go into a slight climb to keep it centered. The missile (AIM-7) fired from the rail on the right wing and swerved under the nose of the airplane. I lost sight of the missile, but Rob saw it guide to a direct hit. Also, about this time one of the MiGs flew right by me. I think my missile hit either the Number 1 or Number 2 aircraft. My theory is that the first section did a wingover reversal to get behind Lou and 'J. C.' That caused my steering dot to climb as I chased them. The second section (at this point, only one plane) continued straight by me. I don't think he ever saw me. The next phase of Lou's tactics was for us to disengage quickly. He was very concerned about trying to turn with a MiG-17 or MiG-19, our most likely foe. We disengaged by lighting afterburner, climbing up through an overcast and rendezvousing. Then we reversed heading, re-established the 'hunter' (search) formation, and went back through the clouds looking for the rest of the MiGs. We saw smoke trails from our missiles, but no MiGs and one parachute."

There was speculation that a third MiG, probably the wingman of the second to be hit, also went down, having ingested debris from his leader's explosion. Shortly after the engagement the two *Freelancer* crews heard that only one of the four MiGs had returned

to base. David Batson had to wait until 1997 for confirmation and declassification of this information when his former CAG, CDR Bob Moore, was able to back up the theory. However, one kill apiece was glory enough at the time, as the two crews made for home:

"By this time we were below bingo fuel, so we headed back to *Midway*. Along the way we were offered refueling, but we declined because if something went wrong we might not make it. We had just enough fuel to land: I was actually showing 400 lbs (about three minutes) at the top of the glide slope. After landing, I taxied by our CO, CDR Bill Franke, who was jumping up and down with his hands over his head. After shut down, Rob came up from the back seat, shook my hand, and said, 'Four more to go!' Lou was escorted up to the Flag bridge where he was congratulated by Secretary Nietze. To get to the ready room we had to pass through the F-8 (VF-111) ready room, where they were yelling and cheering. Our own ready room was packed. Someone handed me a coffee cup. It was full of scotch! Lou quieted everybody down and gave a brief review of what happened. Then we were taken to Air Intelligence (I still had my 'coffee' cup), where we told the Admiral and others what happened. It was an amazing experience. We were told to get some clothes and get ready to go to Saigon, where we were to participate in the daily Press briefing; the Five O'Clock Follies."

Such media treatment, including a stay at General Westmoreland's private house, demonstrated the importance of the Navy's first official MiG kills in lifting morale. The old tradition of fighter pilots as "knights of the air" still held good in an age of radar and missiles. All four received the Silver Star after a brief hiatus when "J. C." was offered a DFC, a lower award, as he was "only an RIO." It would be a few more years before Navy philosophy would abandon the idea of the RIO as a "technician assisting the pilot." By June 1972 Fred Staudenmayer could become the first RIO to command an East Coast F-4 squadron, VF-33.

Ironically, among the seventeen combat losses to Air Wing 2 on that cruise was a sole F-4B (BuNo 152215), crewed by CDR Bill Franke and Rob Doremus. Both men spent seven and a half years as POWs after a SAM hit during an attack on the Than Hoa bridge on 24 August. Lou Page, a "quiet, competent leader who led by example," in David Batson's estimation, took over the *Freelancers*.

Mk 82 SE bombs fill the TERs on a VF-154 F-4B from USS *Coral Sea* in 1966. The ship was commanded by Captain Frank W. Ault, whose investigations into the F-4's armament were to have profound consequences. (Terry Edwards)

Spook Strike

"After completion of the transition we picked up approximately eight other pilots, all second tour and experienced in the F-4, plus a USAF exchange pilot, Captain Tom Rush. I believe it was shortly after 1 January 1965 that we found out that we were scheduled for a Westpac cruise in June. All that time the squadron began dropping some practice bombs and making preparations to fly the aircraft in the attack mode. I was scheduled for release from the Navy in Summer 1965. The squadron didn't want to waste training flights on a future civilian, so I did not get any bombing practice. About mid-April 1965 the Navy postponed all scheduled releases for a year. I had just enough time to go to Survival School and get my carrier landings brought up to date before we sailed aboard the USS *Independence*. As we arrived in the Philippines we picked up a junior pilot, ENS Ralph Gaither, right out of flight training. Of the fifteen pilots all were second-tour (i.e. over four years flying experience), apart from Ralph and three ex-Crusader pilots. By any comparison we were a very experienced squadron. VF-84 had won the Atlantic 'E for Excellence' in its last year with the F-8C and first year with the F-4B.

Meeting MiGs was very much the exception for many of the squadrons deployed to Task Force 77 in the Gulf of Tonkin. VF-84 Jolly Rogers transitioned from the F-8C, received F-4Bs in September 1964, and joined CVW-7 aboard USS Independence for her only war cruise. The A-6A Intruder made its combat debut with VA-75 on that 1965 cruise, supported by VF-41 and the MiG-hungry, ex-Crusader drivers of VF-84. LT Grover G. Erickson was among the F-8 veterans:

"There were only five F-8 pilots who made the transition; myself, Bob Johnson, Ray Herzog, Stan Olmstead, and our CO Jack Waits." (Bill Knutson also flew the skull and crossbones-adorned F-4Bs of VF-84 and remembered Jack Waits as, "very good at leading us, at relaying what combat would be like and preparing us mentally, as well as with the physical skills to succeed. He was one of the few remaining aviators from WWII and Korea junior enough to command an F-4 squadron.")

We started flying in-country missions over South Vietnam around 10 July 1965, and after about five days moved up to Yankee Station and started operations over North Vietnam. About a third of the missions were textbook F-4 CAP or BARCAP. Throughout the six months we operated the North Vietnamese kept their MiGs at home. Another third were leading strikes as flak suppressers. We dropped some 1,000 lb bombs on the flak sites to disrupt their ability to really zero in on the A-4s and A-6As coming in behind us with bigger payloads. That was a mission I never really anticipated as a fighter pilot, and it certainly wasn't the best environment for the Phantom. As scary as it was at first, we never did lose any planes during the actual strike. We did lose two planes en route to the target when a SAM alert caused the strike group to go in at an altitude where the flak could get them."

Although none of CVW-7's thirteen combat losses were to SAMs, its A-4E squadrons were responsible, on 17 October, for the first strike against a SAM site, near Kep airfield. The AAA was murderous, however, and five of the Air Group's F-4Bs fell to it and to small-arms fire over Laos. Among the losses were ENS Gaither and his RIO in F-4B BuNo 151494 as POWs, and LCDR Stan Olmstead (F-4B BuNo 151515) killed in action. Grover Erickson himself fell victim to the flak: "The remaining missions were smaller strikes and recce flights looking for targets of opportunity. That is the type of mission I was on when we were hit. My plane was part of a midnight three-plane strike at some torpedo boats operating out of Bac Long Island." After his successful ejection and recovery LT Erickson wrote a letter to the Weapons Systems Evaluation Group at Arlington, Virginia, describing the sequence of events leading to the loss of F-4B BuNo 151505, 208 AG *Victory 208* of VF-84. It read:

"I was launched from the USS *Independence* about 0100 on 26 October 1966. Weapons loading was as follows: two AIM-7E (station 6 and 7), centerline tank (station 4), one TER 4 with two LAU-3A rocket pods (each station 1 and 9), PMBR with six Mk 24A flares (station 2) and two AIM-9C Sidewinders (station 8). I expended the six flares over the target on the first pass from 6,000 ft.

This *Jolly Rogers* F-4B blasts off with a warload of Mk 83s, AIM-7s, and AIM-9s. (Boeing/McDonnell Douglas)

One of the *Black Knights* F-4Gs trapping on USS *Kitty Hawk*. The 300th Phantom to leave the St. Louis line, BuNo 150642 reverted to F-4B configuration in October 1966 and flew with VF-121 and VF-171. It was rebuilt as an F-4N, flying with VF-51 (1974), VF-201 (1976), VF-202 (1983), and VF-171 in 1984 before being struck off on 22 February 1984. (Boeing/McDonnell Douglas)

On the firing pass I fired all four rocket pods in the ripple mode from about 2,500 ft in a 25-20 degree dive at 450 KTAS. Just after firing I broke hard right to clear the area. After 20-30 degrees of turn I felt the aircraft receive a couple of hits. At first I thought they were near misses, as I had experienced similar sensations before with no damage.

The following events took place over a period of 2-3 minutes that we remained with the aircraft. I heard and felt muffled explosions from the aircraft. I lost complete control of the rudder: full throw of the pedals was available with no reaction. I had the distinct impression that the landing gear extended, indicating down and locked, then retracted. At any rate, after this impression the starboard gear did indicate 'barber pole' [an instrument state showing landing gear extending or retracting]. The aircraft was becoming increasingly harder to control. At this time the fire-warning light on the starboard engine illuminated and I secured the engine. The fuel gauge became very erratic. My RIO reported numerous circuit breakers popping. The cockpit then started to fill with smoke, and we had a complete electrical failure. As the cockpit became dark I noticed flames coming from the port intake. I pushed the battery-powered 'eject' light, heard my RIO eject, and ejected myself.

"After the aircraft was hit we flew about fifteen miles, climbing to about 19,000 ft. We were losing altitude and ejected as the aircraft descended through 15,000 ft. My wingman and his RIO said they saw the aircraft descending in flames. I believe that the aircraft took two hits, probably one in the tail section and one in the wing-root. I consider us lucky to have been able to remain with the aircraft that long and still be able to get out safely. As for the ejection sequence and survival gear, everything worked to complete satisfaction."

The rescue of Grover Erickson and his RIO, LTJG John Perry, recorded in a series of SITREP flash messages between the USS

Independence and the Rescue Center at Da Nang, made their pick-up sound almost routine:

SITREP 1 37 MILES FROM SCENE AT 260225H, F-4B CALLSIGNS VICTORY 200 AND 218. THREE LIGHTS SEEN IN WATER. DO NOT KNOW WHETHER JUNKS OR PILOT.

SITREP 2&3 R. K. TURNER, UH-2A HELO ON SAR SCENE 260245H. PICKED UP ONE SOUL AT 0325H. PROCEEDING TO RECOVER SECOND MAN.

SITREP 6 260510H Victory 208 RIO LTJG JOHN. H. PERRY PICKED UP BY H>B>WILSON. PHYSICAL CHECK BY DOCTOR INDICATES CONDITION GOOD. SAR FORCES RETURNING TO STATION. INCIDENT CLOSED.

The messages ended with one from VF-84's skipper to Cubi Point the same day requesting a replacement for the lost F-4B.

Extracts from a diary which Grover kept during the early part of the cruise show how the F-4B units were heavily committed to strike (and BARCAP) missions of this kind, and the unremitting nature of daily combat flying:

JULY 1st 1965. Flew Strike mission. Dropped 2 x 500lbs bombs on hamlet with VC concentration. John Perry saw light AAA.

JULY 2nd. Flew strike. Dropped 2 x 500 lbs bombs against VC concentration in jungle vicinity 49 m east of Saigon.

JULY 4th Flew BARCAP for 10-12 plane Strike against Sam Dinh POL storage 30m S of Hanoi. Heavy AAA at 12-15,000 ft over target. Good results. No enemy aircraft. One A-4E with small flak hole in wing slat.

Mon JULY 5th. BARCAP for road recce. RA-5C returned with 1 inch flak hole in wing.

JULY 6th BARCAP for road recce. CAPped 10 miles of coast S. of Hanoi. No opposition. Small flak directed at recce. One A-4 came back with numerous holes.

JULY 7th 1003-1254 H. Road recce with CO. Dropped four Mk 84 and four Zuni at small dam.

JULY 8th 1600-1730 H. Escorted ECM aircraft over North VN.

JULY 9th 2200-2400 BARCAP during strike ops over North.

JULY 10th Photo-escort mission over SVN, Laos, and NVN.

JULY 11th 1400-1630 BARCAP south of Hanoi during strike on Than Hoa bridge.

JULY 12th LT Don Boecker/Don Eaton (VA-75) ejected from A-6A over NVN when bomb prematurely exploded. Spent night in jungle hiding from search parties. Air America picked them up following day.

JULY 13th CDR Denton/BN LT Tschudy (VA-75) shot down over Than Hoa bridge. Believe both ejected and captured [in fact another premature ordnance detonation].

In country mission—dropped 4 x Mk81 on VC concentration 15 m W of Nha Trang.

JULY 15th Escorted ECM plane over NVN.

JULY 17th Escorted photo recon over NVN.

JULY 18th 0001 hrs. Launched with 12 x F-4B and four A-6A for strike on Than Hoa bridge. (During launch VF-41 F-4B rolled into catwalk, vicinity of arresting gear). Target was obscured. Half of F-4Bs dropped in formation with A-6A ("systems" bombing), rest milled around, couldn't find target, dropped at sea. Meanwhile, back at the ranch Tilley [mobile deck crane] got F-4B out of catwalk and stalled in landing area, fouling deck. All F-4Bs diverted to Da Nang, landed there at 0230, some with about 800 lbs fuel. Refueled at Da Nang, departed 0530, landed aboard at 0600. Took off 1230 orbited MCAS Chu Lai. Escorted SecDef R. McNamara to CVA 62 [USS *Independence*] and landed back aboard 1400.

JULY 20th 1145H took off with 2 x 1000 lbs bombs for flak suppression on a strike to Bon Xom Long barracks area. Two F-4B (XO and self) rolled in on target first, followed by eight or ten A-4Es. Was supposed to drop on AAA site—got steep and fast—dropped long. A-4Es rolled in at same time. Two got hit (Bill Wheat) and started losing fuel. Were able to meet tanker and take on fuel all the way back to the ship. I could not drop bomb from port wing. Had to divert to Da Nang where Marines were more than happy to get a bomb. Back aboard ship 1615H.1800H RA-5C from RVAH-1 [BuNo 151619] with CO CDR Matula and LT Gronquist got No 1 cross-deck pendant on recovery. Wire broke after 100 ft roll-out. Plane went over angle deck with insufficient flying speed. No survivors.

JULY 21st Road recce Hanoi-Nam Dinh. Dropped 2 x Mk 81 on four barges in river NW of Than Hoa.

JULY 23 0030-0200 Flew Force CAP. Two bolters. Blew two tires, hard landings. Fatigue—got to get more sleep.

JULY 24th Flew aircraft to Cubi Point for corrosion control.

CVW-7's Ordnance Load Plans show that in Summer/Fall 1965 strike sorties VF-41 and VF-84 F-4Bs normally carried eight Mk 82 bombs with M139(I) fuses or four LAU-3 rocket pods, each with nineteen 2.75 inch rockets and two LAU-10 pods each with four 5-inch Zuni rockets. Despite the general absence of MiGs the Ops Officer published a daily MiG Alert key-word. On 11 November, for example, it was "Blue Boy," which would have got the heads turning in the cockpits and the RIOs peering into their scopes.

By the beginning of 1966 ten Navy F-4B squadrons had made combat cruises with Task Force 77 and another nine Fleet squadrons had transitioned to the Phantom. In all cases the emphasis on the F-4's attack capability in a war whose success was measured in bomb tonnage and sortie rates meant some rapid changes in the training routines. The Navy had three specialist attack aircraft available. Its Korean-vintage A-1 Skyraider was ideal for the "in-country" campaign in South Vietnam, but the murderous defenses in the North made its use there too costly by 1967. The excellent A-4 Skyhawk was the principal Navy attacker during *Rolling Thunder*, but it carried half the bomb-load of an F-4 and lacked the afterburner-generated speed to escape from trouble. USS *Bon Homme Richard* lost ten (virtually a whole squadron) on its 1967 cruise, though its F-8 squadrons did score nine MiG-17 kills in response. The heavy-weight, expensive A-6A Intruder was best at solo night and all-weather radar bombing, but it too was flown in daylight Alpha formation attacks. Air Wings were tasked with generating the largest possible attack packages to match the devastation created by the Air Force's large F-105D and F-4C formations. Smaller strike packages of about twenty aircraft could be launched several times during a twelve hour period (followed by a twelve hour standdown). More often, Phantoms operated in pairs against targets in the south. Usually these were directed against smaller targets in the "lower" Route Packs (RP II, III and IV). An Alpha involved the entire Air Wings from up to three carriers in a concerted assault on targets in the Route Package VI area; Hanoi and Haiphong. Unlike the Air Force's practice, USN tactics called for high-altitude ingress to the target and dive attacks.

Inevitably the Phantom was included in these battle formations in a number of roles. As a strike bomber or flak suppresser it could carry a substantial ordnance load and retain at least a pair of Sparrows for "fighter" duties. Other F-4s provided BARCAP, or TARCAP (target combat air patrol), in which they would bomb with the strike but carried more missiles to cover the retreat of the attackers. As MIGCAP, one or two elements of F-4s with a full complement of missiles patrolled off the coast near Haiphong or flew a lateral transit across the Route Pack VIb area in which most Navy targets were located, awaiting the call to chase MiGs off the strikers.

As strike aircraft F-4Bs (usually six aircraft) launched first, because external ordnance increased their already high fuel consumption, and headed for a tanker. They then approached the target

On return from its 1966 cruise F-4G BuNo 150642 had VF-121 *Pacemakers* markings added to its salt-scarred green paint. (Author's collection)

area at around 300 kts and 15-18,000 ft. Nearing the "feet dry" point, speed would be pushed up to 400 kts at 11,000 ft, depending on the cloud ceiling. Navigation duties fell usually to the strike leader and was done with old-fashioned charts and timing for the most part. The Phantoms attacked first, either as flak suppressers or against the main target to make best use of their superior speed. Given good visibility over the target, bombs were usually released in a 45 to 60 degree dive at about 5,000 ft, and the strikers then regrouped quickly to egress the area and head for the carrier.

As a bomber, the F-4 had the inbuilt lifting capacity which MAC had left in the structure from its AH-1 origins, but its bombing system was optimized for tactical nuclear delivery rather than conventional attack. When it became clear that the F-4B was going to war as a strike fighter there was some urgent work at squadron level to improve the aircraft's ordnance delivery. Fred Staudenmayer described how VF-41 *Black Aces* dealt with the problem:

"In 1964-65 VF-41, who had a long turn-around period between deployments, started working to prepare bombing tactics for our upcoming deployment to S. E. Asia. First came bomb testing by the NATC boys at Patuxent River and multiple ejection racks (MERs) and triple ejection racks (TERs) to adapt to the Phantom's weapons stations. For the pilots, an existing bomb sight was adapted off the shelf to allow 'mil' adjustments for lead angle (not lead computing) for dive bombing. Our skipper, Bob Gormley, was very keen on getting the fighters into the air-to-mud business and was interested in using our APQ-72 air-to-air radar for level bomb drops using time, speed, and distance dead reckoning to drop on mark. For instance, our radar had a ten mile scope and we would 'paint' targets, put them on the nose, and fly at 360 kts. At six miles distance we would start the clock and pickle at a fixed number of seconds. Bob even had our radar technicians, the AQs, devise a horizontal line across the scope at exactly six miles to give us a gauge to start our timing.

The radar was outstanding for ingress navigation and was OK for ship targets, and even junk-sized floaters if the seas were not too extreme. The tactics were devised with coastal defenses, surface combatants, and other targets of opportunity. We trained at this extensively, but of course it was terribly crude with no computer; just 'Kentucky windage.' The radar was terrific for navigation and coast-in to the target area, and I personally felt confident to lead in other Air Wing strikes when flying from the sea. We dropped a lot of practice bombs and adjusted the delivery procedure empirically, then we dropped a few Mk 80 series bombs. Our preference was for TERs on the outboard wing station so we could keep missiles on the inboard wing stations and the ever-present centerline tank. I think VF-41 was the only squadron that pursued this level-bombing tactic. After all, we were fighter jocks! Once the squadrons got into Vietnam, however, almost all bombing was dive bombing, during which RIOs were indispensable for calling off altitudes and speeds so that pilots could adjust their runs. We did more level runs in Vietnam with the Marines calling the drops using TPQ gear [see Chapter 4]."

CAPT John Nash was also involved in the groundwork for introducing bombing tactics to F-4 interceptor crews:

"We didn't have an air-to-ground syllabus in the RAG [in 1964]. Between November 1964 and our 1965 cruise [with VF-213 on USS *Kitty Hawk*] the Navy developed an air-to-ground syllabus for the F-4, because it was obvious we were going to do some of that. Initially, we borrowed MERs and TERs from the USAF at Hill AFB, and we had about two weeks to get ready for the cruise. Our CO and XO in VF-213 had been A-1 'Spad' drivers in Korea, had flown F9Fs, FJ4s, etc., and had done bombing. The syllabus was largely self-taught. We basically took a USAF 'Dash 1' [handbook and technical manual], pulled out some 'mil' settings [for the bombsight], started doing some low-level navigation, which you couldn't do in Vietnam because of the jungle cover and AAA, and started practicing low-angle (10 degrees and 30 degrees) bomb deliveries. It turned out that the low level deliveries were almost impossible over there. We delivered napalm and CBU, but the big problem was that when you were at real low altitude in level delivery you couldn't see your target until it was under the nose because of the jungle cover. Delivery of napalm in this way was almost impossible down South because the hootches and other targets were usually in small clearings, and you couldn't see those. We went to a minimum 10 degrees dive delivery for CBUs and napalm. I used a 30 degree dive with a 3,000 ft release point. Everyone had their own techniques. On my first cruise I wasn't very good: when you get AAA in your windscreen your technique starts to degenerate. I extended my tour rather than doing shore duty, as I had qualified as a section and division leader. We had a great CO, Jim Wilson, and I said that if he let me lead flights I would do another cruise. They were happy to get experienced guys back. I had one wingman who was inexperienced, and he lay awake at night asking me how to do this and that.

When I went back for the second VF-213 cruise I was much improved. I could hit targets and concentrate better. In the old 'iron sight' days everything had to be perfect—airspeed, dive angle, and g at release. Later, when I was flying F-4Js with VF-142 I flew an A-7E Corsair II and I had a better CEP on my first mission than I could probably have ever achieved in the Phantom with several thousand hours on it and a lot of bombing." As for munitions, "We

Don Willis noted in his diary on 16 May 1965; "CDR Lee, VAP-61, escorted by an F-4 (LTJG Marv Johnson and ENS Tex McTigue in Switchbox 107, BuNo 149457) was forced to conduct his photo recce of Highway 1 at low level today due to bad weather. Both got hit by small-arms and flak. Talked to Tex—he thinks 37mm is what got their stabilizer." Don's aircraft refueled the mission. This F-4B later became a MiG killer in the hands of LT Winston Copeland and LT Don Bouchoux in June 1972. (Don J. Willis)

Seconds before a successful trap, an F-4B returns from a CAP mission in February 1967. (Frank A. Curcio)

had CBU early on and a reasonable amount of napalm. After fires on the USS *Enterprise* [14 January 1969] and USS *Forrestal* [29 July 1967] they took napalm off the ships. In 1967 during the bomb shortage we were delivering 125 lbs high-fragmentation, high-drag WWII vintage bombs, or the 250 lbs version. Even in 1967 we delivered a lot of 750 lbs high-drag bombs; we could carry ten of those on an F-4. We flew escort for the VA-113 A-4C Skyhawks early on, but they were just too slow. They couldn't make much more than 350 kt with a bomb load aboard, and that load was less than we could get on one wing of a Phantom. They were real experienced air-to-ground guys, but the problem was that they just couldn't survive in that environment." (RADM "Smoke" Wilson remembered the A-4Cs, "at full bore on a hot day, with full ordnance, climbing all the way to the target to get enough altitude for a comfortable roll-in").

Flattening Flak
"The most compatible airplane we flew with was the A-6A. We could carry a load of 2.75 inch rockets as the flak suppressers with an A-6A strike group and bomb with VT fuses. We could fly right with those guys. The only limit was the frangible heads on the 2.75 inch rocket pods. If we got over 500 kts they would blow off, and then you really had a drag problem as you had the flat face of the pod sitting there. For flak suppression we would fly with the A-6As and just dash ahead, pop up, and roll in. The suppression was only good for about a minute. I've rolled in, and on the pull-out from the dive I've seen the guys running out of their little tin tub [shelter] and manning their guns again. They were pretty gutsy.

With the A-4s we would run maybe two or three miles behind them and then catch them when they were trying their pop-up. They would pull the nose up and float through. Frequently they could use a negative g push through to maintain the airspeed and wind up in a 45 degree dive. We would have to accelerate from behind them, pop up to 10-12,000 ft, roll in and deliver our flak suppression weapons while they were doing this. That was really hard to do, as the A-4Cs slowed from 350 kts to maybe 200 kts at the top of their pop-

up. In fact, they had a half-flap delivery mode at one time which was good for stability but not good for survivable speed in a combat situation. On big strikes we took all the tanks off the A-4Cs, loaded them with three 1,000 lbs or 2,0000 lbs bombs, and then had a 'ready deck' for them, as they had about forty minutes of flying time."

John Nash, who became one of the initial cadre of Top Gun instructors, also commented on the other aspect of the A-4 in its later A-4E/F and A-4M versions: "It was the best VFR day fighter I've ever flown. I've never seen an airplane that can beat one in a good dogfight. It has a tremendous thrust-to-weight ratio. Even an F/A-18 has a hard time against the A-4E." As it passed out of service as a bomber the Skyhawk's performance as a MiG simulator made it a vital instrument in ACM training of Phantom crews.

Ken Baldry suggested that Alpha strikes were regarded with mixed feelings:

"These were the most likely places you might find MiG action, which all fighter pilots were salivating about, and of course they were in the most heavily defended areas, which caused all of us a certain amount of heartburn. Mostly, they were made up of a number of A-4 primary bombers, and some other Iron Hand A-4s armed with big AGM-12C Bullpup B guided bombs or AGM-45 Shrike anti-radiation missiles to attack SAM sites. They had a large degree of autonomy from the strike group. Then there were the F-4 flak suppressers and F-4 MiGCAP or TARCAP. The flak suppressers were usually armed with twelve Mk 82 500 lbs bombs with electric fuses to cause air bursts over the offending flak sites. Most of us hated the electric fuses, as they were notorious for not going off, or for premature detonation, so most of us set our switches for a contact burst and long ripple mode of release (giving maximum distances between releases), figuring that covering the maximum amount of territory with ground bursts would be at least as effective as keeping their heads down. We all conjectured that the loss of Major Russ Goodman [USAF exchange pilot KIA in F-4B BuNo 150413, *Showtime 614*, on 20 February 1967 with RIO ENS Gary Thornton, POW, during strike on Thin Lingh Dong rail complex] was very likely due to bomb-to-bomb collision or influence. There was no way to prove it, but the flak in the target area was not too severe and Russ was one cagey guy.

The briefing for an Alpha would be about two hours ahead of launch and would cover all the usual stuff about join-up, frequencies, how much fuel the F-4 TARCAP would take on (in case of MiG engagements), the ingress route, the flak sites to hit, when the flak suppressers would accelerate ahead of the strike group, known SAM sites, general flak plots, egress routes, and 'squawks.' Also, we needed to know who else was on our radio frequency, the MiG code calls of the day, and the myriad of small details that could save your butt. We in the F-4s were not very crazy about flying these strikes with the A-4s, as they were slow with bombloads on and we had to weave to keep our speed up. It was a relief to push it up to gain speed and altitude. Later in the cruise we configured the strike groups with all A-6A/F-4B strikers and A-4 Iron Hand support, and this was a much better evolution, as the A-6 could carry bombs at 500 kts with ease and it carried an enormous load.

The more memorable strikes were around Hanoi (the Van Dien vehicle storage area) and Kep airfield, where all the fighter jocks were praying for MiG encounters. I and my normal 'wingie,' Steve Amman, were assigned TARCAP for that one, to cover two strike groups. Steve was the first Ensign in the U.S. Navy ever to be sent to the Fleet as qualified on the F-4. He joined us during the ORI in Hawaii and became a fine F-4 pilot and a very tough wingman. Ingress for the Van Dien mission was south of Thanh Hoa and then up the mountains until we got out over the plains south of Hanoi. This particular strike was all A-6 and F-4 with A-4 Iron Hand, so we were moving along at a great rate of warp (500 kts or better) at about 10,000 ft. As we came out over the plains the SAM alerts started going off, and it looked like the whole formation was a school of fish just rolling and going for the deck. Our assigned CAP station was along the Red River, and we just kept going lower and faster, pulling many g and keeping a rough and eccentric pattern to the south of Hanoi. I remember seeing several classic 'star' pattern SAM sites, most of which seemed unoccupied [missiles and launch equipment were often moved from site to site overnight to frustrate Iron Hand attacks on them]. The main big stuff and SAMs being hosed off were more up in the target area, but if you stuck your nose very far above 1,500 ft AGL you would certainly get all the SAM alert tones you wanted. I remember my RIO, Jim Stillinger, was literally standing in the back seat at 7g, yelling 'Pull harder, pull harder!' and I obliged. Steve Amman in the meantime was literally hanging on for dear life, and about all I could see of him was the occasional look at his nose when the g load lightened momentarily.

The strike leader called, 'Off target,' and we all headed for the mountains to the south. There was a hell of a lot of smoke, flak, and SAM trails still coming out of the target area, and we were still jinking hard at low altitude until we got over the relative safety of the mountains. We followed the strike group out to 'feet wet,' then our assignment was to cover the egress of another strike group of A-4s and F-8s which had gone in behind us. I got tanked OK but had to pull Steve off the tanker, as an A-4 with battle damage was streaming fuel from his wing and the A-3 tanker was going to have

to wet-nurse him back to the ship. The rest was sort of anti-climactic, as we just bored holes over the mountains 'til the other strike group egressed. We were under the control of an E-2A Hawkeye and just orbiting over a huge hydro-electric dam that the North Vietnamese were so certain we would never hit that they didn't bother to place any defenses anywhere near it. I always dreamed of hitting that dam, like the English squadron the Dam Busters in WWII.

The strike on Kep, another A-6/F-4 evolution, was later in the cruise, and the powers that be were willing to use the A-6s in daytime. We were getting some more interesting targets to hit besides bamboo foot bridges and going down to the DMZ for *Milky* Control to put us on a heading, altitude, and true airspeed to salvo our bombload on suspected troop concentrations north of the DMZ. (*Milky* loved F-4s because we had a true airspeed gauge and the A-4s did not). The Kep strike group was eight A-6s (an amazing number to have up at one time), and since it was day VFR they didn't require full systems. There were sixteen F-4Bs from VF-96 and VF-92 armed with electrically fused bombs. We coasted in over a low mountain range to the north of Haiphong and were doing much warp. As we closed on Kep one SAM alert was noted, and immediately one was fired from the area of Kep. As the SAM climbed you could see a smoke trail from a Shrike missile almost collide with the SAM and then impact with the ground. The missile promptly went stupid, climbed straight up, and detonated far above the strike group. No more SAMs were fired during that particular strike, and we were pretty sure the Shrike had hit the missile's control radar. There was still plenty of flak of all calibers in the target area, but since we were a very fast-moving strike group it didn't appear that much of it was particularly well-aimed: mostly barrage fire. It was heavy enough that I lost sight of the F-4 ahead of me and only regained sight of him as we went 'feet wet.' I let my bombs go at a string of revetments at the east end of the runway, but didn't have time to look back and see what the effect was as we were engaged in jinking to avoid the flak and heading out over the water as fast as we could go. The A-6 was fast enough that one of them was keeping up with me all the way to the coast.

The BDA photos showed cratering of the runway and several MiGs blown apart in their revetments, but the North Vietnamese had tons of trucks covering the holes in no time, and since the MiG was designed for rough field ops it was really no big deal. The only airborne MiG was spotted by the VF-92 XO, 'Skenk' Remsen, who had a good Sidewinder tone and was ready to fire when he was hit in the leg by a flak burst. Fortunately the wound was not major, but it spoiled his aim and the MiG got away. Skenk was really pissed! He also said the flak batteries seemed to be shooting at the MiG as

A *Tophatters* F-4B (BuNo 152327) with the Project *Shoehorn* ECM update. (Author's collection).

much as they did at him. There weren't many MiG sightings during the 1966-67 cruise."

Data Link

John Nash's first squadron assignment was to VF-213 in October 1963 on USS *Hancock*, where they were still flying F3H Demons off the coast of South Vietnam. On return to the USA in December the *Black Lions* began to transition to the F-4B. They then received a dozen F-4Bs, modified on the production line to F-4G standard. The first of these (BuNo 150481) had flown on 20 March 1963 after development work by McDonnell using a number of earlier aircraft. F-4A BuNo 146820 began testing data-link as early as March 1960, and this work was continued in 1961 by F-4As 146821 and 148252-4. The data-link principle originated in the USAF, where interceptors could be guided to their targets without the need for a direct radio link between the fighter and its ground controller. Air Defense Command's interceptors from the early 1950s onwards operated within the semi-automatic ground environment (SAGE) with considerable success. The Navy requested a system using similar ideas, but providing a direct information network linking its ships and aircraft. Initially the F-4 Phantom and E-2 Hawkeye AEW/airborne control aircraft were seen as ideal linked components in that the interception process could be facilitated by linking the Phantom's autopilot to a two-way UHF data-link between the fighter, its carrier, and/or the E-2. Information on the F-4's position, fuel, and armament status, plus information on targets could also be relayed directly back to a controller. Data for the pilot's attention appeared in both F-4 cockpits. The pilot had an "acknowledge" button to indicate that information had been received. One penalty was that housing the system's black boxes behind the RIO's cockpit required a 600 lbs fuel reduction in the capacity of the Number 1 fuel cell. An added bonus in Naval use was that the ship's AN/ASP-10 radar and AN/USC-2 data-link system could be used to control a "hands off" automatic deck landing. A small, fold-out radar reflector was housed ahead of the nosewheel door, enabling the

USS *Franklin D. Roosevelt's* catapult officer prepares to launch a VF-14 jet on 6 September 1966 during the fourth week of the carrier's only war cruise. AB 105 has five strike missions chalked up and many more to follow. (PHC Neal Crowe/U.S. Navy)

ship's radar to trace the F-4 accurately enough for a successful "trap." The complete AN/ASW-13 data-link kit, plus a bolt-on radar reflector was installed in F-4A BuNo 148254. In production F-4Gs this was replaced by the RCA AN/ASW-21, which connected to both cockpits and supplied information on the F-4's fuel, oxygen, and weapons condition. Also, an automatic approach power compensator system (APCS), or "auto-throttle" was included to make rapid engine power corrections in response to the F-4's angle of attack during the final stages of a deck landing. (This was also aboard the F-8 Crusader and improved the F-8's safety record).

VF-96 *Fighting Falcons* (as they were later known) were the first owners of the F-4Gs in July 1963. Ten aircraft were received, the other two (BuNos 150489 and 150625) remaining with NATC at Patuxent River for automatic carrier landing system (ACLS) development. In fact, several problems were experienced with the early versions of ACLS: on occasion the aircraft was thrown into a dive when auto-landing was switched in, requiring the pilot to over-ride the autopilot rather fast. "Smoke" Wilson was Project Pilot on the ACLS and flew 150489 (the other aircraft went to VX-4 for missile trials):

"That test work was the classic 'hours of boredom with moments of stark terror.' When the ACLS worked the Phantom would make an absolutely perfect pass hands off, but 'hard over' signals at 200 ft while approaching the blunt end of the carrier was something else! At the Carrier Suitability Branch we had F-4A BuNo 145311, the record breaker, and BuNo 148368. This F-4B had been in the carrier suitability program for years. Finally, it was sent to the rework facility, having experienced more arrested landings than any other F-4B in the inventory. While in Yokosuka, Japan, with VF-14 I observed my sister squadron VF-32 take custody of a replacement F-4B: BuNo 148368. I went to their Maintenance Officer and suggested he should personally check the log-books for the airplane's history. He nearly fell off the chair when he saw the number of arrestments on the records. As I recall, the plane only flew three or four sorties before becoming a 'hangar queen.'"

VF-96, preparing for another war cruise, reverted to stock F-4Bs and passed the F-4Gs back to VF-121, from whence they moved to the *Black Lions* in January 1964. The decision to designate the "data-link dozen" as F-4Gs was taken by BuWeps in March 1964, and they briefly belonged to a "phantom" Phantom squadron, VF-116. In keeping with the tradition of associating squadron and CVW numbers, the move of VF-213 to CVW-11 led to a belief that they would become VF-116. The new squadron number had been painted on the aircraft before the Chief of Naval Operations (CNO) rescinded the change on 16 September 1964. VF-213 got out the paint pots again and continued their pre-deployment training ahead of a Westpac voyage on USS *Kitty Hawk* in October 1965.

John Nash flew the F-4G and believed it was simply too advanced for its time:

"The problem was that it was pretty much a hard-wired, all analog system. The E-2 Hawkeyes were hard-wired to the point where they had all kinds of bogus information in their systems which, for example, assumed that the F-4 could go from Mach 0.9 to Mach 2.0 in 30 seconds, pull 12 g, etc. When you did intercepts with the E-2's control you would as frequently roll out *in front of* the bogies

VF-14's AB 101 before the addition of the red fuselage flash. Its assigned RIO, LT Gene Blair, later flew as "Smoke" Wilson's back-seater. (R. Besecker via Norman Taylor)

as behind them. Since we were a Fleet squadron and getting ready to do air-to-ground in Vietnam nobody was really interested in data-link. The important part was that the AN/ASW-21—the 'mother board' of the data-link, went on, was changed, and became the backbone of the data-link in the F-14 Tomcat.

Of course, one nice thing was the auto-throttles. The F-4G was the only Phantom with these. I used these religiously; there's nothing better than to be able to put both hands on the stick and have your airspeed stay the same. Some guys wouldn't use them. They'd say that if the auto-throttles quit and you had to use manual throttles you'd be out of practice. Well, normally when you flew auto-throttles you'd have your hand on the throttles anyway, and you'd be sensing the very fine power corrections the system was making. [Some described this process as, 'like milking a mouse'].

I didn't use the data-link during the Westpac cruise. The biggest problem was that the airplane was much more capable than any of the Navy systems. There were some data-link equipped destroyers and cruisers, and you could couple up, select autopilot, and they could drive you around a little bit, but we seldom did that because one part of the system was typically down. In those days data-link was ineffective; a good idea, but well before its time. They put the AN/ASW-25A one-way data-link in the F-4J, but for hard-wired systems you need solid state digital systems where you can change things and then they become more reliable. I don't think they ever came back with the two-way data-link."

The only externally visible difference between the F-4G and F-4B was the temporary tactical paint scheme applied to the aircraft in a 1966 experiment to reduce the visibility of Fleet aircraft in their usual Light Gull Gray and Gloss White schemes. A sample of all the aircraft types in CVW-11, together with CVW-9's Skyhawks and CVW-15's Intruders, Skyhawks, and Vigilantes received a variety of green or green-and-tan/gray schemes on their upper surfaces, with USAF-style 15 inch insignia and only BuNos and Modex numbers applied. VF-213's F-4Gs were given a coat of FS 34102 Olive Drab on their upper surfaces. Capt. Nash felt that the benefits were dubious:

"I'm not sure it did any good, as we weren't in a real air war over there. The paint was only effective when you were looking down on an airplane over the jungle. Ironically, we had some of our airplanes with white radomes, and we were instructed not to paint

the radomes, so you'd have the camouflaged airplane and you'd see a white dart shape, and then you'd pick out the rest of the airplane. It was a good deal for me, as I was sent to Clark AFB in the Philippines for two or three weeks where the USAF painted the airplanes with an overspray of rubberized-type, removable paint. It was different and exciting at the time, but I don't think it did much good. Visibility of the aircraft on deck in night operations was not a particular problem [for the deck crews], as the deck was kept totally dark in those days and you couldn't see anything anyway. They could have painted them black and it wouldn't have made any difference."

Blacked-out decks, ordered because the threat of attack from the coast seemed likely at the time, made night operations even more demanding, particularly in the recovery phase. A different attitude soon prevailed:

"In 1967 the *Kitty Hawk* adopted a 'moonlight' deck lighting scheme so that you could see when you were manning the airplanes. Landing back aboard ship you could make out the flight deck features from about a quarter-mile away, which pretty well cured vertigo. In the early days it was like flying into a barrel. You could just about see the running lights and the mirror and that was it. You could get vertigo so badly you couldn't stand it, feeling like you were 300 ft high as you were crossing the ramp because they had toed-in the lines of 'runway' lights for perspective effect and you couldn't see the deck or the tower. That was terrible. Once they started lighting the deck it made a lot of difference."

CAPT John Brickner moved to Phantoms from the F4D Skyray, which he flew in 1958:

"Back then, night carrier landings were unforgettable. There were only line-up lights and the 'meatball' [mirror landing device], no angle of attack indexers, no 'moon-glows' to light the deck for at least some depth perception. On a black, moonless night it truly was a surprise when the tailhook grabbed the wire. What an exhilarating feeling and a gigantic relief! As the threat of WWII submarines became second priority to night carrier landing accidents the 'brass' slowly installed red 'moon-glow' floodlights that illuminated the deck. Even later there were white floods, which greatly reduced the accident rate."

Despite the hazards John Brickner made 1,044 deck landings (including 312 at night), flew 4,250 flight hours (2,500 in the F-4) and 200 combat missions, eventually commanding VF-111 in 1974-75. In all that time his only mishaps were a few blown tires.

Ken Baldry, flying with VF-96 in 1966, found things rather different on his carrier:

"The USS *Enterprise* used full white floodlights at the masthead to illuminate the landing area, and it was great! You could read a magazine back in the Fly 3 area (aft of the island to the fantail) at night during recovery. Most of the F-4 jocks allowed as how they would come aboard without a 'meatball' fresnel lens operating if a decent night horizon existed, which it never did in Vietnam." Failure to get aboard for any reason usually meant a diversion ashore to Da Nang if the fuel reserves were adequate. However, "everyone hated to 'bingo' to Da Nang. My wingman, Steve Amman, was the Assistant Ordnance Officer and knew his switchology off cold, but he was cursed with repeatedly having a 'hung bomb' and being forced to 'bingo' to Da Nang to have it removed. It was invariably the last recovery, so they would end up spending a good portion of the day at Da Nang. Finally, Steve put out the word that if he had hung ordnance from then on the offending MERs and TERs were going into the water. These bombracks were always in short supply, so the 'ordies' really worked his plane over and prayed that he would be assigned nothing but BARCAPs that needed only missiles. On the last Alpha strike of the cruise to Haiphong, the CO, CDR Sheldon 'Lefty' Schwarz (a former *Blue Angel* with a reputation as a 'tiger' of the first order: if you were in formation with Lefty and something was moving it had to be you!) and his wingman, LT Charles 'Righty' Schwarz both had a hung CBU. Since it was the last recovery prior to departure for Cubi Point and the States they did salvo off two TERs apiece. I thought the ordnance officer was going to have apoplexy, but Lefty just said, 'I'll sign the Survey Report,' and that was that!"

Westpacs for All

The ever-increasing demands of the Vietnam War meant that Atlantic Fleet squadrons were sometimes deployed alongside their Pacific equivalents. Usually, this was for a single cruise, and only one Atlantic squadron, VF-74, made a second visit. The team of VF-41 and VF-84 were first to go West, and their experience was eagerly sought by members of the three East Coast units—VF-11, VF-14 and VF-32—who followed them in 1966-67. John "Smoke" Wilson was Ops Officer with VF-14 *Tophatters* when they boarded the USS *Franklin D. Roosevelt* in June 1966 for a six-month voyage. He had flown two tours with the F3H-2 Demon on USS *Ticonderoga*, flown the F8U-2N Operational Evaluation with VX-4 (including tests with the AIM-9C/D missile), and spent three years conducting carrier suitability trials on a wide selection of USN types, including the F-4B and F-4G, plus SATS arrestment trials on the F-104G, just one of 106 types he had flown by the end of his career. With "Smoke" Wilson at the Naval Test Pilot School was Dick Adams, who was CO of VF-14 when "Smoke" joined the squadron on 3 January 1966:

"The squadron had just moved from Jacksonville to Oceana with the implementation of 'Base Loading' (all F-4s at one base; all A-4s at another). My 'scope' was LT F. E. 'Geno' Blair, a very experienced RIO. He and Jim 'Hummer' Hayden were the old hands among the radar operators. When we started preparing for the deployment we picked the brains of the VF-84 and VF-41 aircrews as to the missions and the tactics they had employed, and then developed the training plan for missions like night road reconnaissance and night bombing under flares that were very new and very challenging."

The emphasis on "attack" training was puzzling to fighter jockeys brought up on "wild blue yonder" intercepts. Reflecting on that situation thirty years later, Chuck J. D'Ambrosia, a *Tophatter* RIO from that cruise said, "Given the short time we had to prepare and the state of readiness we were at, it made eminently good sense to concentrate on the bomb-dropping part of the business like we did. As a bright, shiny young ensign with no experience it was hard to see that at the time." However, bombing under flares was, "probably the least productive and most dangerous type of mission we

flew. Because of the haze the world would light up like a giant illuminated milk bowl. It was impossible to keep from getting disoriented. Also, it was impossible to see anything on the ground when diving at 450 kts in a 40 degree dive angle. After pull-off it was always a scramble to keep from having a mid-air with your wingman. Nobody had lights on, of course."

Pre-deployment training still included some interception practice. "Smoke" Wilson:

"I started out as training officer under Jerry Riendeau as Ops Officer. During the work-up we were doing a 'snoop' exercise off the Carolinas. Because it was a real black night, poor weather, and a pitching deck the CO said he wanted only the most experienced guys standing the Alert. He took the first watch. Shortly afterwards he was relieved by Jerry Riendeau and me, though we didn't normally fly together. We were told to 'man up,' there was a snooper in the area. They were going to launch only one aircraft and put the other on five-minute Alert. When Jerry saw the plane I was assigned was on the starboard cat (that was the 'go-bird') he exercised his seniority and swapped aircraft. He launched, intercepted his snooper, and was back overhead at his 'Charlie' time; no small accomplishment considering all the ship's navigation aids, radio, radar, and lights were off to hide us electronically and visually from the snooper. Jerry then got a wave-off and two bolters and was then at low fuel state. The failure to get aboard was principally due to the night and the pitching deck. An A-4 tanker was launched, and after three plugs without getting any gas he and his RIO, Jon Bertrand, ejected. Jon was uninjured, but Jerry wound up with a compression fracture of his neck. To quote Jerry, 'The moral of this story is never trade airplanes with 'Smoke' Wilson.' He was transferred to hospital for surgery, and I became Ops Officer. The aircraft he took from me was AB 104; I remember that number even today."

Before going to S. E. Asia the *Tophatters* had to go through Survival, Evasion, Resistance, and Escape (SERE) training at Brunswick, Maine.

"The squadron was split into two detachments to be 'indoctrinated.' The skipper, Dick Adams, the CO of VF-32 and I were in

VF-84's aircraft 206 poised for a trap. When this F-4B was assigned to VF-21, LCDR Duke Hernandez and LTJG S. L. Vanhorn ejected from it after a AAA hit over North Vietnam. (U.S. Navy)

the first group. I was designated Head of Escape Committee and escaped twice. Our group was agitating our 'captors' almost as much as they were us. Of course, we realized eight days (with a guaranteed release in reasonably good health) wasn't the same as years in the Hanoi Hilton.

Two weeks later the other half of our squadrons took their turn in the barrel, led by our XO, Jack Koach, and VF-32's XO, Buddy Burt. I got permission from Dick Adams and VF-32's CO to lead the 'Brunswick Strike.' We printed up about a thousand leaflets that read something like: 'We are The Yankee Air Pirates (a favorite NVN name for naval aviators), and unless you release our fellow airmen immediately we will return with devastation, aimed at the lowly members of the Communist Party running this camp (here we listed the names of the people running the camp that I had collected when our exercise was over). You must be especially considerate of CDR Koach and CDR Burt, as harm to these brave warriors will result in grave consequences.' We carefully placed 250 of these leaflets in each speed brake well of two F-4s, and while the Plane Captain held them in place the speed brakes were closed. Knowing the camp routine we launched from Oceana on the first day the new 'POWs' were in the prisoner compound. We came in as low as possible, given the pine trees surrounding the compound. We were slow; 225 kts with flaps and slats down. Just prior to the camp perimeter we went, 'Speed brakes, burner now!,' pulled up as hard as possible without stalling, and headed back for Oceana.

Everyone agreed it was a super coup—even the captors. The guys who were not quite so appreciative were Koach and Burt, who had received increased 'attention.' There were four coffin-like boxes (undersized, of course) in the center of the compound into which un-cooperative prisoners were placed for punishment. When my cruise-mate Gus Watters got back he said, 'J. C., how low were you? I could feel the heat from your afterburners, and I was in the box!'"

During our Operational Readiness Inspection (ORI) off Gitmo (Guantanamo Bay, Cuba) we had a missile shoot. Three shooters were in about a two-mile trail with me as number three. We made a head-on run at a high-flying supersonic target. The XO, Jack H. Koach, was in the lead. As number 3 I had to wait to snap up until the others had made their moves, which put us in a near-vertical attitude as we fired. I shot a Sparrow and rolled over to pull back to the horizon—and there I was, eyeball to eyeball with the missile. It had been ejected, but the rocket motor failed to light. Here I am, pointed straight up at 55,000 ft, running out of airspeed and ideas, and if that's not enough, behind and below me in close proximity is the chase F-4 also approaching an out of control situation. I managed a rudder roll to miss the missile and then fell—literally—about 20,000 ft before gaining enough airspeed to start flying again. Everyone said the contrails from that evolution were very interesting. They should have seen it from where Gene and I were!"

Also during the work-up period, on 10 May 1966 there occurred a famous incident in which an F-4B took off with its wings folded after some confusion on a crowded deck.

"We were doing some refresher landings. Greg Schwalbert, the CO's wingman, was on the catapult ready for the final spot when he was given a 'wings fold' signal to let another airplane

LT Grover Ericksen (left) and LTJG John Perry (with RADM J. R. Reedy, Commander of Task Force 77) after their recovery from the sea in October 1965. (USN via CDR G. Ericksen)

pass. He unlocked the wings mechanically and hit the switch to 'fold.' The F-4 had an occasional habit of folding only one wing when 'fold' was selected, requiring the pilot to cycle the switch. In this case, although Greg had put the switch to 'fold,' *neither* wing folded. The passing aircraft cleared, and the director put the F-4 on the catapult. Greg and RIO Bill Wood's plane (AB 112) was tensioned and launched.

Shortly after the plane started down the catapult the wings folded up and were all the way over center at 180 degrees when he cleared the ship. Something didn't feel right to Greg, so he lit the afterburners and literally rocketed into the air. The Air Boss, Joe Elmer, screamed over the radio 'Don't turn. Your wings are folded!' Because Greg was light on fuel and had lit burner he was flying, but to where? He couldn't land back on the ship, and his current fuel consumption made the sixty miles to Gitmo 'iffy.' One by one, he retracted the landing gear, then half flaps, then all the way up. He jettisoned his racks and centerline tank, at which point he had to make a forty degree turn for the airfield. Gas was looking better, so he pressed on. He dropped the gear and flaps and came over the fence at 200 kts plus. Once on the deck he shut the engines down and picked up the arresting gear.

Next day Rolf Noll, the Maintenance Officer, manually spread the wings, put two bolts in the wing-fold locks, and flew the airplane to the rework facility at Cherry Point. They replaced the actuators, and the plane was returned to the squadron as good as new. Courtesy of Marv Seay, our squadron practical joker, we had two sets of clean skivvies and two pairs of Navy Wings—with their extremities properly folded—to present to Greg and his scope on return to the ship."

The *F. D. R.* sailed for the battle area, and en route "Smoke" and his fellow *Tophatters* survived the usual Shellback ceremony ("My hands and knees were raw from crawling around on the non-skid surface of the deck"), and then Jungle Escape Survival Training (JEST) in the Philippines. Heading straight for Yankee Station on 7 August the carrier became the first to, "Do not pass 'Go,' do not collect $50": previously carriers spent a week or two on Dixie

Station flying over South Vietnam before heading north to Yankee station and more ferocious action.

"Just prior to deployment we picked up a couple of new aircrew. The skipper got one crew as his wingman and I the other—put the new guys with the oldsters. Bob Cross and his 'scope,' Dean Hutchinson, became my wingman. Though the squadron was organized in divisions of four aircraft we seldom flew in anything but pairs except for Alpha strikes. On 10 August 1966 Bob and I flew our first combat hop; a BARCAP twenty miles off Hanoi. Two days later we had our first 'over the beach' go, a night recce near Vinh. Because of the vintage of the USS *Roosevelt* and its arresting gear, and the comparative frailty of the F-4B when subjected to these landing loads, we were restricted in how much fuel we could have aboard when we landed. On night missions we carried two Sparrows and two Sidewinders, a centerline tank and three bombracks, two for bombs and one for flares. All this weight subtracted from the maximum 'come aboard' fuel to the point where the first pass was made with 1,900 lbs, enough for one more try, and then you were in extremis with about 900-1,000 lbs. The F-4B burned about 120 to 150 lbs per minute. There was no way to divert to Da Nang once you had committed to the ship. Picture a first-cruise 'nugget,' 22 or 23 years old, with 200 hours in type and maybe 350 hours total flight time, doing a night recce, bombing under flares, being shot at, and faced with his tenth night carrier landing. I was impressed!

We had one enterprising pilot [the XO] in the squadron who would land 'heavy,' with more fuel aboard than permitted. Just before securing the engines he would push the fuel gauge test switch, which ran the dial indicator and counter down to 1,000 lbs (the correct indication), then immediately secure the electrical power which locked the indications slightly over 1,000 lbs. No-one was any the wiser that he was landing heavy? Wrong! CO Dick Adams found out and really had a piece of him."

Though they were initially charged with flak suppression on Alphas, the CAG switched this role to the A-4s and the F-4Bs became strike bombers. "Usually, four F-4Bs would launch (with an

empty centerline tank and four 1,000 lbs bombs plus four 500 lbs bombs) at three or four knots above minimum flying speed and then proceed to the tanker to fill up the centerline tank. Then the strike group would go to the "plate." The Air Wing code-word to arm everybody up was "Batter up," which the strike leader called about the time we crossed the beach and headed to the target in "Indian country." Flak suppression was a particularly distasteful role, since to kill a flak site required the suppresser to fly down the gun barrel. A bummer! Actually, the A-4s were better off, since they could use their Bullpups and stand off to some degree." To increase the effect of Mk 82s used for flak suppression they were sometimes fitted with a three-foot metal tube screwed into the bomb's nose, with the fuse attached to the front of the tube. These "daisy cutter" fuses ensured the bomb exploded just above ground level, projecting its blast and fragments outwards.

During the pre-Vietnam work-up, VF-14's aircraft were modified to utilize the loft bombing computer for delivery of conventional ordnance, as opposed to nuclear weapons. The radar was also modified to improve the resolution of ground targets. RIO Gene Blair explained:

"This included circuitry to give us an additional range on the scope (ten miles) and gave finer control of the RF gain. With the gain way down we could pick out points, such as small streams, bridges, mountain peaks, etc. which otherwise would have been hashed up. The improved RF gain control was a handy little feature for navigating, join-ups, and I believe it could have been useful in a heat-seeking attack. It gave a good control over a close-in target that otherwise would have 'bloomed' all over the scope in bearing. I could find buoys, boats, the ship's spar target [for ordnance practice at sea]—about anything with it. The radar also had a so-called JetCon Labs modification to it so that the crew could attack a target which had no radar significance of its own by using an offset radar initial point (IP). An attack profile would be strictly level flight with conventional delivery, and the crew would plan a run-in course and speed to the target, spotlighting an offset IP, and when the IP was at a certain slant range and bearing, start the count-down to weapons release, using our nuke release mode."

"Smoke" Wilson noted that, although they had several night strike missions planned they never received permission to execute them. However, he did attempt to improve the lot of the flak suppressers:

"CAG Smith asked me to figure out how far we could loft (toss) proximity-fused 1,000 lbs bombs. After doing the computations he sent me out to try it. (The idea was to loft these bombs into the target area by surprise—and they sail a long, long way! With the correct timing the bomber aircraft could get into their runs while the AAA gunners had their heads down and ringing ears). I ran in at 5,000 ft and 500 kts, pulled up to 15 degrees, and the bombs came off at one-second intervals. After the fourth 'thump' I rolled over to execute a break-away, and there I was, flying formation on 4,000 lbs of proximity-fused bombs. Fortunately, they had not yet armed, or Ron and I would have been history. Another very rapid 'bank and yank' and we were out of there and more than ten miles away when the bombs went off. CAG decided this was not a very accurate method of flak suppression."

Major events, such as MiG kills or, in this case the 1,000th bomb loaded by VF-84's ordnancemen, were usually celebrated with cakes. The squadron XO, CDR Alex Waller, prepares to slice this monster. (USN via CDR G. Ericksen)

Target photo of a North Vietnamese radar site on Bach Long VI Island, used for a September 1965 attack by CVW-7 Phantoms. (USN via CDR G. Ericksen)

Chuck D'Ambrosia tried loft bombing on one mission, too:

"We had six Mk 82s, but could not get over the beach because of a low overcast layer at about 3,000 ft. Off-shore, we could get down and see the beach well. There was a flak site marked on my map behind a stone fence, just up from the shoreline. It was very distinctive and good for radar ranging, so we lofted those six bombs off all at once at the site. Dave Brown, my pilot, made a very smooth pull-up through the overcast, and the bombs came off right on cue. As we rolled 90 degrees for the exit maneuver we were flying wing on these six bombs with their little fuse propellers spinning their little hearts out. It was an astounding sight. We pulled off and dropped down below the overcast again over the water and made a turn so we could see where the bombs hit. After an eternity the area just over the stone fence erupted with a huge, black, dirty-looking explosion. By luck or whatever we had put those bombs exactly where we had intended."

On 23 August "Smoke" had, "the questionable distinction of being the first in the Air Wing to be fired upon by a SAM."

"I was escorting an RF-8G photo run over Haiphong. We set up a CAP between him and the two major MiG airfields to the north and north-west of Hanoi. We picked him up and covered his 'six' as he came out of Haiphong. About the time we went 'feet dry' my wingman demonstrated his cool by reporting over the radio, 'Jesus Christ, it's a f****ing SAM at 2 o'clock. Break right!' We did have code words, but it was really unnecessary under the circumstances. I got the message. I have never seen anything move so fast!"

Geno Blair was replaced by a new RIO, Ron Bird, between line periods, and after three "fam" flights the pair did a coastal recce from Than Hoa to a point South of Haiphong.

"No-one overflew those two places unless they had to. Our hop was a dusk-to-dark mission, and I had great hopes of finding barges moving south. On the way north it was as quiet as a church at midnight, but when we turned south it seemed like not a minute went by when someone wasn't shooting at us. When we got to the end of our run we were about out of the fuel allotted for the recce

and still hadn't found anything to shoot up. We each had four 2.75 inch rocket pods. There was an island a mile or so off Than Hoa that had a known radar and AAA site on it. That seemed to me as good a dump spot as any, so I called for Bob to climb to a roll-in position in 'Washington' (to the north-west of the target), and I went 'Florida' (south-east). As I let go my salvo (they were really spectacular at night) I watched the rockets burn out through the gunsight. It was like watching a movie being played in reverse. Up through the sight came these orange balls of fire, like my rockets coming back. It was three or four 37 mm [Soviet M38/39] AAA sites shooting at me! That was all Bob Cross needed. He was already in his dive, and with a slight adjustment he put all fifty-six of his rockets right in their lap! Blewy—no more AAA! For that we got an 'Attaboy' from our Admiral, and the next day a 'Dumbo Sierra' from Seventh Fleet for 'dueling with a flak site.'

I recall another dusk flight in the same area. We had just installed a radar warning receiver in our airplanes, a real 'Rube Goldberg' set-up, but it worked. In order to get the capability to have a warning that a SAM radar had locked on, the troops fastened an aluminum bracket to the RIO's glare shield [on his helmet] and installed a navigation system antenna on it. The equipment was plugged into the 24 volt utility receptacle, sort of like a 'fuzz buster' you'd put in your car's cigarette lighter. With a phone jack patch the high-pitched 'deedle deedle deedle' that said 'Look out, there's a SAM coming your way!' would be heard by both crewmen."

John Nash also experienced the so-called 'Little Ear' SAM detecting device on his second cruise with VF-213:

"That lasted about two weeks. They wanted to put it on the RIO's helmet and have him turn his head and scan. They looked pretty queer in the back with those things on. You'd look back and say, 'Are you scanning?' Their egos got the better of them, and they said, 'We're not doing this anymore!' It was effective, but strictly directional, and the RIOs were too busy to do that."

"Smoke" was on a barge hunt near Than Hoa on the evening of 28 November 1966 when, for some reason, he couldn't recall Ron Bird testing the APR-27 SAM warning receiver before they launched:

"He responded, 'Yeah, sure did, and it checks OK.' Then, 'deeedle, deedle, deedle.' My first reaction was that he'd pushed the test switch again. This was quickly dispelled by the 'telephone pole' lifting off at 2 o'clock, about eight miles away. I yelled 'Hot dog' over the radio and headed for the deck. As soon as the first SAM went off its auto-pilot at booster burn-out, its first command was to go down, since I was now at fifty feet rather than 5,000 ft where they had initially fired. The SAM nosed over and smartly ran into a karst hill between me and him, with spectacular results. I immediately climbed out of the weeds, where I could have been hit with a rock, only to view another SAM coming off the rail. The high-g barrel roll beat this guy, too." A third SA-2 Guideline [Soviet S-75 *Dvina*] menaced the F-4B, and 'Smoke' out-maneuvered it, too. "I watched it all the way. Ron and I actually heard the roar of the rocket as it went overhead, and we watched the characteristically orange and black fireball as it exploded a hundred yards to our right."

Making contact with his wingman, who had wandered ten miles out and was already "feet wet," "Smoke" released his rockets at a likely target and "beat feet."

Another early SA-2 warning device used in VF-14 F-4Bs was the AN/APR-24, a variant of the USAF's APR-25 Vector IV system. Chuck D'Ambrosia: "APR-24 had a little radar scope on it. The antenna was mounted on the tail, but was subject to significant rain erosion and never worked." SAMs took a heavy toll of strike force aircraft. Chuck:

"I went on the 13 December 1966 Alpha strike on a truck park on the outskirts of Hanoi. Dave Brown and I were assigned a race-track pattern just south of the target to serve as 'blockers' for any MiGs that might show up. We had entered 'feet dry' just north of Than Hoa, flew north and set up our pattern. I started counting SAM trails as they lifted off. I stopped counting about half-way through the mission at sixty, though I may have double-counted some, as we were doing a lot of maneuvering. Most seemed to be launched on a ballistic trajectory because most went to a high altitude and detonated. Some aircraft were undoubtedly targeted, but for the most part they looked like they were being fired in panic. As the strike group departed to the south we were the last aircraft out to cover the withdrawal. Talk about a lonely feeling. I saw another missile lift off and detonate at about 8,000 ft, and I saw pieces of an A-4E Skyhawk [BuNo 151068] tumble out of the sky. It was so hazy I couldn't see the plane before the missile hit it. The pilot was LT C. D. Wilson of VA-72 [KIA]. The wreckage landed near a karst spine. We didn't see a chute, but we made a low-level pass around the wreckage anyway. By this time everyone else was long gone, and we were anxious to be out of there. We were pushing 600 kts at 4,000 ft, jinking continuously. Just prior to 'feet wet' we jinked right and then back left. As I looked back behind the aircraft for missiles I saw the distinctive orange cloud of a SAM detonation about 100 yards behind us, right in our wake. The same SAM site that had gotten the A-4 almost got us, too."

SA-2s were controlled by a large but road-mobile *Fan Song* guidance radar unit, which emitted a distinctive demodulated "warble" and was effective at distances of up to seventy miles. Targets were initially tracked by a P-12 *Spoon Rest* radar, or the earlier P-8/10 *Knife Rest* or PRV-11 *Side Net* height-finder and P-35 *Bar Lock* search radar. In the ever-advancing electronic war other emissions could be picked up from the truck-mounted *Long Talk* GCI radars used at MiG airfields in conjunction with *Two Spot* azimuth/elevation trackers. Optical tracking by a PUAZO-3 rangefinder, or radar fire control by a SZON-4/9 unit was used for heavy 85 mm AAA batteries up to their range limit of 30,000 ft. Most common was the 37 mm M38/39 gun, which relied on an optical lead-computing sight and could destroy an aircraft at 10,000 ft, while the heavier 57 mm S-60 gun, usually operated in batteries of six weapons with a single fire-control system, had twice that range. The heaviest gun in the North was the 100 mm KS-19, which could bring down an aircraft at more than 35,000 ft. Equally deadly to low flyers below 6,500 ft was the ubiquitous DShK-38 12.7 mm machine gun firing 600 rpm.

Life on the "boat" offered little time for relaxation, either, as "Smoke" recollected:

"The hours were long during the line periods. We would fly twelve hours at a whack, and when we were not flying there were two F-4 crews in the cockpit, all strapped in and ready to be airborne in five minutes. There were also two manned up during routine ops. The crew was able to sleep anywhere, anytime. When we were the 'BLUE' carrier (on duty 6pm to 6am) it was breakfast, breakfast, breakfast. We were either flying, sleeping, standing Alert, or doing paper work."

Towards the end of the *Roosevelt's* 95 days on line a major Alpha strike was flown:

"There were two rings around Haiphong (20 miles diameter) and Hanoi (50 miles) inside of which JCS and Mr. McNamara had prohibited any strikes. On the morning of 2 December we conducted the first raid inside this. It was a three-carrier co-ordinated 'go' with USS *Roosevelt*, *Ticonderoga*, and *Constellation*. The target was the Van Diem vehicle depot, or 'bicycle works,' where Russian trucks were de-mothballed and prepared for movement of men and materials to the south. Going after the truck park was particularly satisfying to me after taking pictures of them on the ships entering Haiphong, then trying to find and destroy them with six 250 lbs bombs under flares at night. There had to be an easier way.

On that strike there were nineteen SAMs fired at us before we got to the target. The strike group from the '*Tico*' had more. They were the new guys, and opted to ingress over Than Hoa in spite of warnings from the other CAGs, and they lost a couple of aircraft. We came in the back door from the south-west; relatively safer. The post-strike pictures show a string of bombs across an open field, not exactly close to the target, the result of another SAM evasion. This guy rolled over to drop, and as he lined up his sight he saw a 'telephone pole' coming at him and grabbed a whole fistful of stick, including the bomb pickle.

On 16 December the CNO, ADM David McDonald, visited the USS *Roosevelt*. The next day I was scheduled to fly one of our birds to Cubi Point for a fresh water wash-down and pick up one that was ashore. As I was approaching the catapult I got the signal to shut down, and I was told to stay in the plane and check the switches were off; they were going to arm us with missiles. Next I

The Samh Dinh POL storage area, 30 miles south of Hanoi, with defensive positions ringed in pen. A simple red arrow shows the F-4 crew their target. (USN via CDR G. Ericksen)

was called to the ready room, where my wingman, Bob Cross, was suiting up. The day before, the airplane carrying the CNO had strayed very close to Hainan Island (China), and our Intelligence types had learned that should they get close to Hainan on the return trip the Chinese would shoot the plane down (á la Yamamoto in WWII). Our mission was to fly fighter escort for the CNO's airplane and then divert to Cubi. Normally, an 800 mile jaunt is no big deal. However, when half of it is flown escorting a Grumman C-1A 'COD' aircraft at about 140 kts it becomes a real challenge. As we were about to start engines CAG Smith climbed up the boarding steps and said, 'Smoke, you know what to do if any Chinese MiGs show up, don't you?' I replied, 'Yeah, CAG. I'm going to shoot their ass down.' He just smiled, slapped me on the shoulder, and departed the deck with a thumbs-up. Fortunately, no problems materialized, and after 2hrs 45 minutes we landed at Cubi with just enough fuel to taxi to our parking spot."

Connie's Kills

After their use by VF-213 the small batch of F-4Gs were converted back to standard F-4B configuration, losing their data-link equipment but retaining the smaller No.1 fuel cell. One specimen, BuNo 150633, was dropped off the winch while being loaded onto USS *Kitty Hawk* at Cubi Point, snapping off its tail section. Another, BuNo 150629, passed to VF-142 on 16 August 1967 and flew with its stock F-4Bs. It was one of these F-4Bs, BuNo 152247, which was used by LT Guy H. Freeborn and LTJG Robert J. Elliott to make the first of USS *Constellation's* four MiG kills on that cruise. The former F-4G made the last kill on its own final sortie. Ex-F3H Demon pilot Guy Freeborn described the background to his kill:

"I had been an F-4B Tactics Instructor in VF-121 for over two years and the Training Officer in VF-142 prior to the cruise. Squadron activity in 1967 was, 'Fly, eat, drink, play cards, man the Alert 5 watch, and curse the politicians for putting us in a war they wouldn't let us win." Things livened up considerably when Guy and his RIO launched as Number Two to LCDRs Bob Davis and 'Swede' Elie with a big, two-carrier Alpha against the Phu Ly trans-shipment area. "The engagement was planned by us based on the environment, i.e. positive radar or radio control and thin cloud layers at 22,000 ft. We were BARCAP for the strike group, and held station just under the cloud layer instead of the normal 15-18,000 ft for the surprise element, which worked just as we had planned." Normally, the MiGs attacked from above the cloud under ground radar (GCI) control, and the Phantom crews hoped to catch them as they broke through the cloud layer in pursuit of the Navy attackers. VF-142's new tactic led to Task Force 77's first MiG-21 kill. Usually, MiG-17s were sent to operate over the Navy's northern Route Packs. "We were very familiar with the MiG-21's capabilities and tactics. We just hadn't seen many. They were mainly into 'hit the strike elements and run for home,' with not much dog-fight capability. The feedback [from USAF crews who had met the MiG-21 the previous year] mostly confirmed this. They seemed to like the three plane formation with one up front and two tailing the lead aircraft. They also liked to run in multi-pass."

After a series of MiG calls to the BARCAP patrollers they were told that MiGs had closed to within fifteen miles astern of their

VF-11 adopted revised tail markings for their brief war cruise on USS *Forrestal* in September 1967. F-4B BuNo 153068, seen here launching with Mk 82 bombs and AIM-9Bs during their five days on the line, was one of the Phantoms which survived the carrier's deck fire on 29 July 1967. (via J. T. Thompson)

section. Both Phantoms turned back sharply and caught the MiGs as they dropped through the cloud base. With the Navy fighters now below their six o'clock the MiGs were unable to see them. Bob Davis and Guy pushed up their speed to close on the prey and set up their missile switches.

"My choice of missile for close-in fighting was always Sidewinder. Bob Davis and 'Swede' Elie fired two Sparrows with no good guidance. Their kill was also Sidewinder. Reliability of the missiles was a major issue then. The gunpod didn't work well and was too limited, so nobody wanted it. What we really wanted were built-in guns like Air Force F-4E Phantoms."

Although this view was shared by a number of the crews interviewed for this book, many felt that the Phantom's lack of agility in close fights would have made the gun a useless burden on the majority of missions. However, on this occasion the unpredictable behavior of the missiles made what should have been a textbook MiG kill into a frustrating process. CDR Davis' Sparrows failed to initiate guidance, and he switched to Sidewinders. Meanwhile, Guy Freeborn went straight for the 'heat' option, and his first AIM-9 burst close to the left MiG, causing enough damage to make it trail vapor. Seconds later a Sidewinder from the lead *Dakota* F-4B also exploded close to the other MiG-21, but a second failed to guide and went 'ballistic.' The Phantom pilots pulled back on the speed to avoid an overshoot, CDR Davis making a high yo-yo to do so. He came back down behind Freeborn's MiG, destroying it with a third and fourth AIM-9. Guy, in *Dakota 202*, turned instead to the other MiG, following into a climbing turn and firing another AIM-9, which remained stubbornly on the rail, but a third shot flew up the slender MiG's tailpipe, disintegrating the fighter. It had taken nine missile shots to bring the two MiGs down, and only three had found their targets with lethal results "as advertised."

Also with VF-142 *Ghostriders* aboard the *Constellation* in 1967 were LCDR Eugene P. Lund and LTJG James R. Borst. On 30 October they were flying an early afternoon MiGCAP in *Dakota 204*, the squadron's former F-4G (BuNo 150629). LTJG Ron C. Ludlow and LTJG Bruce L. Hardison were on their wing, flying abeam at one mile separation. Geno Lund's debrief tape captured the sequence of events leading to their MiG kill. In retrospect, Geno noted that it showed the crew of *204* were 'both pretty excited' as they told their tale in the noisy conditions back aboard ship:

"We were on the northern MiGCAP station, which is over the ridge north-west of Haiphong. We went 'feet dry' just west of Cam Pha at 270 degrees at 18,000 ft, and we tracked down the northern ridge. About abeam of Haiphong we got our first 'Bandits' call, not from *Harbormaster* [USN radar controller in a destroyer], but on Guard [frequency]. About a minute later *Harbormaster* came up and said he had an 'unknown,' not a bogie, at 270 degrees and about 50 miles. Shortly afterwards he called it a bogie and commenced a vector from 270 degrees to 330 degrees. Shortly after that he confirmed that we were cleared to fire on the bogie, which was now at 353.60 degrees. He called them at 'angels ten,' and I suggested that they were climbing. He then said that there appeared to be a pair of bogies and continued to give us vectors. Things got down to about ten miles, and we got a lock-up on our radar at about fifteen miles. It appeared to be two bogies, from the scope. I more or less neglected any other vectors from *Harbormaster*, and we drove on in.

At 3 to 4 miles I already had it in range, and I got a tally-ho [visual sighting] on *four* MiG-17s flying a finger-four formation in two sections; one section which we were locked onto, and another about 2-300 ft astern, all at about 18,000ft. My range was about mid-range [for the AIM-7], the missile [AIM-7] commenced to guide absolutely beautifully and impacted on the Number 2 MiG, who was on the left wing of Number 1. It hit just aft of the cockpit, and he blew up and entered a flat spin. I never saw a chute." ['Biff' Borst added later in the debrief; "Geno said 'We nailed it,' and I looked out and saw a MiG-17 with its tail on fire spinning into the ground."]

At that juncture the second section broke left and the first section broke right. I came up in a high yo-yo to the right. My wingman was on my right, and he crossed over. I came back down and committed towards the second section with my wingman hanging in there. A couple of times when I lost sight of them he was able to take them on a lot better than I was. They appeared to be pretty aggressive, but I don't think they fired anything at us. I didn't see any guns or missiles. Shortly after a couple of high-g turns [with the MiG leader] I think I picked up one section of them dead ahead [two aircraft circling at 5,000 ft], slightly down at (estimated) 3 to 4 miles. I told James Borst to 'go boresight,' which he did and he locked up immediately at about three miles. The dot was in the center [of his scope], and they were above mid-range—about a mile or so—so I fired again. The missile (AIM-7E) left the launcher, and about one to two hundred feet from the airplane it exploded and broke up into all kinds of pieces. I felt a jarring sensation in the airplane; I didn't know if it was jet wash or what. Apparently, the right engine had stalled, and I didn't know it.

From that time on the airplane was sluggish. I'd gone into full burner on both engines and another high yo-yo to get back on [the MiG]. I zipped past one, canopy to canopy at about 20 ft, and this one I determined to be a camouflaged, olive drab MiG-17. I didn't get a shot at him because he was head-on. I almost got a shot on one more run-through. I picked him up again dead ahead at about three

Snakeye-toting VF-154 F-4Bs cruise towards their target from USS *Coral Sea* in 1966. F-4B BuNo 152974 was shot down by AAA on 4 January 1967, but its crew, LT A. M. Van Pelt and ENS R. A. Morris, were recovered. (Terry Edwards)

miles, but I had about 80 degrees [of turn] to go to get on his tail. I was coming down at him at 30 degrees down, but I couldn't get a 'growl' on my Sidewinder because of the aspect. It was a good Sidewinder, but we couldn't get a lock-up on him.

We met a couple more times, and that's when I looked down and figured out why the airplane was so sluggish. The right engine was at idle, and the TOT was about 300 degrees with fuel flow about 500 lbs though the throttle was at full afterburner. I figured I had a stall, so I pulled it back to idle and started moving the throttle up to 70 to 80 per cent, but I got a heavy rumble in the airplane and the rpm dropped off, so I figured I had a fire in the right engine. Utility pressure was down about 1,200 lbs, so I figured I had taken a hit from the fire or from the engine itself and I had better get out of there, so I put full afterburner on the left engine. Unfortunately, my wingman's radar malfunctioned, so he could not fire an AIM-7, and since he was a good wingman supporting me he was unable to position himself for a Sidewinder shot.

My wingman had reported that the remaining three MiGs had descended to the south east towards Phuc Yen. We passed Kep airfield and still had the three bogies ahead of us at about ten miles, but they descended in a south-easterly direction. I wanted to go out as quickly as I could, as I wasn't sure what else was going to happen to the airplane."

As he proceeded out at 400 kt and 18,000 ft his wingman gave Geno's BuNo 150629 a look-over and reported hydraulic fluid all over the underside of the Phantom, which confirmed his cockpit readings: zero hydraulic pressure, a speed brake warning light, and, "the full utility failure routine." [On most Navy F-4s the hydraulics powered around twenty-seven crucial systems, from the aileron power controls to the chaff dispenser door.]

"I came out of burner, started heading south and called for a tanker. I had about 6,500 lbs of fuel, barely enough to make it, but I could take another 1,000 lbs. Then I realized I couldn't stick my probe out and couldn't have tanked if I had wanted to. I determined I could make the ship with about 3,500 lbs, going 80 miles out, single engined at about 300 kts. My wingman looked me over again and reported several holes in my starboard, forward Sparrow missile well. He didn't say whether they were gunfire; they seemed to be ragged holes which could indicate they came either from the missile motor or fire from the missile as it exploded back down the airplane. I called the ship and requested a straight-in approach. I'd 'dirty up' at about five miles, and I would need assistance after the arresting gear since I wouldn't have any brakes.

At eight miles from the ship I attempted to lower the landing gear. Nothing. I pulled both the gear and flap circuit breakers, pulled the gear handle out, and the indicator remained 'Up, up, up.' My wingman checked, and the doors hadn't even cracked open."

Repeated attempts and recourse to the emergency pneumatic flap handle failed, suggesting holes in the pneumatic lines also, though the gauge showed 3,000 psi. The pair were advised to eject. Fuel was down to 600 lbs, and at one mile from the ship and 5,000 ft, "Biff and I went through pre-ejection. We lowered our seats, knee-pads off, sat flat in the seat, rudder pedals fully forward, visors down." Biff went first, watched by Geno in the rear view mirror. "I pulled the handle and got quite a jolt—a definite shock, then further jolts from the drogue 'chute. I could see a big splash in the water where the plane had gone in." Removing his face mask, Geno decided to wait until he was closer to the water before dropping his seat bucket, and then fell among the parachute lines. Looking up, he saw the *Connie's* rescue helicopter above him. Both men survived the experience unharmed and fought on, but 'Biff' Borst was killed in an A-7 ACM accident after transfer to an attack squadron.

Over thirty years later, Geno concluded that the rogue AIM-7 which downed his F-4B was ejected at the wrong angle, and when the motor ignited the missile came back up and struck the belly of the aircraft, flew out a short distance in front and then exploded, showering debris down the starboard side and into the right intake. CVW-14 scored a further victory on 26 October when LTJG Robert P. Hickey and LTJG Jeremy C. Morris of sister squadron VF-143 brought down a MiG-21 despite being almost shot down by a missile from another F-4B. Cleared by *Harbormaster* to fire on bandits without the customary visual "tally ho," Dave Grossesch in the lead F-4B lost his radar just as he was about to fire an AIM-7 and, according to Guy Freeborn, "had problems with his switchology

and couldn't get his act together." Hickey and Morris attempted an AIM-7 shot at the "MiGs," but it failed to guide, and an AIM-9D shot also failed to acquire the target. As they got within visual range the two *Pukin' Dogs* realized their "bandits" were actually two other F-4B *Dogs* who were unaware of their lucky escape. Re-vectored to another contact, Hickey and Morris launched an AIM-7 which destroyed a MiG-21.

VF-143's curious nickname derived from the "winged griffon" insignia which appeared on its aircrafts' flanks, and which was irreverently compared to a vomiting dog by one observer. Perversely, the idea was adopted, but it did have the unwanted side-effect of getting the squadron's name deleted from press releases, as the "puking dog" image was not considered "good news."

In all, five F-4Bs were lost by the *Constellation's* Air Wing on the 1967 cruise, and the other losses, including both COs, typify some more of the hazards facing their crews. Guy Freeborn felt that the losses stemmed partly from the F-4 squadrons' heavy involvement in strike and flak suppression, which brought them within the lethal envelope of the AAA: "Skipper Robin McGlohn, VF-142, and skipper Bill Lawrence, VF-143, were both shot down by ground fire. Tom Sitek, VF-142, was shot down by SAMs." CDR Lawrence was at 10,000 ft, traveling at 500 kts when he was hit by a lucky 85 mm round. Realizing that his aircraft had been struck, and hydraulic pressure was failing, he nevertheless continued with his flak suppression run, dropping his CBU on target. As the hydraulic pressure faded to zero he had great difficulty in pulling out of the dive, but managed to reach 10,000 ft before the F-4 went into a flat spin. It had fallen to below 2,000 ft when the crew ejected. CDR McGlohn and his RIO LTJG McIlrath were recovered, but CDR Lawrence and LTJG Jim Bailey, his 6 ft 6 inch RIO, became POWs. Tom Sitek had survived an earlier SAM hit to BuNo 151406, which had its right stabilator torn off and huge gashes in its wing. The SAM hit to BuNo 149498 on 23 August 1967 proved fatal to Sitek and his RIO, ENS P. L. Ness. The other loss, also to VF-142, was during the June line period when LCDR F. L. Raines and ENS C. L. Lownes (in BuNo 150439) fell to AAA but were picked up. On the following cruise in 1968-69 LTJG Mark Gartley and LT Bill Mayhew were victims of a "blue on blue" accident in F-4B 151404 of VF-142. They had flown thirty missions, mostly over North Vietnam, attacking coastal targets with 500 lbs and 1,000 lbs bombs (one of which "hung," causing a Da Nang diversion: their worst problem before 17 August), and then hitting the tanker and flying a MiGCAP off the coast. "Usually," as Bill explained, "there was no trade. MiGs never came far enough south before turning back." On 17 August 1968 they had, "the first MiG engagement for some time. We were told to intercept two MiGs. We pickled off the centerline tank, got behind them into a firing position, and then couldn't get the missile to launch. One MiG went into afterburner as it turned away and would have made a great target. We saw another "MiG" coming at us in the other direction with a smoke trail behind it." Four years later he found out that the "MiG" was actually an F-4B, possibly his wingman, which had launched an AIM-9D in their direction. The missile impacted their Phantom, and they both became POWs.

Mk 82 bombs fall from a VF-154 Phantom onto artillery positions above the DMZ. F-4B BuNo 153006 has both APR-30 fin-cap antennas and a third beneath the radome. The *Black Knights* made five war cruises on USS *Ranger*. (USN/ J. L. McDowell via Norman Taylor)

Aboard USS *Coral Sea* at the height of the *Rolling Thunder* campaign were the *Black Knights* of VF-154. The carrier had made its first war cruise in 1965 with VF-151 in F-4Bs partnered by VF-154, which still had F-8D Crusaders at the time. There were twenty-one combat losses, including six Crusaders, but the F-4 squadron emerged intact and claimed the only MiG kill of the cruise. LCDR Dan MacIntyre and LTJG Alan Johnson used an AIM-7D Sparrow to knock a MiG-17 from a three-bogie formation, and were lined up for a second kill when they realized their "wingie" was in danger from another MiG and broke off to go and save him. For *Coral Sea's* second cruise VF-154 joined VF-21 with F-4Bs and departed for the Gulf of Tonkin on 29 July 1966. The *Black Knights* were led by Ken Wallace, and the XO was Will Haff, a former F-8 driver with VF-111. CAPT Haff looked back on the cruise, which he considered:

"Far from perfect. We took some heavy losses, and both Wallace and myself lost wingmen. LCDR Nels Tanner and backseater LT Ross R. Terry [BuNo 152093, lost on 9 October] were bagged and spent many years (1966-72) in the POW camps of North Vietnam. LT Al van Pelt [and ENS R. A. Morris in BuNo 152974, 4 January 1967] was hit [by AAA], but ejected over the Gulf and was recovered OK. I took a good hit while escorting a RA-5C and had a nice hole between the left intake duct and canopy. I guess it was a non-VT fused round or just a plain steel projectile, as it just passed through the aircraft. An explosive head would probably have destroyed the F-4, but there was not even a hydraulic leak. The F-4 could take a good shot and still make it home."

Of *Coral Sea's* sixteen combat losses, most were to AAA. Four were downed in the "safer" Route Packs II and III in an eleven-day period. In contrast, the 1967-68 cruise on USS R*anger*, for which Will Haff took command of VF-154, was:

"A near flawless show with no combat losses to the squadron, but a few scars. LCDR Lew Mitchell led a flight of two F-4Bs into Laos for a radar-controlled bomb release above overcast. Returning to Ranger, they apparently homed in on the Northern SAR destroyer's TACAN for the let-down through the overcast. When they broke out underneath the clag there was no carrier, only a destroyer and some shoreline, and they were over a hundred miles

north of USS *Ranger* with minimum fuel. The KA-3B tanker was over *Ranger*, where the recovery was under way. A fiasco followed; both crews ejected, and two F-4Bs [BuNos 151499 and 151506] were lost from fuel exhaustion due to aircrew error and poor radar control from the destroyer and *Ranger*. All crew were recovered OK. Later, the new XO CDR Don Pringle and his RIO ejected after take-off from NAS Cubi Point, and the F-4B [BuNo 150644] crashed into an ammunition storage bunker. A fire in the wheel well was apparently the culprit, but there wasn't much remaining for the Accident Board to investigate. It was suspected that the ejection may have been a bit premature. The control tower yelled 'Fire!' to Pringle just after the gear doors had closed, and he had never hit the brakes after raising the gear to stop wheel rotation. Some foreign material from the runway may have stuck to the tire and ignited in the well from the friction."

Both combat losses on the cruise were VF-21's, and ironically both involved the same pilot, LCDR Duke E. Hernandez, "a class act and a real credit to Naval aviation," in Will Haff's view. Shot down by AAA with his RIO, LTJG S. L. Vanhorn in BuNo 151492 on 16 December 1967, Duke was recovered but was struck a second time by flak on 28 April 1968 with RIO LTJG D. J. Lorscher in the back of BuNo 153014.

Fuel starvation losses were an occasional price to pay for a breakdown in the difficult business of finding and landing on a comparatively small carrier deck in a vast sea area, particularly in bad weather or darkness. VF-96 lost two F-4Bs in the same month, December 1965, during their first USS *Enterprise* cruise. On return from a strike mission LTJGs Robert Miller and "Duke" Martin in BuNo 149468 made a series of "bolters" and then could not link up with the A-4C "buddy" tanker, electing to eject when the fuel ran down to 300 lbs. It gave the *Enterprise's* UH-2A SAR helo detachment (HC-1, Det 1) their first business, but they were called out again on 28 December to pull out LT Dean Forsgren and LTJG Robert Jewell, whose F-4B (BuNo 151438) ran out of the vital fluid after failing to trap in bad weather. On USS *Roosevelt*, VF-14 had to land at absolute minimum fuel to achieve the correct weight for the ship's short-span arresting gear. A "bolter" meant immediate in-flight refueling, though F-4B BuNo 151018, flown by "Frenchy" DuCharme, didn't even manage that. Preferring his own judgment to the aircraft's auto-throttle, DuCharme failed to get aboard after a night sortie and was unable to plug into the tanker for the requisite 800 lbs of gas for a go-around. His Phantom flamed out, and both crew returned by helicopter instead.

Sights for Specters

As the Navy took stock of the Phantom's performance in Rolling *Thunder* it was clear that the aircraft's overall outstanding success was tempered by a number of understandable problems. Many of these originated in the conditions and Rules of Engagement of the war itself, but others arose from the unforeseen emphasis on missions other than interception which the F-4B was required to perform. Even before it arrived in S. E. Asia many squadrons had realized that they were likely to use the F-4B, like all other "fighter" types in the war—including the F-8—as a bomber. As the F-4B's nuclear toss-bombing computer was totally unsuited for precision

F-4B BuNo 152980, flying from USS *Forrestal* when the ship returned to its usual Mediterranean operating area. (USN via Angelo Romano)

attacks with conventional bombs it had been necessary to jury-rig a more effective bomb-sight for the first squadrons to enter combat. Like VF-41, VF-96 *Tiger Stripes* found their own solution. In August 1964 when the squadron was designated with a secondary bombing mission the CO, CDR William Frazer, realized there would only be time for installation checks on the new MER and TER bomb-racks on an aircraft, which was essentially devoid of an effective bombing system.

En route to battle the USS *Ranger* called at Subic Bay in August 1964, and some rapid modifications were made to the F-4Bs. The fixed reflector sight-plate on the pilot's cockpit coaming was replaced by the sight from the A-1 Skyraider, which had a movable plate, enabling the pilot to crank in lead angles. Metal collars to hold the "Spad" sight-head were specially cast at Subic's foundry. The sight then had to be calibrated, using the "zero" setting on the plate as the boresight for air-to-air missile, and for the zero position to calculate a correction factor for lead angles, depending on speed, type of ordnance, and dive angle. The control knob which set the "pipper" of the sight had to be re-adjusted for each mission, as it was often displaced by the shock of a deck-landing. VF-96 then tested the system at Subic Bay using Zuni rockets and Mk 81 bombs, the weapons they expected to load for their intended flak suppression role. They became the first USN F-4 unit to develop air-to-ground techniques in this way, and the delivery methods they evolved remained virtually unchanged throughout the Rolling Thunder campaign.

VF-114, aboard USS *Kitty Hawk* in 1965, came up with a similar modification to their gun-sights. Fritz Klumpp joined the squadron when they returned from their first Westpac deployment, remaining an *Aardvark* until the end of the 1966 deployment.

"The F-4Bs that the squadron flew on the first two cruises had no air-to-ground capability. Between cruises we received new aircraft, capable of air-to-ground, but we had to borrow MERs and TERs from the Marines at Cherry Point. Since we had no delivery tactics or sight settings Dave Coker and myself calculated sight settings and developed delivery methods for our squadron. Initially, we calculated settings for 10, 30, and 40 degree deliveries with release altitudes of 3,000 and 4,000 ft. I believe we later added a 45 degree dive. We soon found that our Westinghouse sights would not maintain the boresight setting during carrier arrestment because of a worm gear design, but our Ordnance Officer, LT Ted Wright, found some Skyraider gunsights at Litchfield Park. We commandeered enough for our aircraft, and by casting a ring we were able to swap our Westinghouse sights. Where the Westinghouse sights also obstructed the forward view when coming aboard ship, the Skyraider versions were completely out of sight."

By 1966, when John Nash deployed to Yankee Station, delivery procedures had been thoroughly tested in action:

"Basically, our attack profile was to run along at 3-4,000 ft. Anything below that was almost uninhabitable because of the small-arms fire and 37 mm. Then you'd pop up to 11-13,000 ft and roll in, wing-over or inverted, and accelerate the airplane back. With FACs we would fly race-track patterns at 10-12,000 ft and roll into a 30 degree dive from that altitude. My recovery altitude was around 15-1800 ft, which was in AAA range. On the next cruise there was still no Navy directive on minimum altitudes. Air Wings did it the way they wanted to. When you left the line you'd tell the relieving Air Wing, 'You get below 3,000 ft and you are going to get a lot of flak.' The first thing you'd hear is that they flew their first strike at 500 ft and lost five airplanes. It was everyone learning their own lesson again and again."

Flak suppression was one answer to North Vietnam's burgeoning anti-aircraft defenses, which by 1967 were the densest and most deadly in any war. AAA was by far the biggest threat to the Navy's F-4s, accounting for 53 out of the total of 71 combat losses. The North Vietnamese rapidly increased their AAA network in response to Rolli*ng Thunder*, using Soviet fire-control radars to direct the guns. As Will Haff put it, "AAA simply became intense around important targets, and our aircraft losses increased proportionately. Repeat bombing passes for Alpha strikes were stopped, and you had better make your first run a good one, drop all your bombs, and start jinking as you exited the area."

The F-4 was no more vulnerable to damage from AAA fragments than other sophisticated 1960s jets. Lacking any armor protection, since none had been designed into this missile-armed interceptor, the hydraulic lines would quickly lose their fluid after a small hit, and the stabilator would usually swing to the "full up" position, throwing the aircraft into an unrecoverable pitch-up angle. There was no real defense against flak except to stay away from it or above it, but many of the F-4's missions permitted it to do neither of these. SAMs were statistically less problematic, accounting for thirteen USN F-4s, but their psychological effect was usually horrifying. After the first USN loss to a SAM, an A-4 on 12 August 1965, Naval strike aircraft began to approach their targets at lower altitudes, as the SA-2's *Fan Song* radar was poor at tracking targets below 3,000 ft. This naturally increased their exposure to small-bore anti-aircraft fire. The Soviet SA-2 *Guideline* SAM was designed to kill B-52 bombers at high altitude, where its fragmenting 280 lbs warhead could cause lethal damage within a radius of over a mile. At lower altitudes it could still penetrate thin-skinned jets at 1,000 ft. Initially, Navy pilots had to rely on a visual detection of the missile and then attempt to out-maneuver it by turning into its track at the last moment, breaking the lock of its guidance system. With the adrenaline hitting the stops at that point it is understandable that there were no standard tactics for that eventuality. As Jerry Houston put it, "All that b.s. about planned maneuvers against SAMs you could see and track; most of those droll pearls came from POW experts." To Guy Freeborn, "The greatest threat were the SAMs. There were so many of them, and they were more 'human' because they could track us. Bob Elliott, my RIO, and I had one pass by so close to us we could see the rivets."

The whole business became much more intense at night. Will Haff recounted a hair-raising incident from his 1966-67 cruise on USS *Coral Sea.*

"Night flying was as common as day flying, and for the F-4s it was generally night road recce for trucks or targets of opportunity. I almost became addicted to taking the scheduled night armed recce flights; they were a real challenge. On one of these, Harley Hall was my wingman, and he was as good as they come in talent and attitude." [He later became leader of the *Blue Angels* and was one of the last losses of the Vietnam war when his VF-143 F-4J was shot down in January 1973, and he died in captivity.] "After launch I took a flight up north towards Haiphong and then headed south along the coast about 10-15 miles offshore at about 7-8,000 ft. I thought we were out of missile range and relatively safe. Airspeed was only about 280-300 kts as we burned down our centerline tanks. The sky was absolutely clear, with a full moon, and complacency had me in its hand. I thought two orange blobs towards the coastline were flares from some A-4 aircraft also on road recce. Boy, was I wrong!

The cockpit 'tweedle' [APR-26] suddenly sounded, and its warning lights flashed. A missile or two had locked onto me and was approaching from the right beam. The orange ball was getting bigger, too. I dropped the nose and pushed the throttles to 100 per cent, but not in burner, and prepared to break right into the missiles. I never called a break to Harley, but he hung with me as I pulled into the SAMs. I recall seeing the chaff flashing and reflecting off the wings in the moonlight. The scene in the rear-view mirror was breath-taking. One missile blew up in the chaff cloud, and the other appeared to lose lock and fell off into the Gulf as it ran out of thrust. Steve Rudloff, my RIO, had made the most out of the chaff located behind the cockpit. I continued right and headed straight at the coast, where smoke and a red glow seemed to illuminate the site for our CBUs. A quick pop-up, and there it was, right below the pipper. We unloaded the ordnance and made a few passes, and except for lingering brush fires and a secondary or two everything was dead quiet on the ground. Results are hard to verify in combat, but it was a very satisfying feeling for myself and Harley."

Their escape was facilitated by two modifications to the F-4B in response to the SAM threat. Chaff was a long-established means of confusing hostile radar emissions, and F-4Bs were retro-fitted with the AN/ALE-29A dispenser in a compartment each side of the upper fuselage above the wing trailing edge. Release of the chaff bundles was controlled by the RIO, and flares were also added later. The other aid to survival was the APR-26 threat warning receiver, which gave an indication of the type of threat, though not its location relative to the aircraft. Ken Baldry had the device in his F-4B in 1966-67:

"The most advanced gadgetry we had was the APR-26, a little mini-scope in the cockpit that showed spokes around a clock code for various radars (*Firecan*, etc.), but I really don't remember using it a hell of a lot. You were not tempted to stick your head in the cockpit very much when 'feet dry,' as none of the warning gadgets we had were better than a pair of frightened eyeballs."

Commenting on a later version of the radar homing and warning set (RHAW), Jerry Houston had similar reactions:

"The RHAW gear gave good early warning, but between the threat sector indicator, the flashing red lights, and the high-pitched warbler, mostly they just accelerated your heart-beat. Besides, you couldn't see the threat sector indicator with all those tears in your eyes!"

Each crew had its own solution to the increasing influx of new electronic devices, as the defense electronics industry sought to keep pace with advances in the Soviet radar technology exported to North Vietnam. Will Haff's approach was very direct:

VF-14's usual sister squadron was VF-32 *Swordsmen*. The squadron made one Westpac deployment in 1966-67 and then moved to the USS *J. F. Kennedy*, where BuNo 149461 was photographed taking the wire in November 1968. It was later selected for QF-4B conversion. (via Peter B. Mersky)

"I always lightened my cockpit load by simply turning off all the things and equipment that bothered me. It is still probably the way a pilot should handle the load. Unless on a vector from an airborne controller or PIRAZ [radar ship], I always shut down almost all electronics, including radar, while approaching the coast prior to going 'feet dry.' That was not the case on the night recce with Harley Hall: everything was on and saying, 'Here we are, come and get us!'—and they tried to do just that. I never feared SAMs with my electronics shut down, as 'skin painting' [by hostile radars] is marginal, especially at lower altitudes. It is the old game of simply trying to out-fox your adversary."

Fritz Klumpp described the *Aardvarks'* tactics to avoid the NVA defenses:

"I believe that in the early stages of the action over the North we lost fewer aircraft than the Air Force because of our attack profile. Since we had no ECM capability to detect SAMs we elected to proceed low-level and climb to 3-5,000 ft crossing the coast. We kept g's on the aircraft all the time, jinking in altitude and heading. Approaching the target we would light burner to clear the smoke trail and pop up to 8-10,000 ft to roll in. We would 'fan' the target so that everyone would be on the target at the same time, rolling in on different headings: 'small aircraft, large airspace.'"

A much fuller response to the SAM threat was Project *Shoehorn*, an initiative which was powered along in 1966-67 by RADM Julian Lake, who continued as an electronic warfare consultant into the late 1990s. So-called because the dense structure of U.S. combat aircraft required the RHAW components to be "shoe-horned" into any available internal space, the system was fitted to A-4E/F Skyhawks and F-4Bs during depot-level maintenance as the AN/APR-30. An antenna was installed in the redundant AAA-4 IR sensor housing under the radome, and another at the rear of the fin-tip. Many aircraft had a further antenna on the front of the fin-cap. Navy and Marine F-4Bs did not carry the external ECM pods which USAF units received from September 1966 onwards when the first deliveries of QRC-160 (ALQ-87) barrage jamming pods were delivered to the F-105 units in Thailand. They relied instead on specialist jamming aircraft. LT COL John Harty, who was McDonnell-Douglas' "Mr Maintenance," working on the Phantom project for 28 years and becoming F-4 Support Program Manager, explained that *Shoehorn* was, "the result of the Navy tasking McDonnell for an Engineering Change Proposal (ECP) to provide both active and passive countermeasures systems for the F-4B. All the design and integration was done in St Louis. McDonnell provided all of the Group A kits. The initial 'trial' installation of the AFC (airframe change) was accomplished in St Louis. It was at that time the largest retro-fit change ever attempted on the F-4. We even had to remove the aft fuselage to install some of the provisions. Displays and controls were installed in the cockpits. The biggest challenge was running the wire bundles from the Group B equipment to the displays. Antennas were installed in the nose, tail, and wings."

However good the warning system, the task of avoiding SAMs devolved mainly onto the skill and ingenuity of Phantom aircrew. Once a SAM had "seen" its target it required a little over a minute to lock onto it, so that the aircraft had a little time and a warning of the threat to enable the pilot to initiate evasive maneuvers. However, it was never clear, until radar emissions were picked up from the *Fan Song* radar, which sites were actually occupied by the transportable SA-2 batteries. John Nash remembered how quickly the game changed over the North with the introduction of the missiles:

"The A-4 had the *Shoehorn* modification, and they said, 'The A-4s are the attack airplanes, the F-4s aren't going to need this. On my first cruise (1966) we overflew SAM sites and couldn't attack them because the Russians were on them. We would overfly and report them. After they had been made operational they said, 'OK, you can go get them now the Russians have gone.' On that first cruise I overflew Hanoi at 25,000 ft several times because there was no AAA to bother you and the SAMs weren't operational. It changed rapidly. An old CO of mine from VX-4, Bill Franke, was in the first F-4 to be shot down by a SAM and was a POW for seven years."

The advent of the Texas Instruments AGM-45 Shrike anti-radiation missile also helped to reduce the SAM threat from 1966 onwards. Carried by A-4, A-6, and A-7 Iron Hand flights, the missile was launched within about five miles of hostile emitters and its seeker homed onto the pre-set frequencies of the Soviet *Fan Song* or *Firecan* guidance radars, either destroying them, or more often forcing them to shut down apart from quick radar peeps at potential targets. A further aid was the introduction of the QRC-248 transponder in Summer 1967. Installed in some EC-121 radar picket aircraft flying offshore, it enabled them to interrogate the IFF transponders in MiGs at low level and long range. When it worked, "Bandits" could be more easily identified among the air traffic, and U.S. fighters could be controlled more effectively without risk of attacks on "friendly" aircraft.

Of all the threats to USN F-4 activity the MiGs were the most publicized, but actually the least destructive. The Navy lost seven Phantoms to MiGs during the war in exchange for forty-one enemy aircraft downed. The first losses were to VF-151 *Vigilantes* aboard USS *Coral Sea* on 19 November 1967. LCDR Doug Clower and LTJG W. O. Estes, with wingmen LTJG J. E. Teague and T. G. Stier, were being vectored to attack a MiG formation near Kien An

airfield when they were jumped by two other MiGs. Doug Clower assured the author that these were MiG-21s, and that the pilot of the one which blew the wing off his F-4B [BuNo 150997, *Switchbox 110*] with an AA-2 missile was named Khoa. (Other sources suggest Nguyen Dinh Phuc or Nguyen Phi Hung in MiG-17s). Teague's aircraft, *Switchbox 115*, [BuNo 152304] may have ingested debris from Clower's, or was hit by gunfire. All four men went to the "Hanoi Hilton" as POWs until 1972. Ironically, Doug was one of the cadre of former VF-84 Crusader pilots who were sent to F-4 units to teach ACM skills using tactics which later became the basis of the Top Gun ACM syllabus. With only ten hours on the F-4 he had been directed straight to the ACM syllabus during RAG training and had to do a "run out" maneuver with one of the VF-121 instructors. In this he was supposed to separate about five miles from his opponent and then return and engage him. Instead, Doug "went high" and came down on the instructor's tail five times out of five. As a result, he was marked "below average" by his assessor for winning with non-standard tactics. Things didn't improve when he took on one of the most experienced F-4 exponents and got the better of him in a series of scissors maneuvers. He was told not to make a fool of his instructor and was transferred to ground attack. Although these same F-8 skills were to make F-4 pilots great dogfighters in years to come, Doug had no chance to test them in combat, as he had no warning of the MiGs which shot him.

Poor rearward visibility was one of the F-4's constant problems in the air combat scenario for which the aircraft was never intended. Mid-1950s fighters were designed mainly for high-Mach speed, using engines which were barely able to provide it. Any aerodynamic feature which might increase drag, such as a projecting cockpit canopy which allowed good visibility, was streamlined into the fuselage contours as much as possible. In the F-4, there were numerous reports of RIOs unstrapping in the cockpit and turning around to kneel on the seat in order to watch the 'six o'clock' in a fight. However, as pilot David Daniels pointed out, "That meant unstrapping from your parachute, and I wouldn't want to do that, particularly with a MiG on my tail!" John Nash discovered that other Phantom users found speedy solutions to that shortcoming:

Loaded for CAP, a VF-154 F-4B is positioned on USS *Ranger* for a 14 December 1967 mission. (U.S. Navy)

"You can't see straight aft in a Phantom at all. Internal mirrors were fitted which had very limited use, but it was quite a while before the Israeli practice of fitting an external mirror on the center canopy bow was adopted. I went to Israel the day after the Yom Kippur War stopped, doing a survey for VX-4. They had put an external mirror on the center of the canopy. They did that overnight. It would have taken the U.S. Navy two years to get permission to do that. It helped the Israelis considerably when they had guys aft of them in gun range. When you get guys outside of half a mile from you you're not going to see them in a mirror. It's like those warnings on your car mirror; 'Vehicles may be closer than they appear in the mirror.' You try to find a MiG at a mile and you won't see him in those mirrors."

One way in which MiGs could fairly easily detect approaching Phantoms was by their long plume of engine smoke, sometimes visible at up to twenty miles unless the aircraft was in afterburner, which combusted all the exhaust particles. To some extent the amount of smoke depended on climatic conditions. John "Smash" Nash: "On a bright clear day you wouldn't see any smoke until you were quite close. In Vietnam, where it was humid and hot you could see the smoke trails for several miles. There was an anti-smoke system using fuel additives in early Phantoms, but we never had anything to put in there, even in combat, so we never could use the system. When the 'dash ten' engines [J79-GE-10 in the F-4J] came in they did away with the smoke." In combat, pilots often tried to keep minimum afterburner in use to eliminate the trail, though the smoke could have its uses in identifying friend from foe to other F-4s: "One of the standard calls, when running an intercept on bogies, was 'We're at 20 miles and in burners'" [to avoid unwelcome attention from other F-4 crews].

In most respects the ultimate MiG or SAM detector was a capable RIO who could keep his head turning, scanning for incoming threats. Increasingly, RIOs learned to combine their radar duties with periods with "eyes out of cockpit," and this technique was to be the salvation of many Phantoms. The problem in 1966-67 was a lack of RIOs. On their 1966-67 cruise VF-213 were, "down to eleven pilots and nine RIOs for a time. We had a pretty high attrition rate for back-seaters—they quit." Will Haff judged that it was, "a real overload trying to operate and interpret radar and still keep your head out of the cockpit looking for MiGs or SAMs. In a high-maneuvering, high-g environment the RIO and his radar were almost useless. However, the back-seater was a wonderful set of extra eyeballs when he forgot about the scope and looked outside at the combat environment. My biggest concern was always the safety and responsibility that I had, when I was in the air, for the RIO in the rear seat. I had always been accustomed to single-seaters where, if I crashed or killed myself at least I didn't carry the burden of doing the same to the RIO behind me. The F-4 undoubtedly made me a bit more conservative in my approach to aircraft safety. A true fighter pilot is probably a bit reckless, and that may be the edge that makes him so good in dog-fighting tactics."

One of the frustrations which made the RIO's task harder was the unreliability of the radars of the day. When it was working well the APQ-72/Aero-IA system was unsurpassed, but vacuum-tube technology did not take kindly to the rigors of life on a carrier.

Ideally, an RIO needed to test his equipment prior to a deck launch and mission, but this required a mobile air conditioning system to be hauled into place to cool the radar components. In flight, this was effected by the airflow, but on a hot, static deck overheating and failure could quickly occur. Maneuvering the few air conditioners around a crowded deck was too difficult an exercise during a large Alpha launch. The result, in John Nash's recollection, was that little pre-launch testing was done:

"The first time you knew whether it worked was in flight. Truly, we didn't have much of an air war at the time [1966]. We always wanted the radar working, but the failure rate in those days was about seven-tenths of an hour. You could take off with a good radar, and the odds were that you would come back with at least a discrepancy, if not a 'down' radar. A few airplanes had radars that would stay up for several sorties, and on others the radar was never up. Some planes just had lemons for radars; as much as they worked on it the Radar Techs (AQs) could never maintain the radar, and weren't helped by the heat and humidity. We had tech reps from Westinghouse who might change all the components. Their knowledge was much deeper than our AQs'. It was an endless battle."

Some crew felt that the F-4's radio provision was inadequate. There was a single two-way radio requiring some deft changes of frequency to monitor the action successfully and avoid missing vital updates in the course of a strike. Controls were on the right side of the front cockpit and on the RIO's left side-panel. Mostly, the communications problems arose from the vast number of conflicting transmissions, often made simultaneously or at too great a length by excited aircrew during a mission. However, UHF calls could be interrupted if the F-4 was turning hard, causing a wing to blank off the UHF antenna. A small, perennial problem with Phantoms of all generations was water ingress into the radio compartment. Key components of the system were beneath the rear seat, requiring its removal to perform repairs. Like all U.S. combat aircraft a Guard frequency was available for emergency use and MiG or SAM warnings. All too often during Alpha strikes it could be blocked by the wail of distress signals from downed aircrews' rescue beepers. F-4s were fitted with an auxiliary "receive only" radio which enabled crews to monitor transmissions from other elements in a strike package. Not until the ultimate naval Phantom, the F-4S, did the radio suite comprise dual UHF radios when a pair of new AN/ARC-159s became available.

When all else failed the inevitable ejection from an F-4 usually went as prescribed. Fritz Klumpp was one of the F-4B crewmen who successfully escaped from their stricken fighters and were recovered. During USS *Kitty Hawks'* first combat cruise AAA crippled his plane [BuNo 152233] on 31 January 1966, his thirtieth mission. "We made it to the water before being forced to eject. I was back in the cockpit two days later and completed 106 more missions by the end of May." Fritz received the DFC for the first successful strike on the Hai Duong bridge, and went on to fly for McDonnell Douglas Aircraft until 1997. Although most of VF-114's losses were to AAA, two revealed another hazard of low-level strike sorties when they sustained lethal damage from their own bomb fragments.

On other occasions the ejection went less smoothly. John Nash described another loss to the unfortunate VF-114 on their next deployment in 1967, though the sequence of events began a few weeks earlier. ENS Jim Laing, rated as an unusually proficient RIO, had scored a MiG-17 with LT Charlie Southwick in *Linfield 210* on 24 April during the first authorized attack on the MiG airfield at Kep. In one of the biggest "furball" dogfights of the war they met three MiGs head-on, climbed sharply, and then came down and entered a "wagon wheel" circle of MiGs, each guarding the other's "six." At a mile range Southwick released an AIM-9 which impacted his target MiG-17's wing and sent it tumbling down. Denny Wisely and Gareth Anderson in another *Aardvark* F-4B saw a second MiG closing on Southwick's tail for a missile shot and warned them in time for *Linfield 210* to evade the MiG's AA-2 missile. Wisely then closed on the MiG and fired an AIM-9 up its tailpipe for another "exemplar" kill. AAA damage to *Linfield 210* (BuNo 153000) before engaging the MiGs had knocked out the fuel system, preventing fuel transferring from the wing tanks, and its crew headed for the coast, just making "feet wet" when the starved engines spooled down and the Phantom started to fall. Their wingman photographed Jim as he rocketed out of the cockpit, and the image was frequently published. Charlie Southwick and another RIO, LT D. J. Rollins, were shot down in BuNo 153001 a few weeks later, this time by a misfired Zuni rocket from their own F-4B. Both men became POWS, and the wreck of their F-4B was exhibited in Hanoi.

Teamed with Denny Wisely, ENS Laing was launched with John Nash on a strike mission on 21 May 1967. John Nash takes up the story:

"We were targeted on a truck park south of Hanoi with ten 250 lbs bombs each. I heard Denny's voice screaming that he had been hit [by small-arms fire]. When I had delivered my bombs I saw him at a distance at real low altitude, on fire, and I joined up with him. They were on fire pretty good because they were in afterburner. I had them come out of burner and the fire went out. Part of the left stabilator and part of the leading edge of the left wing had been blown off. Fuel had been streaming out of there into the afterburner and burning. He wanted to go back to the ship, but I said, 'No, we're going to Thailand,' because he was in pretty bad shape. I had them start to climb."

Wisely had been on a TARCAP and evaded three SAMs, but in so doing had come down within reach of the ever-present AAA.

"I watched a missile following their plane and asked if they had ejected one—they had not." [The missiles on 213 NH had started

The F-4B assigned to Geno Lund and Biff Borst (though not the one used for their MiG kill) during a diversion to Da Nang. In later life it became a QF-4N. (J. T. Thompson)

firing themselves off] "At 8,000 ft, heading over the mountains, their airplane started to roll. I said, 'Are you rolling [deliberately]?' and he said 'No.' Then I watched the back-seater, Jim Laing, eject. They must have been doing about 450 kt when they bailed out. His chute opened immediately on the way out of the airplane, which it should not have done. His seat hadn't even detached from the ejection rod when the chute started to open, and I watched his arm twist around the seat. He went through the tree canopy and hit the ground." The tremendous deceleration of the premature parachute opening broke Jim's arm and severely wrenched his other limbs. "I had word from Denny on the emergency radio, and he said, 'I'm hanging in the top of a 200 ft tree! We can hear voices.' Some of the local tribesmen were out looking for him." It took two hours to recover them in an operation for which John Nash acted as airborne controller. "The USN helo sent to pick them up got shot up and his hoist didn't work. They wound up having to take that helicopter, land it in a valley, and destroy it. Then a Jolly Green Giant helo with 'Sandy' A-1 escorts came in and rescued them."

Can-Do Phantoms

Navy F-4s became involved in quite a number of other missions in S. E. Asia, including escort duties for OP-2E *Igloo White* sensor-dropping aircraft and EC-121M intelligence gatherers. There were occasional operations with EA-3B "Whales" of VQ-1 and VQ-2 based at Da Nang. Ken Baldry described a night with the "Queer Whale":

"LT Rich Wilson approached me in the ready room and asked me if I would mind carrying napalm on the night road recce I was scheduled for that night (a useless mission if ever there was one!). It seems the ship's ordnance types had been digging around in the depths of the ordnance storage lockers and come up with four filled napalm bombs. They wanted desperately to get them off the ship, as our commitments in RP VI and RP VIA definitely did not call for napalm. I was to follow one of those sensor-equipped EA-3B Queer Whales around, and if he found a likely IR target my wingman would go 'One potato, two potato' and pop a flare. Then I would commence the 15 degrees glide release that Rich Wilson and I had concocted and release the napalm. All this was going on in radar trail on a moonless night that was blacker than Uncle Ho's heart, and the adrenaline level was right up there! For once, things went pretty well as briefed: the Whale driver said, 'I got a good return,' Steve Amman popped a flare, and I saw a road. Where there was a road at night in North Vietnam there were generally trucks along it, so I let fly with the whole load and broke left for the beach, calling for Steve to break off and join me offshore. When the napalm hit, the whole area for about a mile was lit up, and secondary explosions started going off all along the road. It attracted a lot of attention from the A-4s that were in the area, and they came in and hit some more of it. I never did find out what I hit, but it sure made a lot of interesting explosions!

The other nasty mission I had was again with the Queer Whale down South of Than Hoa. Again, Steve and I were to follow the Whale in five-mile radar trail with me as the tail-end Charlie. Conditions were about the same as far as dark and adrenaline were concerned, with the added gambit of not being real certain of the height of the terrain all through the area. We decided on a pickle altitude of 6,000 ft to be sure the pullout area wasn't stuffed with rocks. The Whale called 'Target' and Steve popped off two flares. I pulled up to start a dive as Steve pulled up high and left to set up his own run. I saw some sort of buildings and decided that was good enough. At about 8,000 ft all hell broke loose in the way of flak; more colors and sizes than I imagined existed! I let fly with my twelve Mk 82s, and Steve dumped his in the same hole. I pulled off in afterburner to get the hell out of there and pulled off the wrong way—towards Laos. RIO Jim Stillinger politely reminded me I was a dumb sh*t and that 'feet wet' was the other way, so we had to exit back through the target area where we were again treated to a great pyrotechnic display. It would have been better than a circus if I hadn't been in the middle of it. After all that all I had to do was negotiate a night carrier landing. It turned out that the Whale had designated a target in the Ha Tinh area, which was a known training area for flak crews. They sure gave us a display!"

Flying escort for the majestic RA-5C Vigilante on its *Blue Tree* and BDA recce sorties was a frequent assignment for Phantoms throughout the war. After a strike on the Kep MiG-nest Ken and Steve were required to:

"...hang around and play nursemaid to the RA-5C photo bird which was coming through for BDA. Steve and I headed down over the mountains to the south of Kep to await the Vigilante. The Vigi and Phantom both produced voluminous quantities of smoke when not in afterburner [both used the J79], and the Vigi was a huge airplane, extremely fast but not very maneuverable. The weather was just about as CAVU as it ever gets in that area, so we could see the Vigi from a long way out and set up an intercept, rolling in about half a mile abeam of him on both sides. The Vigilante RIO was fairly new, and of course had limited visibility in the back seat of that monster. He started a panic call for help: 'MiGs, MiGs!' and my RIO, Jim, keyed the mike and said, 'MiGs don't smoke.' Nothing more was heard from the Vigi, but I would have loved to sit in on that debrief.

Vigilantes were fun to escort, but they could sure run you out of fuel in a hurry. I remember one flight from Vinh to the DMZ at Mach 1.2 and about 2,000 ft. I was on fumes by the time we got to the tanker, and that SOB was still loaded with JP5.

Another Vigi flight I ended up on was one of those luck-of-the-draw things where I was spare BARCAP and only meant to launch if the BARCAP went down, but the Vigi escort went down instead. 'Climax Tower' said, in essence, 'Brief on Guard frequency.' We got a common frequency, and the only question I really had to ask was, 'How much fuel should I take now, and will there be a tanker after we go feet wet?' You knew you were going to go fast; it was only a matter of how long and how far. I just about sucked a KA-3 tanker dry, and then we coasted in north of Vinh and headed for an airfield I had never heard of to see if there were any MiGs there. We didn't find any, but we sure stirred that end of North Vietnam with sonic booms. When we coasted out I got another 6,000 lbs of fuel just to make Charlie time with 'max trap fuel' [about 4,000 lbs, the limit for carrier arrestment]. We tanked so often it was second nature to us, and the F-4 was a nice, stable machine to do it with."

A pleasing study of VF-96 F-4Bs. (via Angelo Romano)

Will Haff expounded on the fuel consumption issue:

"Photo escort was primarily another set of eyes for the RA-5C. It had plenty of speed and was sometimes tough to cover, especially if you were S-turning or jinking above and behind him. Once the RA-5C started a photo run in burner it was a real fight just to stay with him, as he was a clean bird, whereas the F-4 had tanks, racks, and Sidewinders. Low fuel state problems would easily occur if you were not careful."

On recce escort Phantoms flew one mile abeam their charge and slightly above it. Their effectiveness and the RA-5C's speed at low level can be judged by the fact that only two escorted reconnaissance types fell to MiGs during the war, though the RA-5C's losses to other causes were heavy: seventeen to SAMs and AAA. Even recce flights in comparatively safe areas produced some unpleasant surprises. Ken Baldry told of an escort mission where one VF-96 pilot found an unorthodox use for a Sparrow missile:

"Major Russ Goodman, who had been a Thunderbird demo team announcer and had lots of time in the F-100, made the transition to the F-4B at VF-121, and he could soon fly rings around a lot of folks. Early in the cruise Russ was assigned as Vigi escort for a coastal recce mission. The weather averaged 1,000 ft ceiling and less than three miles visibility. The RA-5C was not supposed to go over land, only to use sensors and sideways-looking radar (which never turned up a damn thing in the whole cruise, as I remember it). Somehow it got closer to land than anticipated and got bagged by flak. Russ saw the airplane explode, heard beepers from the 'chutes, and started a search for survivors while he called for SAR. The crew [CDR D. H. Jarvis and LTJG P. H. Artlip, RVAH-7] were in the water, and the CO called Art on the PRC radio, directing him overhead. Russ saw a North Vietnamese PT boat headed in the direction of the downed crew, locked it up on his radar, and fired a Sparrow at it. He said he couldn't tell if he hit it, but the PT boat turned tail and ran. About that time the SAR helo showed up and Russ vectored him onto the downed crew. We all enjoyed the case of Scotch the Vigi CO provided."

Flying escort abeam the photo ship was certainly preferable to an earlier tactic of flying three miles astern. Chuck D'Ambrosia, with VF-14 in 1966, once chased the photo RF-8 over one of the hydro-electric dams in North Vietnam:

BuNo 152316 of VF-161 gets a top-up. (via Angelo Romano)

"He passed over the dam without drawing a shot. When we went over I looked down and saw the shock waves coming off the gun barrels and laying down the trees and foliage for a good hundred feet around each gun as half a dozen very large guns (85 or 100 mm) fired at us. I even saw one projectile go past the canopy. The angle on it was just right. Fortunately, it did not detonate."

Meanwhile, the limited air war also continued, though the majority of the MiG kills still went to the F-8 squadrons who had shot down twenty MiGs by the end of 1968, sixteen of them MiG-17s. Paradoxically, the "last of the gunfighters" scored only three of these kills with its 20mm guns, and in two of those combats either a Zuni rocket or AIM-9D was also used to finish off the MiG. The twelve MiG kills by F-4s in the same period included four MiG-21s. Nine kills were with the AIM-7 Sparrow, including a couple of Antonov An-2 biplanes which were downed by CVW-11 Phantoms with that missile. John Nash explained that those two unusual encounters arose out of an Alert 5 aboard USS Kitty *Hawk* on 20 December 1966:

"We flew our sorties every day, and at nights there was always Alert 5 [two F-4Bs with crew strapped in ready for take-off at five minutes notice]. On my first cruise the longest period of sleep I had was four and a half hours. You could fly two or three times a day and, with the limited numbers of pilots we had, you'd have an Alert 5. This was two hours in the cockpit, in flight gear, on the catapult. I went up one night to relieve a guy called Dave McCrea. Our stateroom was close to the forward wardroom, so I said to my RIO, 'Let's stop and get a coffee and a couple of doughnuts.' We spent about ten minutes there, and then we went to the airplane sitting on the cat. I beat on the side of the Phantom, woke up McCrea, and said 'OK, come on down now.' No sooner had I said that than the Air Boss came up on the speaker and said 'Launch the Alert 5.' In those days we had a contest to see who could get the most night traps (we were young and didn't know any better). Anyway, I said 'Come on down,' and he gave me the finger and said 'The heck with you, I'm going to take off.' So they got the airplane started, and the Air Boss came on the horn again and said, 'Your bogie is at 350 degrees at 25 and you're cleared to fire.' At that point I realized I had made a big mistake with the coffee and doughnuts."

So LT Dave "Barrel" McCrea with ENS Dave Nichols in F-4B BuNo 153019 of VF-213 and LT Denny Wisely and RIO LT David Jordan in 153022 of VF-114 were launched from the carrier with marginal visibility against two bogies. As there were no other U.S. aircraft in the area the slow-moving plots were assumed to be hostile. "They were two An-2s calibrating the surface-to-air radar off Than Hoa in the clouds. Possibly they were trying to drag out the Alert guys, too?"

In other combats LT "Squeaky" McGuigan and LTJG Robert Fowler from VF-161, and LCDR Dan McIntyre and LT Alan Johnson from sister squadron VF-151 scored MiG-17s. McIntyre was another former *Blue Angel* with a long career in F-8s and earlier jets. His kill was not publicized for several years, as there was some possibility that the MiG belonged to the Peoples' Republic of China. Its formation leader was seen heading over the Chinese border at the end of the engagement. Possibly the MiG was merely escaping to safety. It was not uncommon for VPAF MiGs to decamp to Chinese sanctuaries when their airfields were under attack, and the A-4 strike for which McIntyre's *Switchbox 107* was providing CAP was close to Kep airfield. Alan Johnson picked up three MiGs on his radar at about eighteen miles, and the crew made the vital visual ID at three miles. At the time, identity via IFF could not be relied upon since the IFF transponder equipment fitted to U.S. aircraft could malfunction or may have been switched off. On

VF-161 devised an extension of the intake warning band to add to their decor in August 1969. (Duane Kasulka via Norman Taylor)

this occasion *Switchbox 107* (BuNo 150634) was out of touch with Red Crown, the offshore radar controller ship, and F-8 TARCAP aircraft were known to be in the vicinity. McIntyre turned in behind the Number Two MiG and launched a Sparrow, which fatally damaged the fighter. *Switchbox 107* then came under attack by the lead MiG-17, which rapidly turned back into him. Evading him, McIntyre went after the third MiG but then saw that his wingman had the lead MiG after him instead. Turning to aid the threatened Phantom, McIntyre saw the skilled MiG pilot break away and head for China "on the deck" and too low for the Phantom's missiles to lock it up. "Killer" McGuigan and Fowler's MiG-17 on 13 July 1966 was the first of six by the *Chargers*, placing them second only to VF-96 as major "distributors of MiG parts." VF-161 had been the last F3H Demon squadron to transition to the F-4B, LT Bryce Thompson having flown the last Demon to storage at Litchfield, Arizona. With VF-151 they would also become the final deployed USN Phantom squadron.

Left: LTJG Biff Borst (left) and LCDR Geno Lund en route to the ready room. (via Geno Lund)

LCDR Gayle "Swede" Elie and LCDR Robert C. Davis after their MiG kill mission with VF-142. (via Geno Lund)

4

CAS Plus:
Marine Phantoms in Vietnam

Having converted six squadrons, three of which had been F3D night-fighter units, to the Phantom the USMC soon found itself drawn into the S. E. Asia maelstrom. From 1 August 1963 the F-4B squadrons became known as VMFA rather than VMF units, the "A" signifying the attack or close air support role for which the Marines were to use their Phantoms predominantly. The first Marine Air Group (MAG-11) with the F-4B assembled at NAS Atsugi in 1963, and VMFA-531 became the first squadron to be shore-based in Vietnam when they deployed to Da Nang on 10 May 1965 with the 9th Marine Expeditionary Brigade, though they only stayed three months initially. Colonel (then Major) Jim Sherman was among that first cohort, as was Marine record breaker Tom Miller:

"Tom Miller was XO of the group that followed us to Da Nang in 1965. He flew on my wing on ten or so missions, but wouldn't lead as he wasn't in the squadron." On arrival they found that, "Da Nang was a desolate, dilapidated airfield. It consisted of a single runway with two parallel taxi-ways. On the east side were three good-sized hangars, the northernmost occupied by the SVAF flying AD Skyraiders. The center hangar was occupied by transient aircraft and had the control tower on the roof. The *Gray Ghosts* confiscated the third hangar and the area next to it where we stored our ordnance."

Another *Gray Ghost* at the time was Captain Michael "Lancer" Sullivan from Beverly Hills, California, who commented on the

short time that the squadron had to prepare for its CAS/attack roles prior to the first Vietnam deployment:

"The aircraft was capable of conducting air-to-ground, but there were no bomb-racks and ordnance testing was still being conducted by VX-5 at China Lake. We didn't drop our first bomb or shoot our first rocket until just before we went into Vietnam from Atsugi. When we got to Vietnam we had only old WWII or Korean bombs, and they were all high-drag. The supply was limited, and on our first few combat sorties we might have only two bombs per aircraft. We had new MER (for the centerline) and TER (for wing pylons) bomb racks which contained an explosive charge to push a 'foot' onto the bomb and help to propel it clear of the aircraft." The squadron collected these when they visited Subic Bay for weapons training prior to deploying to Da Nang. "If we were carrying three bombracks and dropping single bombs the system dropped one bomb from the center, one from the left, and then one from the right to keep the load symmetrical. However, if all stations weren't loaded you had to manually by-pass the empty station to one that was loaded. Needless to say, we dropped as many bombs accidentally as we did intentionally.

I was on the first F-4 napalm strike, and we dropped twelve Mk 79 Mod 1, 1,000 lbs 'napes'; six per aircraft. Eleven were duds, the reason being that the fuzes were held into the canister by wooden pegs, and of course when the canister hit the ground the impact had

The *Gray Ghosts* insignia adorns the tail of BuNo 148373, one of the squadron's first batch of Block 6 F-4Bs. (via J. T. Thompson)

Gray Ghosts **F-4Bs receive attention from the maintainers in the squadron hangar at Da Nang. (via Peter B. Mersky)**

VMFA-531's later markings included additional yellow flashes and canopy sills. (via J. T. Thompson)

the fuze separate and ignite in the air, but the napalm didn't ignite. So little was known by the ground FACs about the F-4 and its capabilities that the FAC asked us to come back and try to ignite the nape with our 20 mm guns, as it was right where he wanted it. We sadly informed him that the F-4 had no gun."

Jim Sherman was also very aware of the temporary ordnance shortage as the Marines set up their F-4 operations in April, 1965:

"There was no ordnance pre-positioned in Da Nang or elsewhere in South Vietnam. The Navy had sufficient ordnance in Subic Bay and had ships re-supplying the two carriers on Yankee station. All available amphibious shipping had just completed the offload of the Marine Brigade at Da Nang, and one LST was dispatched to Subic to haul aviation ordnance for the *Gray Ghosts*. So for the first ten days the Navy used R6D four-engined transports to fly in ordnance. They preferred to haul in rocket pods rather than iron bombs because of the weight versus volume. (Rockets also had three times the accuracy of bombs). When the LST returned to Da Nang it was loaded with loose 500 lbs bombs in the well-deck. Napalm was in short supply because the Navy didn't use it on carriers. Our first delivery of napalm tanks were WWII external fuel tanks which we filled ourselves. Napalm tanks were extremely fragile and made of very thin aluminum so they would break up and scatter napalm when they hit the ground. They could easily be punctured in loading and were loaded individually. With ordnance generally in the first few months over fifty per cent failed to drop successfully. The napalm was essentially 'home-made,' from JP-5 and a thickening agent that looked like plain jello powder. Every time an aircraft was brought into the hangars it was defueled. The fuel was considered contaminated and could not be used in another aircraft. We stored this fuel in a 10,000 gallon black rubber bladder, mixed some of it with the right proportion of dry powder, and pumped it into the empty napalm containers. A lazy man's way was to put 'x' amount of thickening powder in the empty canister, hang the canister on the airplane and then, with a refueler truck, fill the container with 150 gallons of JP-5. The movement of the aircraft on the ground and en route to the target would hopefully mix the two ingredients. There were two

LTCOL J. R. Sherman, USMC, as CO of VMFA-334. (via J. R. Sherman)

Maintainers at work on a *Gray Ghosts* F-4B in 1965. (via Peter B. Mersky)

FLT. LT. James Sawyer (seated) and SQN. LDR. Ian Hamilton, the first RAF officers to fly Marine Phantoms with the *Gray Ghosts*, after a night trap aboard USS *Forrestal*. (USMC via Group Captain. J. Sawyer)

fuzes—nose and tail—which were phosphorus and produced the high flashpoint needed to ignite the napalm. I have seen two or three gallons of napalm which had leaked out underneath an F-4 being blown away but not ignited by the aircraft's engines as the F-4 taxied over it.

There were also some personal equipment changes to be made for the new environment: "We arrived in-country wearing orange flight suits, and it didn't take a genius to figure out that if we had to use evasion tactics after ejecting over the jungle we wouldn't get very far. So the powers that be ordered us camouflaged flight suits from somewhere. When they arrived they were fit for operations in the arctic tundra, as they were about a quarter of an inch thick. I ended up flying in my old herring-bone utilities, which turned out to be perfect, as they had pockets and we wore them when we weren't

flying, so to fly all you had to put on was a g-suit, torso harness, and your helmet. Of course, they weren't fireproof, but no-one cared because if they ever thought they'd need a fireproof flight suit they'd never go flying anyhow!"

F-4B pilot and author John Trotti commented on the sartorial situation in VMFA-314 *Black Knights* in 1966 during their first Da Nang deployment:

"VMFA-314 wore camouflaged flight suits in early 1966. They were made out of something approximating to duffel bag material—a great idea for crashing through the underbrush, but hotter than blue blazes to wear. We had them made up in Iwakuni by girls from the 'Green Door'"

Manfred "Fokker" Rietsch, also a *Black Knight* at the time, remembered the squadron having colorful, tailor-made squadron

Refueling an F-4B at Da Nang. This was usually the first call after landing from a mission. (via Peter B. Mersky)

party suits, but there were also black flight suits which John Trotti estimated were not the best idea due to the Viet Cong preference for black combat gear, "and after the night of 22 February when the Wing area at Da Nang was infiltrated, we put them away except for a squadron photo."

For the first few days at Da Nang the squadron under its CO, Oklahoma-born LT COL William C. McGraw, was tasked with providing air defense for the base area, but there was no North Vietnamese aerial response to the Marines' arrival and the *Gray Ghosts* quickly moved into the mud-moving business. Initially this was with four-aircraft flights, but after several near collisions over the targets two-ship sorties became standard. Jim Sherman's first mission followed an initial "show of force" flight by the squadron; a ten-aircraft local tour in formation on 13 April, "to let both friend and foe know we were there."

"Shortly after 1700 we were scrambled to contact an airborne FAC on the east end of the Cua Viet River. We flew just off the coast at 15,000 ft and 350 kts. Each aircraft was loaded with six pods of 2.75 inch rockets. The pods were about 20 inches in diameter and produced extremely high drag. I contacted the FAC, and he said the weather was 2,000 to 2,500 ft broken overcast with 2 to 3 miles visibility. There had been an engagement between ARVN

LT COL Tom Miller (left) with CWO John D. Cummings at Da Nang with VMFA-531. In 1972 "Lil" John Cummings and "Bear" Lasseter became the Marines' MiG killers. (courtesy of John D. Cummings)

and Viet Cong, and the VC had destroyed an ARVN personnel carrier, which was still burning on the north side of the river. South of that were three small hamlets, one with grass huts, and the VC had disappeared into them. The FAC radioed, 'I'll drop a smoke flare in the hamlet, and I want you to make a positive ID pass over the target.' All this time we had four F-4Bs and an observation plane trying to avoid one another and visually pick out what the FAC had said. Fortunately, F-4B numbers Three and Four decided to do what I should have told them at the beginning and climbed above the overcast out of the way.

I informed the FAC that I had his smoke-flare in sight and would make my identification run south-to-north. I asked how many VC were there and he said, 'About a hundred.' I pulled the nose of my F-4 up about 30 to 40 degrees, and in the overcast reversed direction and came back in a shallower dive over the thatched huts at 200 ft and 350 kts. As I pulled up the FAC said, 'Here they come. They are running from the huts to the north into the paddy fields where they are hiding.' I did another yo-yo turn into the overcast and came back to the south to see them fleeing into the rice. I had my switches set for 'rockets—salvo,' turned on my master armament switch, and when I was less than 300 ft from them at 50 ft altitude I fired all the 108 rockets at once. My wingman asked the FAC if he wanted him to hit the same area, and the FAC replied that there was nothing left to hit."

Since Viet Cong activity tended to happen mostly at night the practice of bombing under flares was introduced as a means of seeing the target and catching troops and vehicles in the open. It was a tactic which was to be repeated in Kosovo in 1999. In Vietnam, some aircrew found this a vertiginous and dangerous experience, but commanders were under pressure to stop the well-organized, nocturnal assaults and supply movements by their adversaries. Sometimes it worked well. Jim Sherman:

"Shortly after our arrival at Da Nang we were asked if we would work under flares at night. There were five pilots who had come from A-4C Skyhawks, and we were comfortable under the flares.

Cockpit view of a successful "plug" with the tanker, vital to virtually all F-4 operations. (USN via A. Thornborough)

Brits in the Marines: (from left) Keith Shipman, Mike Shaw, Jim Sawyer, and Ian Hamilton with VMFA-323's CO, Norman Gourley (center). (Group Captain J. Sawyer)

VMFA-122 F-4Bs in 1966 at El Toro prior to the unit's first deployment to Da Nang. (Steve Kraus collection via Norman Taylor)

After a series of lectures the 'attack guys' would take a 'fighter pilot' wingman and go out into the dark. Some of the RIOs had flown under flares in Korea. In general, the RIOs didn't mind it, because they felt that what you can't see won't hurt you! The results were often very productive. One night, two of us former 'attack types' were diverted to a location about fifty miles south of Da Nang to contact a FAC aircraft. It turned out that the FAC was on an Admin flight from Saigon, and all of a sudden he had received massive small-arms fire from a valley below. When we were overhead he flew his O-1 Bird Dog over the valley with the lights on to draw their fire. He then proceeded on to Da Nang while my wingman and I set up a race-track pattern at 8,000 ft and dove on the valley. When they opened fire on one of us we aimed at the greatest concentration of ground-fire and dropped a pair of 500 lbs bombs. The wingman used the lead's bomb-blast to roll in on and refined his aim point on the heaviest ground-fire. On each run there was less and less ground-fire, and by our sixth and final run it was minimal. They didn't realize that if they hadn't fired on us we would not have known where to drop.

About two months later the same wingman and I were directed to a location 80 miles south of Da Nang to work with a 'Spooky' AC-47 [gunship], which had a Vietnamese talking to ground forces on FM radio, and they reported a sizeable force crossing a major river. 'Spooky' found a major target area and dropped a string of flares. We each dropped six 500 lbs bombs and departed the target area as instructed. As expected, in two or three minutes the ground forces informed 'Spooky' the enemy were back in the river picking up their dead and wounded. 'Spooky' returned to the area and dropped another stick of flares and we expended our remaining six bombs on them. This was the first confirmed sighting of uniformed NVA troops in South Vietnam."

The air-to-air mission for which the *Gray Ghosts* were so well-prepared never really materialized, and by 1968 the training syllabus was, "terribly light on air-to-air," according to former VMFA-531 member, Richard Tipton. "We had the rudimentary work of intercepts, but almost no ACM." He did a brief attachment to VMFAT-201, a Beaufort-based F-4B training squadron in 1968 and "got a taste of ACM." (Back with VMFA-513 he happened to fly

A true "flying nightmare," complete with 19-shot rocket pods and centerline Snakeye bombs. (via Peter B. Mersky)

The front half of a J79 engine on a transport stand in the Chu Lai engine shop. (Frank Shelton)

The VMFA-323 hangar at Chu Lai. BuNo 149427 (left) is unusual in having a red turbine warning band. It was severely damaged by AAA while with VMFA-314 in 1968, but appeared at Wright Patterson AFB in 1980 as a BDRF airframe. (Frank Shelton)

some vintage F-4Bs, including BuNo 148368, which still had an F4H-1 nameplate and the stick housing in the rear cockpit for the detachable control column.) "Our ACM training was minimal to say the least, and later, when I was training VMFA-212 new pilots, they had considerably more training and ACM work."

The build-up at Da Nang continued with VMFA-314 *Black Knights* replacing the *Gray Ghosts* (who returned to Cherry Point to pass on their experience to new crews) in June 1965. VMFA-323 *Death Rattlers* followed in December, having acquired their F-4Bs the previous August and moved to Cherry Point. The squadron had also done a Cuban Missile Crisis duty at Key West in December 1964 under their CO, LT COL Norman Gourlay. In April 1965 they were put on alert for possible operations in the Dominican Republic and flew armed recce missions as a show of strength. The *Death Rattlers* were at NAS Roosevelt Roads, Puerto Rico, at the time (the first Marine F-4s to visit the base), and they indulged in some energetic CAS and missile-launching practice. Unusually, they were permitted to fire eight Sparrows, hitting supersonic targets with every one. On 25 October they began the long haul to Da Nang via Atsugi.

The *Death Rattlers'* first loss, a week after arrival, was their XO, Major John Dunn, with CWO John Frederick. On a night armed recce over North Vietnam their Phantom (BuNo 152261) was the only Marine F-4 to be felled by a SAM during the war. MAJ Dunn spent seven years as a POW, though his RIO died in captivity. The squadron nevertheless racked up 407 sorties in their first month and embarked on an exhausting variety of missions, including escort for the KC-130F refuelers of VMGR-152 (though most refueling tracks were over-water), *Firecracker* CAP for VMCJ-1's reconnaissance aircraft, and many CAS and attack sorties during Operation *Harvest Moon* and in the *Steel Tiger* area. By the end of February 1966 they had already flown 1,567 combat sorties, and in March they moved to Taiwan to provide Air Defense Alert before returning to Da Nang in July.

Shortly after their return a six-aircraft detachment was sent to the newly-constructed Marine Corps airfield at Chu Lai, and in October the whole squadron moved there, transferring to MAG-13

in the process. Chu Lai, on a coastal site some 45 miles south of Da Nang, was built to relieve the increasing congestion of USMC and USAF units, all using a single runway, at Da Nang, and it became the Marines' own base with short airfield for tactical support (SATS) facilities. Devised at the USMC Development Test Center, Quantico, this innovation transferred the essentials of a carrier deck onto a land foundation and could be air-transported and erected within 48 hours maximum. A 2,000 ft runway of AM2 aluminum mats, complete with mobile catapult and arresting gear (MOREST) was established at Chu Lai and later supplemented by conventional runways. Elements of MAG-12 and -13 combined as MAG-22, operating four Marine F-4 units there by 1969. Like Da Nang, the base was under constant threat from NVA troops, particularly in its ear-

"Phantom Phixer" patch. (Peter St. Cyr)

A VMFA-542 *Bengals* F-4B at the Da Nang fuel pit, 1968. (Frank Shelton)

A VMFA-314 *Black Knights* F-4B with a unusual side-number, 1B. (C. Moggeridge)

liest days. The personnel lived in traditional Marine "field" conditions, comprising tents for the majority.

Also involved in the 1st MAW deployments to Da Nang and Chu Lai were VMFA-115 *Silver Eagles*, an ex-F4D Skyray unit which flew an incredible 35,000 combat sorties during its three periods in the war zone, the highest total for any Marine F-4 unit. They were based at Iwakuni between deployments and were well-placed to return to Da Nang in August 1970 for two further deployments, followed by a final period at the new Thai base at Nam Phong which kept the *Silver Eagles* engaged until August 1973 after almost eight years of war.

In skilled hands a Phantom could operate from these basic, short runways with considerable payloads. "Lancer" Sullivan once

took off from the Expeditionary runway at Twenty-Nine Palms, California, with eight 1,000 lbs GP bombs underwing. "I made it with plenty to spare. The field is at 2,000 ft AGL, and it was summer in the desert. The only glitch on post-flight was a small crack about a foot long in the bottom skin of one wing. It was caused by the undulating effect of the "washboard" AM2 matting." Reflecting on Chu Lai over thirty years later, MGEN Sullivan considered it "a good place to operate from. All we had there were Marine tactical squadrons. The F-4 squadrons were at the northern part of the airfield , the A-4s and A-6s on the southern part. The Wing [HQ] was up at Da Nang and left us alone. We had a great thatch-covered "O" Club with beer costing ten cents. Our aircrew huts were air-conditioned. Life was good. The A-4s initially operated

Wall-to-wall bombs on VMFA-323's BuNo 151459 at Chu Lai. (Frank Shelton)

By 1976 VMFA-323's *Death Rattler* markings had become rather more elaborate, as this F-4N at Yuma shows. (James E. Rotramel)

from the SATS strip and then moved over to the 10,000 ft concrete runway when that was completed. In mid-1969 I remember sitting outside the Club and watching one of the newly-arrived F-4J Phantoms land on the SATS strip [in error]. I guess it was a rough ride. The guy bought lots of beer at the Club later. No damage, except a bruised ego."

The *Black Knights* had island-hopped to Da Nang in 1966 and, like the other based units, began bombing operations which frequently used the TPQ-10 radar guidance system and sometimes involved crews in up to six sorties a day. Established at a series of fixed points in South Vietnam by Marine Air Tactical Control units, the TPQ system was similar to (though not compatible with) the

Air Force's more powerful MSQ *Combat Skyspot*. A radar controller, using an AN/APN-154 augmented receiver in the F-4, guided aircraft to a pre-determined ordnance release point, enabling fairly accurate deliveries in bad weather or through cloud, though it did tend to break lock at low altitudes. Richard Tipton, who was assigned to VMFA-513 *Flying Nightmares* in 1968 later flew TPQs with VMFA-542 *Bengals* from Da Nang and also spent four months as an Air Liaison Officer, controlling TPQ bombing from the ground at ranges of up to 50 miles from the aircraft. He described it as, "harassment fire as opposed to CAS. We filled in with TPQ-10 bombing at night to hit roads and truck sites, etc. Rarely was it used for CAS." Paul Fratarangelo flew 485 combat missions, serving

The *Gray Ghosts* also used the "canvas" of their vertical stabilizers more fully in the mid-1970s. F-4N BuNo 153914 was photographed at El Toro in January 1976. (James E. Rotramel)

The *Black Knights* color scheme on its F-4Ns in January 1976. (James E. Rotramel)

with all four Marine aircraft Wings and eventually retiring as a General in 1997. In the F-4 he qualified to fly as a NFO (RIO) and then as a Naval Aviator (front seat). While at Chu Lai he undertook, "quite a few radar bombing missions." With VMFA-333 later in the war he also recalled MSQ missions in which Marine Phantoms dropped ordnance with Air Force ARN-92 LORAN-equipped *Pave Phantom* F-4Ds from Ubon, Thailand:

"We dropped when we saw his bombs start to fall. Some Marine squadrons, VMFA-542 for one, carried eighteen 500 lbs bombs, six on the centerline MER and three each on TERs hung on the inboard and outboard stations. We called our TPQ/MSQ missions *'Penlights,'* compared with the B-52 *Arc Light* missions. Most of these missions were flown below the DMZ. Marine TPQ controllers acquired a lock-on visually, aided by our TACAN cuts, and

A Chu Lai based F-4B with radome swung back (revealing the radar antenna) and canopy open, showing the pilot's instrument panel. (Frank Shelton)

guided us to a set speed and altitude. We got a 'Standby - standby - mark - mark' signal to drop. Aided by Intelligence we avoided flying these missions inside known SAM rings. They were fairly common in 1967, particularly when the weather was too bad to fly CAS, but they comprised a very small proportion of our missions by 1972-73."

Ordnance-dropping with a Forward Air Controller (FAC) was more usual and more accurate. Richard Tipton:

"I flew with many different types of FACs, from O-1 Bird Dogs, O-2A 'Mixmasters,' OV-10A Broncos, A-1 'Sandies,' TA-4 Skyhawks, and F-4 'Fast Facs.' The main difference was the obvious element of speed. With the slower-moving airplanes you had to keep them in sight so you didn't get in their way, and they worked at different altitudes, also. All of them were great to work with, and most especially when they were in contact with 'friendlies' on the ground and you were on CAS. Most of the time the high-speed FACs (TA-4 and F-4) were used 'out of country' [not in the South] and in the *Steel Tiger* area [Laos]. The biggest advantage in working with a high-speed FAC was that your bombing run was more like a coordinated multi-plane attack rather than a 'spotter' plane plus your aircraft. Many times you joined the FAC in formation and made your pass together. Most of the time it was only a single pass because of AAA, and a second run would have been a death trap. We flew with TA-4s from our own MAG, and quite often they were other pilots from your own squadron flying them. Many times when we were flying *Steel Tiger* missions we used the TA-4 Skyhawk as 'bait,' looking for gun sites, and when the NVA came up on the lone TA-4 we were following in trail and pounced on them with our ordnance. Often an NFO with a hand-held camera was in the second seat of the TA-4, taking photos of gun sites to bring back and target for night drops that night. Nothing like your own photo-bird intelligence!"

BuNo 150442 of VMFA-542 with seven BLU-77 Mod 5 napalm canisters. Each held 63 gallons of jet fuel and 43 lbs of imbiber beads to form the napalm gel. (Frank Shelton)

Another typical *Bengals* warload; M117 750 lbs bombs with M131 fin units. (Frank Shelton)

The pressure on reconnaissance resources in the war area to provide up-to-date target information was severe, and many potential targets had moved on by the time adequate information and clearance by higher authority could be provided. The Marines preferred to do all that with their own resources as far as possible. "The *Bengals* delivered an immense amount of ordnance. We would fly around the clock, and with a whole squadron complement of eighteen aircraft we would fly about 100 to 150 sorties a day. Many of these were short-duration on the CAS hot-pad, but others were sometimes 2-3 hour BARCAPs off Haiphong or 1-2 hours into Laos on Direct Air Support (DAS). We worked on a rotating schedule with the day divided into four to six segments, and as an individual pilot you would catch two or three segments. Most of the time we would work for about 10-14 days continuously and then get one or two days off. We mostly slept when that happened, although we did party a little. The squadron would get a one-day stand down about every three months. Pilots would be worked into the schedule so that for a while you got only day missions, then a mixture of day

and night, then an emphasis on night, etc. I did crew scheduling, and it was quite a trick to move the crews around and adjust for aircraft."

Warloads and Weapons
VMFA-542 transitioned from the F4D Skyray to the "double burner" F-4B in November 1963 and arrived at Da Nang in July 1965. In keeping with Marine tradition they flew their first combat sorties within hours of arrival. Their second combat tour at Da Nang began on 1 March 1966, and the third from Chu Lai on 10 October that year. Between these duties the squadron deployed to Iwakuni for training. By November 1966 they had already passed the "5,000 combat sorties in RVN" milestone and delivered no less than 10 million pounds of ordnance. All of this was loaded aboard the "Spooks" by the squadron Ordnance Team, who referred to themselves as "The Real Tigers." Often it was handloaded using the traditional "hernia bar," or "idiot bar" method, and sometimes assisted by the flight crew. The "bar" was a three-foot metal rod

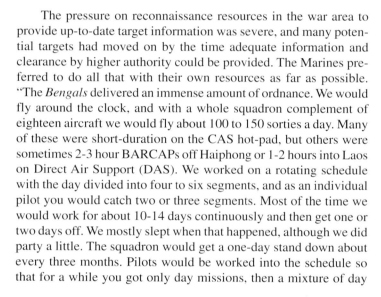

Yet another combination: triple rocket pods on the inboard pylons and Mk 81s on the centerline. (Frank Shelton)

VMFA-323 markings in an early version on F-4B BuNo 150442, which was preserved (as an F-4N) at NAS Memphis twenty-five years later. (via J. T. Thompson)

By October 1972 BuNo 150442 had received revised *Death Rattlers* markings when it was photographed at Kelly AFB. (Norman Taylor)

VMFA-115 flew Phantoms for almost twenty-one years. Their "double nuts" plane, BuNo 153036, carries their distinctive *Silver Eagles* logo in 1971. (Mike Wilson via Steven Albright)

screwed into the fuze sockets at each end of the bomb. The process was eased by the arrival of SATS mechanical weapons loaders which could upload MERs and TERs pre-loaded with bombs in the ordnance staging area. Fuses and arming wires were already in place. This saved considerable time and energy, and the SATS loader could also hang individual bombs, though as Jim Sherman noted: "Experience of the SATS driver was a big factor in the time to load the aircraft. Initially it could take a driver a frustrating 15-20 minutes to get a single bomb rack to lock into position. Within a month the same driver would lock up on the first attempt fifty per cent of the time." Ideally, SATS loading reduced the time to arm a Phantom from around four hours to fifteen minutes.

The range of ordnance available expanded, too, though delivery was made with the same primitive "iron cross" bombsight used by Navy Phantoms. MGEN "Lancer" Sullivan:

"The bombsight was just a drum you turned to get the aiming "pipper" to the proper mil setting." The pipper was an aiming dot projected onto the gunsight glass, and its position would be set in 'mils' according to the ordnance carried and the chosen dive angle,

Maneuvering F-4B WH 7 from its shelter at Chu Lai in 1968. (Frank Shelton)

Caught here at Misawa AB, Japan, on 17 August 1974, BuNo 153036 retains its VMFA-115 scheme but is re-numbered as VE 1. (Norman Taylor)

Revetments at Da Nang provided some protection from Viet Cong rockets, but aircraft were still vulnerable to saboteurs. (Frank Shelton)

Nose marking details of VMFA-314's BuNo 149416. Note the black and white radome. (J.T Thompson)

so that the aircraft would be at the correct attitude when the weapons were punched off their racks. The pilot also had to arm his ordnance. This system remained in the F-4B throughout its career and was carried over into the re-manufactured F-4N series in 1972.

On my second tour we had adapted LAU-33A/A two-tube Zuni rocket pods, and each aircraft carried two pods [on the inboard LAU-17 pylon]. It was a great weapon to get the gunners' attention in North Vietnam or Laos. We'd have our 110-135 mils cranked in on the sight (via the drum) for the bombing setting, but prior to take-off we'd enter in 35 mils on the sight (this was the boresight setting for the F-4 and for the Sidewinder) and then take out a grease pencil and put a mark where the 35 mil pipper was. Then we'd return the drum to the proper bomb-sight setting we were going to use. The tactic was that at the top of your bomb-run (as long as you could get the grease pencil 'pipper' where you thought the AAA was) you'd fire two or four Zunis via the trigger and still have your 'pickle' switch for dropping bombs later in the run.

It was highly effective, since I don't know of a single Marine F-4 loss on a pass in high-AAA territory when they used the Zunis to commence the bomb-run. You can imagine that, for a good AAA gunner, a couple of five-inch rockets going about Mach 3 when they hit the ground certainly would get his attention and cause him to get behind with his tracking. They sound like a freight train coming at you. As far I know the Marines were the only ones to use LAU-33s on F-4s, and not all squadrons did so. I was Maintenance Officer, and we didn't fly air-to-ground without them in VMFA-314."

The LAU-33A/A and its companion LAU-35A were used primarily by F-8s on their fuselage Sidewinder stations. Jim Sherman agreed that Zunis were the weapon of choice for high-threat areas, though he employed a rather different tactic for them: "You couldn't drop bombs and then get your nose back down to where the gun emplacements were. It was best to follow a bomber, offset about 30 to 45 degrees and 3,000 ft above, and then fire rockets on the guns when they opened fire on the bombing F-4." However, Jim was not so happy with the 2.75 inch rocket pods in general use. "The pod

had the drag of a large safe. The rockets had to be fired on 'salvo' to get them all off at once, but they could hit the turbulence of other rockets and go erratic, which wasn't so good for CAS. If you fired them on 'ripple' you had to start shooting at a greater distance, lobbing the rockets to the target and then pushing the Phantom's nose down on the target as you got closer, which caused the rocket blast to go down the engine intakes, and the heat from the rocket motors caused the aircraft engines to go apesh*t."

Paul Fratarangelo had the chance to observe USMC F-4 attacks from the FAC's angle, too:

"Because of our lack of a bombing system, almost all our missions were under FAC control. DAS missions were usually controlled by FAC(A) [airborne]. I recall being surprised to be controlled by FAC(A) up to 50 nm north of the DMZ. We relied on a mark, or the FAC would talk us onto a key terrain feature. Flight lead would usually drop one bomb, and FAC would adjust for the remainder of our runs. I flew missions in the back seat of an O-1E

The *Black Knights* insignia on BuNo 149416, with the helmet plume and squadron number in red. (J. T. Thompson)

VMFA-314 was sometimes called the "Volkswagen Squadron." (J. T. Thompson)

BuNo 150651 shows the location of VMFA-314's red lance marking. This F-4B was previously on loan to the USAF as 62-17177, and it also flew with VMFA-323 as a F-4N, ending its days at AMARC as 8F0089 from October 1983. (J. T. Thompson)

Bird Dog FAC(A) in September 1967. The Army pilot would make the engine backfire, and we would watch the entire human population under us disappear into holes leading to underground tunnels. Because of the 'iron cross' bombing accuracy of the F-4Bs, Marine delivery tactics relied on reducing the slant range to the target at weapons release. Before the Air Force dictated a minimum altitude of 3,000 ft it was commonplace to deliver Snakeye and napalm at 100 ft AGL. I believe multiple runs were also prohibited at the same time that the minimum release altitude was raised. We liked 20 degree dive runs, releasing at 1,800 ft AGL, immediately followed by a 4 to 5g 'pull' and a 'bottom out' at around 600-800 ft.

When I came back to the Tonkin Gulf in 1972 with VMFA-333 the minimum Mk 82 fuze setting had been raised from 2 seconds to 4-6 seconds. When the cloud level went down to below 4,000 ft AGL there was no way we could deliver our ordnance with a time of flight sufficient to arm the bombs. After going through the disgust of dumping a few loads in the water I would re-set the fuzes myself during pre-flight when I knew the weather would be a factor."

Some crews compensated for the lack of an accurate delivery system by using dive angles approaching 90 degrees, a tactic which worked well for slower-moving WWII dive-bombers but took courage to try in a Phantom. The advent of the Naval Weapons Center-developed Mk 82 SE Snakeye retarded bomb removed some of the risk of dropping at low level since the bomb's own extending "petal" airbrakes slowed it sufficiently to explode after the bomber aircraft had cleared the target. The four airbrakes were held closed by a retaining ring and a wire attached to the rear arming solenoid on the TER or MER. When the pilot chose to arm the bomb at both nose and tail the wire would release the retaining ring and deploy the "petals" when the bomb dropped, causing it to fall more directly downwards and allowing safe drops at altitudes as low as 150 ft. Jim Sherman: "On the other hand, if the pilot armed his bomb 'nose only' the arming wire remained with the bomb and the 'spades' [airbrakes] would not deploy to slow the bomb, which

would then go off under the aircraft." Snakeye required a 450 kts upper speed limit for dropping, the same as for napalm, to avoid damage to the retard "spades," but faster drops did occur. Manfred Rietsch: "I dropped them at 550 kts with good success in CAS missions with lots of AAA, and Mk 80 series bombs very fast; a few times transonic. Many of the ordnance limitations were based on a peace-time test environment. In Laos or up North where people are shooting at you, you wanted to be steep, fast, and unpredictable. You improvised on text-book solutions. There was a certain amount of Kentucky windage involved in order to hit the target successfully with a manual bomber like the F-4. That came with experience. I watched a Major from another squadron (a former test pilot) do perfect bombing runs; pipper on, on speed, on dive angle, release at the right altitude. The guy was so perfect his backseater would say his bombing runs were like flying a GCA. The problem was, he couldn't hit his ass."

Pulling chocks on VMFA-542's Number 7 in readiness for another CAS mission. (Frank Shelton)

The "welcome" notice at the entrance to Chu Lai, a base built on sand. (Frank Shelton)

A napalm-laden *Bengals* F-4B eases out for the runway and a CAS sortie. (Frank Shelton)

Snakeyes required slightly different delivery settings, but, like so many other aspects of bombing in the "manual" F-4, they were open to interpretation via experience. COL Rietsch:

"The TAC manual gave you an approximately 110 mil setting for 30 to 45 degree 'slick' bomb delivery at 500 kts. The good book also told you to use a 150 mil setting for Snakeyes, a 10 degree delivery and 450 kts. I never delivered Snakeyes at less than 500 kts (speed is life, specially when you are close to the ground), so I

used 110 mil for everything. An advantage of using the same mil setting all the time was that you dropped by 'sight picture' rather than by numbers. The long-short error is considerable if you drop at 8 degrees instead of 11. The gyros precess and are not very accurate. The bottom line was that most good bombers used sight pictures instead of relying on 'numbers.'"

Also available was the 750 lbs M117 bomb of Korean War vintage (but still in use in 1991), some of which were donated by

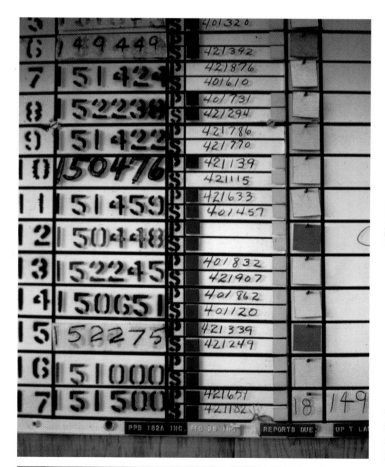

Time for towing back to the flight line at Chu Lai after refueling at the "pit." An EF-10B stands with wings folded in the distance. (Frank Shelton)

The aircraft status board for VMFA-542 at Chu Lai. Each F-4B BuNo is accompanied on the right by its engine serial numbers for port (P) and starboard (S) J79s and red squares indicating aircraft which the power plant shop needed to get back in the air. (Frank Shelton)

F-4B WH18 rolls out from the VMFA-542 revetments at Chu Lai. (Frank Shelton)

Plane captain and pilot do pre-flight checks on aircraft 10 (BuNo 150476) before "pulling chocks" for roll-out. (Frank Shelton)

the USAF units at Da Nang in 1967. It was favored as a blast bomb for destroying light structures or clearing areas of jungle, and it had a retarded version, the M117R. Low altitude delivery of any ordnance, particularly napalm, which required straight-and-level delivery had its attendant risks, as Paul Fratarangelo explained:

"Most of our losses were due to small-arms. We rarely experienced AAA south of the DMZ during 1967. Multiple runs using release altitudes of 100 to 1,800 ft over intense automatic weapon fire was the reason for most of them. But Army and Marine ground forces were all over the 1 Corps area [the Northern half of South Vietnam], and when we were needed for CAS we were never deterred by ground fire. Perhaps that is why Senior HQ changed the delivery Rules of Engagement."

A new weapon in 1968 was the Mk 20 Rockeye II, also developed by the USN at China Lake and used extensively in Vietnam and, much later, in Operation *Desert Storm*. It used the ISC Technologies Mk 7 dispenser to hold 247 Mk 118 dart-like sub-munitions which had both armor-piercing and anti-personnel capability. Although highly efficient and popular as a flak suppression weapon, the Mk 20 posed certain handling problems, as Jim Sherman described:

"The bomblets had no safety devices, and the Navy had great apprehension about storing them aboard carriers. In case of a hung canister they would not bring the aircraft back aboard ship in case the Rockeye came off on arrestment and scattered live bomblets on a flight deck loaded with aircraft. At Da Nang and Chu Lai they put armor plate on two crash trucks. When a hung Rockeye fell off upon landing the two trucks hooked onto the overrun anchor chain and drew the weapon backwards down the runway to sweep and detonate the bomblets. It raised havoc with the runway lights! Also, with a maximum four-second delay, this allowed release at 8,000 ft with the canister opening four seconds later at 1,500 ft. If you had no target altitude information there was a fifty per cent chance of a dud when the canister hit the ground."

Although there was a low risk of meeting the Vietnamese Peoples' Air Force over targets in Laos and near the DMZ, missiles were carried as a precaution. The F-4 was, after all, a fighter! Rich-

ard Tipton: "Missiles were normally carried on every mission. When we flew CAS we carried one or two Sparrows and maybe two Sidewinders. When we went on DAS up North we carried all we could, especially Sidewinders. We had fired them in the USA during training, and actually got a chance to go to the Pacific Missile Range from Cubi Point to do it again while we were in-country."

One of the most controversial extensions of the F-4's armory was the external gun pod. Because the USAF's Phantoms prior to the F-4E lacked internal guns the Air Force developed the SUU-16 and SUU-23 gunpods using the M61 Vulcan rotary cannon. As it was best carried on the centerline pylon the Navy ruled it out for their F-4s, since the hardpoint was needed for the 600 gallon tank. USN F-4s could catapult-launch with two 370 gallon wing-tanks, but they tended to cause unfavorable center of gravity situations on launch and were hard to upload to the aircraft on deck. However, some units did train with the weapon. Bill "Burner" Beardsley flew a combat cruise with VF-102: "We practiced with the gunpod for possible close air support missions, but all our flying was up North

Skilled hands safety-wire components in place after fixing a fuel system problem on a *Bengals* F-4B. The six internal fuel cells were "piped" together in the center fuselage, on top of the cells, and accessed by removing a large cover plate on the upper rear fuselage. (Frank Shelton)

VMFA-312 flew F-8E Crusaders during their time at Da Nang, transitioning to F-4Bs on their return to Cherry Point in 1966. BuNo 152281, seen at Kelly AFB in December 1971, was assigned to Major Hadley and LT Derringer. (Norman Taylor)

A line-up of VMFAT-101 *Sharpshooters'* first batch of F-4Bs in 1969 at El Toro where they trained Marine F-4 crews. Codes changed to "SH" in March, 1973. (C. Moggeridge)

and we never used them in combat." In the USAF the gun pod was used with some success from May 1967 as a close-in, air-to-air weapon where it was responsible for the destruction of ten MiGs. Initially, it had been used as a strafing weapon, though accuracy was reduced by the lack of a lead-computing sight and the difficulty of aligning the 2,000 lbs pod exactly on its belly mounting. It had to be bore-sighted every time it was hung on the F-4.

Deployed Marine squadrons began to receive the Hughes Mk 4 HIPEG pod with Mk 11 gun, but it had mixed success. John Trotti pointed out that HIPEG had originally been conceived as a back-up weapon for the Marines' Ontos 106 mm rocket launcher vehicle, and Hughes sought another application for it after Ontos was canceled. It appeared beneath a variety of USMC types:

"I first fired one from an A-4 Skyhawk, and at a ten-degree dive the little 'Toot' damn near stopped dead in its tracks. Luck prevailed, and the gun jammed after about two seconds. Next time we saw the 'pistol' it carried a one-second burst limit to keep the barrel inserts from burning up." It is likely that VMFA-115 *Silver Eagles* were the first to use the pod in Vietnam, combined with other ordnance. The ammunition used for the gun included proximity fuzes screwed into the nose of HE 1 rounds. "While the combination had a great history of success in two previous wars, it was over-matched by the HIPEG whose pistol-like cylinder, in revolving in a counter-clockwise direction at 6,000 rpm, managed to un-

seat the fuzes (or the aluminum caps used in their place), allowing the barrel rifling to finish the job of disengaging them. Low on mass, the fuzes became intake fodder, dinging some engines pretty badly. I believe one of VMFA-115's pilots had an inflight engine failure leading to the hasty withdrawal of the HE 1/proximity fuze combo.

In Spring 1966 VMFA-314 had a gun pod explode, ripping the rack and attachment fittings right off the bottom of the F-4B. While the aircraft was repairable, it had to be flown back to Nippi in Atsugi. HIPEG's main component was an eighteen-inch, 500 lbs cylinder revolving at 6,000 rpm; a pretty potent gyro. Things are OK at 1g, but the minute you go below 0.8g or above 1.2 g the gyro begins to precess. At 2.5 g the 'right-hand rule' wins out over the sway braces on the pod's mounting rack and the pod wants to slew sideways and pitch up."

For Manfred Rietsch:

"The advantage of the gun pod was its great accuracy. You felt comfortable using it close to troops, as there was little lateral error. The gun with 750 rounds gave you three to five passes—a long time to keep the bad guys' heads down. It was also an excellent weapon to go after targets on steep hills or on the face of sheer cliffs. I once had a target near the Laotian border which had a bunch of NVA hiding in a cave half-way up a cliff. A perfect target for the gun or Zunis but, alas, the pod again went 'burp' and jammed. I

F-4B BuNo 151436 from VMFA-314.
(via Angelo Romano)

A *Silver Eagles* F-4B displays the unit's tasteful markings. This aircraft was destroyed by fire in October 1974. (via Angelo Romano)

'accidentally' jettisoned that particular pod over the water, since this was the third or fourth time it had jammed on me while going after juicy targets.

Obviously the biggest drawback to the pod was operational reliability. In 1968-69 I recall a 30 to 40 per cent reliability with the pods. It got to the point that the pilots didn't want to carry it, and it was viewed as useless junk taking up a weapons station. During the summer of 1969 Hughes sent a new 'Mk 4' tech rep to the Group, and the reliability of the pods improved considerably. In retrospect, I'm sorry I jettisoned that pod, but I was mad!"

With VMFA-334 *Falcons*, Jim Sherman also found that the pod could evoke some strong emotions:

"The normal probability of the gun firing all its ammo was less than 25 per cent. Often it would jam on the first burst. The *Falcons* had a fire-out rate of 80 per cent. Why?" The squadron had experienced a three-month delay before deploying, and during that time Sherman arranged for one of his Corporal Ordnancemen to attend a special course on the gun pod at the Los Angeles factory where it was made. "While in Vietnam all he did was work on gun pods all by himself. He was there observing when you took out *his* gun pod, and he was there when you taxied in to find out first-hand how *his* pod worked. I made a mistake one day of responding to his query with, 'Your G— d—n gun jammed on the first burst!' As I walked back to the hangar I noticed he was sitting on a bench, crying. As I consoled him he explained that he felt he had let the squadron and the ground Marines down, because it was a matter of life and death to them that *his* gun pod worked perfectly every time."

This devoted Ordnanceman would not have been pleased by the reaction of "Lancer" Sullivan's CO in VMFA-314 to the gun's reliability problems: "It jammed most of the time, and on several sorties you couldn't even get one round out of it. Our CO said to check it over the water on take-off, and if it didn't work, use it as a bomb and jettison it on the enemy. The Hughes tech rep was the most despised man at Chu Lai!" Richard Tipton's squadron carried the gun, mostly with napalm outboard. "If you used it in short bursts it worked fine, but if you held the trigger down it would overheat and jam." In other squadrons the weapon was more popular, and VMFA-122 were unique in re-wiring some of their F-4Bs to carry

pods on the outboard wing stations, also. This three-gun installation was impressive, but the pod's unreliability made it a rather hazardous configuration to fly, since a jam in one outer pod while the other two were blasting away could cause recoil to slew the Phantom off course. Some of these "over-gunned" Phantoms found their way into other units, and their unorthodox wiring for that and other ordnance combinations baffled the maintainers. Manfred Rietsch:

"We had two aircraft in the squadron that were configured as 'superbombers.' They could carry twenty-four 500 lbs bombs. These aircraft could go to the DMZ, give the FAC ten minutes on station and return with minimum fuel to Chu Lai. Only experienced pilots flew them. The reason for this configuration was that we had intermittent insoluble fuel transfer problems in the outboard stations of these two aircraft. We finally threw in the towel and took the tanks off. They became of limited utility, but we had to do something. We had received these Phantoms from VMFA-122 where the CO, Col John Verdi, apparently was experimenting with a variety of unorthodox configurations, for example, three gun pods. His maintenance people changed some wiring to the outboard stations, and we could never figure out what they had done. Hence, the 'superbomber.' Even the depot maintenance electrical experts at Atsugi were unable to cope with these random modifications, and it was left to individual squadrons to try and sort out the mess."

Ordnance loads were varied according to the mission, one of which was Hot Pad Alert which, according to Richard Tipton, was similar at both Da Nang and Chu Lai:

"Normally two aircraft were assigned to the pad. The typical load was either 'Snake' [Mk 82 SE] and 'nape,' with maybe daisycutters on the snakes, or perhaps 'nape' and gun pod. All flights were controlled by the 7th Air Force, and they also had USAF F-4Es of the 366th TFW across the field from MAG-11 in 1969-70. They had concrete revetments at the south end of the runway, and we had open sheet-metal revetments on the north end. Ordnance loads were set up by 7th AF in an overall manner so that they could meet any emergency situation with the appropriate ordnance. Crews were on Hot Pad for eight hours. Four of us would sit in the ready room fully suited up and play acey-deucey or cards, or doze until the bell went off. The Phantoms were pre-flighted at the beginning of the period, and the ground crews were assigned over the same period. When the Alert bell went we would run to the aircraft and climb in.

The RIO would be getting the target and DAS controller to contact us, and also sorting out the radio frequencies, while I would be starting the aircraft. All this time the plane captains would be strapping us in at the shoulders. With one hand I was starting the engines and with the other I would be fastening the seat belt. Last, as we taxied we would be hooking up the leg restraints [which pulled the legs back against the seat on ejection preventing injury from flailing limbs] and closing the canopy as we rolled. Meanwhile, the RIO would be getting take-off clearance from the tower. It was one continuous roll from the chocks to the runway and full power, then afterburner. We were armed in the chocks as we started. All of this could be accomplished in less than five minutes with practice.

At Da Nang there normally wasn't that much wind, and if there was it was right down the runway. Seventh Air Force would scramble both the VMFA-542 birds and the 366th TFW F-4Es on the other side of the field. We would try to get airborne before they did. We would roll from the north end on the western runway and they would roll from the south end of the parallel eastern runway. We would invariably pass each other head-to-head at mid-field! If the target was north of Da Nang they would get there first, and vice versa. This is the only time I ever saw jets taking off from the same airfield in opposing directions as standard operating procedure for the Hot Pads. Da Nang was the busiest runway in the world, with a take-off or landing every thirty seconds around the clock, on average. When a crew returned their aircraft was hot-refueled and re-launched if needed as soon as new ordnance could be loaded. After eight hours the next crew came on. We only pulled the Hot Pad three or four times a week, but the squadron maintained it most of the time.

Sometimes you would have a five-minute Alert section standing by with aircrew strapped in, but most of the time you would be on fifteen-minute alert in a trailer with flight-gear on, or the one-hour alert, napping in the trailer. As soon as a section was launched the pad had to be reconstituted, ordnance loaded, and aircraft pre-flight. The thirty-minute or one-hour crew would then move up and become the fifteen-minute pad. The Hot Pad was good when things were happening on the ground. You got launched often, and they were usually good hops, with troops in contact, etc. When it was slow you got lots of naps."

Manfred Rietsch added that the Hot Pad, in his experience, rotated the squadrons every week.

"At times you had an airborne section on Alert for immediate response. In this 'soft ordnance' section one aircraft would carry six Mk 81 or Mk 82 SEs on TERs and the Mk 4 gun—which didn't work most of the time—on the centerline. The other F-4 had wall-to-wall napalm. Certain pre-fragged missions also carried that load. The 'soft ordnance' would be delivered at low angle (from 0-20 degrees) and was particularly useful with troops in contact (TIC). Since the low angle delivery took out most of the chance of lateral error you could get pretty close to the troops. During one mission when our 'grunts' were in deep trouble getting overrun I delivered 'nape' within 30 to 40 feet from the friendlies, running parallel to their front line."

The Marines occasionally shared the Navy's offshore BARCAP duty, and the two Marine bases operated an air-to-air pad, partly for point defense early in the war. However, in Manfred's estimation, "The air-to-air pad was a joke because it rarely or never got launched. Occasionally they might take a launch at night to go to Yankee Station if the Navy couldn't or didn't want to take the CAP. Since the day air-to-air pad never got launched I remember one of the sister squadrons towing a 'hangar queen' [unserviceable 'parts locker' F-4] to the Hot Pad spot!" These moments of relief for the Navy's BARCAP chore sometimes arose due to a lack of ready decks off-shore, but they could lead to some unexpected hardships for the Da Nang crews. John Trotti once found himself stuck on a BARCAP orbit off Haiphong, expecting a four-hour mission with one tanker 'plug.' When his relief failed to show up on two consecutive occasions the patrol dragged on for over eight hours.

VMFA-531 *Gray Ghosts'* **BuNo 152275. (via Angelo Romano)**

Manfred Rietsch observed a change in the pattern of activity after the 'bombing halt' from November 1968 when the number of carriers on Yankee station was reduced:

"We ended up flying more late night CAPs as a result. These were usually terribly dull missions, since the MiGs wouldn't fly at night and would rarely come out over water. We suspected that this was the reason the Navy pawned off these missions on us, though the carriers did need some down time. Even so, it would have been nice to fly some day-time hops when there was some activity. The joke among our aircrews was that the 'O' Club at Kep airfield had a coin-activated radar scope, and that late at night, after many beers the MiG drivers would put a few coins into the radar machine and watch us dumb Marines orbit off Haiphong. In all seriousness, these missions required considerable assets. It took 45 minutes to get to the CAP station from Chu Lai. Normal on-station time was one hour—unless your relief didn't show (which happened frequently). Of course, we needed a new KC-130 tanker every couple of hours. The weather was usually not that great, and tanking at night from a 210 kts KC-130, tobogganing downhill with a heavy F-4B and dodging weather was not a lot of fun.

The squadron's average air-to-air effort was at most 20 per cent. Since the emphasis was on sortie generation and tonnage dropped it was difficult to keep the air-to-air weapons and radar working. The F-4B's tube-type pulse radar required a fair amount of maintenance and didn't like the heat. When I returned to the States in 1970 we insisted on 'up' radars in VMFA-531, and with a good Avionics Officer who was also a RIO (Jack Bardon) we had good systems availability."

At least BARCAP offered the slight chance of a MiG engagement to fighter crews, though there was also that possibility on some of the escort sorties they included in their menu. "Escort" offered some varied missions: "Lancer" Sullivan even found himself flying the first Marine helicopter escort, leading four F-4Bs riding shotgun on a gaggle of forty-four helicopters. Another was to accompany Marine A-6s on their nocturnal strikes in Laos. Richard Tipton remembered *Steel Tiger* A-6A escorts in 1969-70, "up and down the country all the way to the Mu Gia Pass at the junction of China, North Vietnam, and Laos. When the A-6s dropped and were out of the way we would go in after the guns that would come up, following their tracers and dumping Rockeye on the sites, as well as bombs. VMCJ-1 [a USMC composite reconnaissance and electronic warfare squadron] was based at Da Nang, and since we were the only F-4B squadron there we would be tasked with escort for their RF-4B Phantoms, EA-6A Intruders, and 'Willy the Whale' EF-10Bs. Sometimes we would escort B-52 *Arc Light* drops." Escort assignments tended to vary from squadron to squadron. John Trotti flew escort on the EF-10B in 1965-66, but they were beginning to be phased out by September 1966. "Thereafter our duties were almost exclusively for our photo-birds, Fast FACs along the Trails and (of course) to protect the mighty BUFFs as they came to collect their flight pay, combat theater pay, and hostile fire pay all in one fell swoop!" By 1970 the VPAF was so determined to destroy a B-52 that new MiG diversionary airfields were built near the Laotian border, and there were a few occasions on which MiGs came very close to achieving a kill despite the F-4 escorts. Manfred

Rietsch also flew "shotgun" on *Arc Lights*:

"When the air campaign focus switched from the North to interdicting the Ho Chi Minh Trail we also occasionally did B-52 escorts over Laos. We would pick the BUFFs up over the water near Da Nang, go into Laos, bring them back 'feet wet' and go home. Sometimes they would exit south over Thailand. These were also very dull missions, unless you got nearer the North Vietnam border." One of these escort flights almost led to a major international incident. "The night after Ho Chi Minh's funeral, during a B-52 escort mission, we ran a night visual ID on the International Control Commission airplane coming out of Hanoi en route to the Cambodian capital. The intercept occurred over North Vietnam about 120 miles south of Hanoi. The Russian Premier Kosygin was on board. However, the Air Force controller agency, *Moonbeam* in Thailand, was unaware of the flight when they vectored and cleared us on the 'bandit.' There was a lot of attention after we landed!

By mid-1969 I estimate that 25 per cent of the air-to-ground effort went into Laos. Some real interesting and challenging missions were night A-6 escorts, looking for trucks on the trail. We would provide flak suppression while the A-6 was rooting around at low level chasing trucks. The single-ship F-4 would carry two 2,000 lbs bombs with 'daisy cutter' fuzes, four Zuni pods, two CBUs with instantaneous fuzes and two with delayed fuzes. With a load-out like that you got a lot of respect. Sometimes we would get more secondaries, going after the guns and staging areas along the trails, than the A-6 did going after the trucks."

Only the more experienced crews flew these challenging and exciting hops. Sometimes the F-4 escort could be less than welcome. Tom Idema, flying USMC A-6s, said that he hated to have them rendezvous with him for MiGCAP:

"They always used their intercept radar to effect the hook-up. Personally, I hated that since that was a fire-control radar 'feeding' their missiles: one mistake and I'd catch one up the tailpipe. On one mission I had an F-4 joker who continued to hone his skills on me even after repeatedly telling him not to do so. He did it one more time from close range and I went into active ECM mode and overpowered his radar, rendering it useless. He had to abort his mission and return to base. He was upset—but so was I."

After the bombing halt we would also do escort missions for our RF-4Bs out of Da Nang over the North. These missions were mostly from Vinh to the south, but occasionally you would go fur-

VMFA-321's star-spangled scheme even extended to its TERs. This F-4B, seen at NAS North Island in 1975, was later preserved in VMFA-323 markings on the USS *Intrepid* Sea Air Space Museum. (Duane Kasulka via Norman Taylor)

ther north. Our F-4 escort was usually configured with a couple of Sidewinders and a couple of Sparrows. In addition, we carried two or four four-shot Zuni pods for flak suppression. These missions were fun: the RF-4Bs would accelerate over the water to 550 kts plus and then go feet dry at about 6,000 ft. We would hang above and behind them at about 12 to 14,000 ft. A slick RF-4B was hard to keep up with, specially when we were carrying Zuni pods. The RF-4B would make a straight run from the water over to Laos, turn around, and make another run from Laos to 'feet wet,' and that was the mission. I got a couple of [MiG] vectors on these hops, but never got closer than twelve miles to a MiG. Usually you could find some 'hostile AAA sites' on the second run across and go after them. We didn't see any MiG activity in Laos. After 1 November 1968 the MiGs would fly, but generally stay inside the North. There was an occasional stray into Laos, but no encounters during the time I was there."

Although the pressures of flying in such hostile climatic conditions and against such formidable defenses were taxing enough, there were still other unexpected hazards. One was the ever-present risk of a mid-air collision in the crowded skies, particularly around Da Nang itself. The KC-130F tanker tracks were off-shore or close to the beach and normally safe. However, on 18 May 1969 a VMGR-152 KC-130F was passing fuel to two *Black Knights* F-4Bs over Phu Bai when a *Bengals* F-4B, *Roughneck 250* flown by LT Charles W. Piggott and CAPT John L. Nalls impacted the Hercules' right wing at the Number 3 engine. As Richard Tipton recounted, "It was Piggott's first or second flight in-country, and Nalls was in the last week of his thirteen-month tour. It was a visual reconnaissance flight in which an experienced RIO took a new pilot up to familiarize him with the ground under him so that he could pick out landmarks to navigate from. They didn't even carry ordnance other than perhaps a missile. We lost the Phantom and its crew, the tanker went down with all lost, and *Thunder 2*, the Phantom refueling on the right wing, was destroyed, though its crew (Maj J. Moody and his RIO) ejected. *Thunder 1*, flown by MAJ. A. Gillespie and 1LT Vern Maddox, managed to get their damaged F-4B back to Da Nang." In all, the *Bengals* lost six Phantoms and five crews during their 1969-70 tour.

Attacks on the two USMC bases by NVA and VC infiltrators also caused severe attrition at times. Chu Lai was hit by some forty-eight 122 mm rockets on the night of 31 January 1968, killing CAPT Arthur Delahoussaye and seriously injuring 1LT Richard Kerr. Two VMFA-314 Phantoms were destroyed and the base's bomb dump was blown up. Undeterred, VMFA-323 launched missions within a few hours of the attack, and despite continued rocket attacks over the following months flew over 600 sorties per month, many in support of the besieged Marines at Khe Sanh. More worrying for aircrew was the prospect of capture and imprisonment after being shot down. *The Death Rattlers'* CO, LT COL Harry T. Hageman, and his RIO, CAPT Dennis Brandon, were leading a flight of "nape and snake" armed F-4Bs against a suspected AAA position on 7 January 1968. During the first low-angle pass their Phantom took several 50 cal. hits in the belly and started to burn. Brandon, with 300 combat missions to learn from, realized their plight and ejected immediately. The CO stayed a little longer, but the Phantom began to hurtle end over end at 100 ft altitude. Hageman ejected, using the secondary handle [on the seat pan], and fortunately exiting at a moment when the cockpit was pointing upwards. His parachute had barely deployed when his heels hit the ground a short distance from the flaming wreck of his aircraft. Both men hid in the elephant grass and evaded NVA troops, who they could hear looking for them. While their wingman held off the soldiers with some low, though unarmed passes, the rescue helicopter slipped in and recovered them.

LT COL Hageman's predecessor, LT COL Edison W. Miller, was not so fortunate. During an attack on two large, tracked vehicles in October 1967 his F-4B took a 37 mm hit, and he with his RIO, 1LT James Warner, were forced to bale out near Cape Mui Lay. His parachute was seen to enter a treeline near the village, while Warner landed just to the south. Warner was reporting his safe landing and arranging a pickup location on his URT-10 rescue radio when three NVA soldiers appeared and took him off to prison. They were both reported MIA, but in fact they spent five years in captivity.

Phixing Phantoms
Keeping the F-4s airborne from the two bases was in any case a tough enough task. Chu Lai's runways were very hard on tires, particularly the narrower F-4B type. Manfred Rietsch: "The F-4J with its 'fat' tires did a lot better at Chu Lai than the F-4Bs with their

VMFA-323 progressed to F-4Ns, deploying on USS *Coral Sea* in 1979-80. (via Angelo Romano)

skinny ones, and you got a lot more landings out of them. The biggest cause of tire wear and failure was heavy gross weight and heat. You had to taxi very slowly with a fully loaded aircraft and avoid making 'square' turns, or you would wear an immediate bald spot on a tire or even blow it. The F-4B nose gear tire was limited to about 180 to 190 kts. At 53,000 lbs a F-4B on a 110 degree F runway temperature was very nose heavy. If you didn't use full aft stick at 100 kts on the take-off roll you could hit 180 kts before the nose gear was fully off the ground." John Trotti added that the tires on the F-4J model were, "prone to aquaplaning because they had lower pressure (215 psi against 235 psi for the F-4B), and the wider footprint reduced the tire-to-pavement contact pressure." On landing a Phantom of any variety it was necessary to hold full back stick to increase that contact pressure as much as possible so that the induced friction could help to brake the aircraft.

Avionics were the main source of maintenance overtime, and reliability declined rapidly with the fairly basic facilities available at the South Vietnamese bases. John Trotti: "When we (VMFA-314) picked up brand new '152 Series' [BuNos beginning 152] F-4Bs in 1965 you could lock up the planet Pluto with the radars. By March 1966 if the Goodyear blimp got in your way at five miles the RIO wouldn't have had a clue." Manfred Rietsch compared the F-4 with its McDonnell Douglas successor in Navy and Marine service, the F/A-18 Hornet. "I flew 800 hours in the F/A-18 during *Desert Storm* and never aborted a flight airborne for maintenance problems. I never had a down radar. In the F-4, every twenty or thirty flights you could count on a utility or PC failure. The tube radios, TACANs, etc. had a meantime between failures of perhaps 30 to 40 hours. The early AWG-10 radar in the F-4J would perhaps stay up for one or two flights. For the F-4 you were looking at forty plus maintenance man hours per flight hour (compared with 8 to 10 for the Hornet). The F-4 was a maintenance challenge. VMFA-314 at Chu Lai would fly about 800 hours per month. We would start the day with about 14 to 16 aircraft—out of twenty—serviceable. By the time the day schedule ended you had perhaps six or eight machines for your night sorties. Then the miracle workers would slave all night and you would be back to fifteen or so by the morning. This was the story every day. It took five elapsed maintenance hours to change the radio. You had to remove the seat to change radios and navigation black boxes. Changing an engine was a sixteen-hour process, including a low-power run, high-power run, and flight test. In the F/A-18 the airplane will do the majority of the trouble-shooting for you. All avionics boxes are accessible at eye level. It takes two minutes to change an INS. We changed an engine during deployments in less than an hour: no engine 'trimming,' just a low-power check for air leaks and—go fly!"

High utilization of the F-4 inevitably complicated the maintenance schedules, but even the small routine jobs were time consuming. Rodney D. Preston served with VF-161 on USS *Midway*, but the squadron was home-based at Atsugi and spent time ashore. His job as Aviation Structural Mechanic (AMS) included the kinds of tasks which were common to all Navy and Marine "Phantom Phixers," for example, dealing with the drag 'chute.

"On short runways our aircrew would use a drag 'chute, and when coming up the taxi way towards the parking space the aircraft would stop and we would have to run behind the plane, grab the 'chute, and secure it around a weapons pylon, enabling the Phantom to finish movement to the parking area. Now, this may not seem so bad until you consider that the 'chute was trailing behind the aircraft and we had to retrieve it while two J79s were trying to cook us like hot dogs at a barbecue. After the aircraft was shut down it was time to re-pack the 'chute into the tail compartment. First you had to untangle the many lines that held the 'chute to the aircraft. Then we would have to replace it into a space that is only about 1.5 ft high, a foot wide, and 1.5 ft deep. A lot of our guys would have cut-off baseball bats to use as a ram for getting the last of the 'chute into the compartment. The next step was to pack into the same space a small, spring-loaded parachute that was attached to the main 'chute and assisted it to deploy into the airstream. To top all this off was the attempt to close the door [which was actually the tip of the rear fuselage]. This was also spring-loaded, and had to be latched, which usually required us to grab the door with one hand and hang onto it with all our weight while using the other hand to latch it. The plane was not built to be mechanic friendly!" Where possible, use of the drag chute was avoided, as COL Sherman noted: "We landed without deploying our drag 'chutes when away from home, as the aircrew normally ended up having to re-pack them."

Phantom Phan Phixers

Crucial to the F-4 operation was the careful nurturing of its dependable J79 engines. CAS missions could be flown without a RIO, radar, or rockets, but two trusty turbines were *de rigeur*. Frank G. Shelton, Jr., worked with VMFA-542 and VMFA-323's F-4s and at the Iwakuni repair facility. The Da Nang and Chu Lai "Supersonic Propulsion Engine Maintenance Specialists" performed J79-8A/B engine overhauls at around 1,000 hours and took care of most internal problems.

"The things that usually had to be repaired were the combustion liners (or 'cans') that would crack due to the high temperature (625 degrees C at the exhaust). The first, second, and third stage turbine nozzles also cracked. Usually, the parts could be welded, but sometimes had to be replaced. It took a crew of three to five mechanics two or three weeks to dis-assemble, inspect, and re-assemble an engine, and almost as long to have a part sent out for welding and repair. In the USA each engine was thoroughly checked by running it in a test cell where bearings were tested for vibration, and the whole engine was tested for oil and air leaks. The fuel control was adjusted so the engine could not exceed 100 per cent rpm and would run properly at 65 per cent, which was considered 'idle rpm.' In Vietnam we didn't have facilities to overhaul engines. [This was usually done at Iwakuni].

During reassembly, several critical adjustments had to be performed and usually re-checked. One of these was the synchronization of the inlet guide vanes to the fuel control. This was critical, because it ensured that the air was directed to the next set of compressor blades at the proper angle for maximum compression as the air moved through the compressor section. It also helped to eliminate compressor stall. Another was synchronization of the nozzle area control with the fuel control, ensuring that the exhaust gas nozzle at the end of the afterburner would be at the correct position for the engine speed and help to eliminate flame-outs. It was very important that every bolt or nut used to re-assemble the engine was torqued to the proper value. One, attaching the turbine to the compressor, had to be torqued to 350 ft/lbs."

Frank also worked on the Phantom's fuel system, replacing fuel cells in several aircraft.

"The Phantom's fuel cells were basically rubber bladders inserted inside the fuselage. I had to replace two of them; I believe they had bullet holes. The first was a Number One fuel cell, the largest. Basically it was one large cell with a dividing baffle of 'flapper' valves (which closed when the plane was inverted) in its center. This compartment of the cell held enough fuel for thirty seconds inverted flight." Frank found the smell of the fuel cells and the leaked fuel a real problem when working inside them. "It didn't take long for those fumes to affect one's ability to think properly, since there was no ventilation inside that part of the airframe, so a gas mask was necessary. The cells were made with rubber loops on the outside, and there were loops or eyelets in the airframe that the 'parachute cord' was routed through to support the cells. You had to stand inside the cells while installing them, and to avoid putting a hole in them you had to remove your shoes."

One Da Nang F-4B pilot came to Frank complaining about engine vibration, and Frank climbed all over the aircraft with the J79s idling, checking the structure. Eventually, they shut the engines down and Frank peered up the tailpipe to see that the three concentric steel rings forming the afterburner flame holder were hanging on by only two of the five supporting struts. Confident that he could refix them in around thirty minutes with a few new pins, he decided to double check with the manual and found that it required complete removal and replacement of the engine; "about an eight hour job if everything went right."

During its first seven years in Marine service the F-4 achieved a formidable reputation for versatility, and its power and dependability impressed those who flew it. One of their real achievements was the successful evolution of "crew coordination"; cockpit teamwork by both occupants. For Richard Tipton the RIO was, "first and foremost a second set of eyes and a second head. In CAS he mainly monitored gauges and called off altitudes and release points. Basically, he let the pilot stay 'out of the cockpit' while he stayed 'in.' He did most of the communications and radio work, and usually kept the pilot informed where the FAC was so that we didn't run him over—a real problem with slow-moving FAC(A)s. The principal difference in the DAS role was that he was actively monitoring the radar, AAA, SAM, and enemy aircraft threats. He had most of the 'black boxes' for passive surveillance of enemy radars and a second set of eyes that could relay, for example, AAA gun information to the pilot while he was making a strike on a target."

Manfred Rietsch considered that there was always, "a special relationship between a pilot and his RIO if they were crewed up together. You couldn't hide a screw-up from your RIO, but there was an unwritten rule that he would never tell another pilot about your mistakes. Flying fighters is a very competitive business. You never want the other pilots to see you sweat. You hide your shortfalls from them, but your RIO is like your mother or wife. I was blessed with flying most of my missions with two superb RIOs, Dick Kindsfater ('K-9') and Frank McDuffee. Rooting around Laos at night with people shooting at you, it was good to have another guy in the airplane. K-9 and Frank saved my tail more than once when I was about to run into the ground or do something stupid."

In 1965-66 the Marines tended to suffer the same shortage of RIOs as the Navy. Unlike the USAF, the USN maintained that training syllabuses should not be streamlined to satisfy the wartime demand for personnel, so the supply of fully-trained RIOs was limited. John Trotti observed that:

"During the early days we flew a much more chaotic flight schedule—not necessarily more sorties, but more 'peaks and valleys' [in frequency]. By the second month (with VMFA-314) we had fewer effective RIOs than pilots, and I actually flew some night *Steel Tigers* (the ubiquitous 'Manual 45 to Tchepone') in March/April 1966 with no-one in the back seat. A typical baptism for replacement RIOs (who, in 1966 were arriving with fewer than twenty hours in the aircraft) was to throw them in the back seat for a *Steel Tiger* after one daytime familiarization hop."

The most accurate impressions of the effectiveness of Marine F-4 strikes came from the Tactical Air Controllers who, day after day, watched sections of Phantoms rolling "in, hot" to deliver their weapons, and heard the results from troops in contact who were all too close to the explosions. One such report was received from CAPT. George E. Buchner, a 20th TASS FAC(A) who had overseen a strike by a *Castor Oil* (VMFA-542) flight attacking enemy troops close to Chu Lai base.

"A Korean Company was receiving fire from a fortified village fourteen miles from Chu Lai field. It was necessary to accomplish an emergency re-supply. I put the number two aircraft (CAPT. 'Woody' Wooldridge and CAPT. Kevin Rick in F-4B WH 00) in with their napalm. He destroyed ten military structures with all napalm on target. A chopper went into the zone, but received fire, even with the number one F-4B (MAJ. 'Happy' Bradshaw and CAPT. John Marshall in WH 7) making dry passes. At this time the company commander (ROK) requested bombs on the village. He was obviously impressed by the accuracy of the lead aircraft, because his company was only 50 meters away across an open rice paddie. The lead aircraft delivered his bombs on target, destroying ten more military structures. While the dust from the last bomb was still rising the Koreans charged across the paddie and took the village. They counted 19 KBA in the village and met no resistance. They also reported that the flight had received ground fire from the village. I want to commend Castor Oil 062 flight for their patience,

A VMFA-251 F-4B at Beaufort in 1970. The *Thunderbolts* flew the F-4B from 1964 and remained with Phantoms until 1985. (Mike Wilson via Steven Albright)

cooperation, good judgment, accuracy, and oustanding results. George E. Buchner, USAF."

When VMFA-542 handed over to VMFA-314 at the end of their deployment in November 1967 and headed off to Iwakuni for "rest and re-training," the squadron lyricist left this verse for the incoming Marines:

To 314
Welcome aboard, Black Knights of the sky,
You've come to replace us, down here at Chu Lai.
This is your hangar and these are your huts,
And this is your roadway, with deep muddy ruts.
It's gonna stay muddy, Black Knights, never fear,
For this is November and the Monsoons are here,
The days are delightful, and the night hops-what fun!
When the weather for landing is a hundred and one
And the showers are freezing and the Club's out of beer
And the mortar shells falling uncomfortably near.
Oh the flares burn so brightly, up there in the sky,
And the Tigers are weeping to be leaving Chu Lai,
Where the sunsets are lovely, o'er the blue of the bay,
And the rain falls so softly (fourteen inches a day)
And that quiet green valley, barely fifty yards long,
Has a welcome committee - a battalion of Cong!
As much as it hurts us (down to the last man)
We'll try to be brave, up there in Japan,
We'll fight those hot showers, those floor-shows so grand,
We'll remember with fondness those walks in the sand,
We'll try to be brave, and we'll fight the good fight,
And we'll dream of Chu Lai on our starched sheets each night,
Yes, we'll screw up our courage and live with our grief,
And wish we were there for that five o'clock brief.
Yes, welcome aboard, Black Knights of the sky,
You lucky replacements, down here at Chu Lai,
Don't ever be bitter - no, just carry on -
And don't bother looking, for the Tiger....is gone! (HDB)

Heading for a target in Cambodia is *Silver Eagles* BuNo 152217 with Rockeyes aboard. (Major J. Vancy Bounds via Steven Albright)

5

Marine Photo Phantoms

McDonnell's reconnaissance variant of the Phantom dated back to the origins of the F3H-G/H proposals to the Navy in the Summer of 1953. The company envisaged an unarmed, camera-carrying version of their design (the Model 98F) just as they had done with the F2H-2P Banshee. Although the "recce" Demon (F3H-1P) was a victim of the type's powerplant development problems, the company's RF-101C Voodoo had replaced the Republic RF-84F as the USAF's principal tactical reconnaissance vehicle, sustaining the design team's capability in this area. Navy interest in the Model 98F was insufficient to warrant further development, but in 1958 the USAF discussed a Model 98-AX variant with externally podded reconnaissance systems. This series of studies culminated in the RF-110A on 3 January 1961. From the earliest stages it was foreseen that camera systems would be used together with infrared devices; originally a short-wave system, but after June 1960 a long-wave system. In the Navy's earliest Model 98-AK proposal the nose forward of station 77 was replaced by a lengthened version housing the reconnaissance equipment. A Litton INS was included, the first inertial system in any F-4, also a stereo TV view-finder and an external ELINT pod. Attracted by its very advanced qualification the USAF received a detailed RF-110A proposal in June 1962 to compare with rival bids from Republic (RF-105A) and Douglas (RA-3C). The decision to adopt the F-4C as replacement for the F-105D clinched the selection of the McDonnell design, and an order was placed on 5 April 1962 with mock-up approval on 31 October that year. The first two prototypes, by then redesignated once again as YRF-4Cs, were adapted from Block 14 F-4Bs (airframes 266 and 268) and given USAF serials, 62-12200 and 62-12201. The first of these became McDonnell's hardest-working prototype, proving the configuration for the YF-4E, the slatted F-4E, the fly-by-wire Phantom, and canard foreplane F-4E. The first aircraft was flown from St. Louis by Bill Ross on 8 August 1963, and testing at Edwards AFB began later that month. The RF-4C's sophisticated sensor systems were fitted, and testing of these began at Holloman AFB in January 1964. In fact, Holloman had tested some of the podded optical systems intended for the earlier Model 98F proposals using F-4A Number 9 (BuNo 145308).

* VMCJ Photo/Electronic warfare logo.

The first RF-4B, BuNo 151975, on an early test flight over Lambert Field, St Louis, in March 1965. The prototype survived in squadron service until 25 April 1980 when it was destroyed in a mid-air collision with BuNo 151983, both from VMFP-3. (Boeing/McDonnell Douglas)

A hard-worked RF-4B-24-MC of VMCJ-1, home-based at Iwakuni from 1966 to July 1975 with many detachments to Da Nang. (Norman Taylor)

The ground-crew of Col McElroy's RF-4B made imaginative use of its "double nuts" Modex, though this aircraft was even more eye-catching in its later Bicentennial scheme. Note the curvature of the drop tanks, less angular than the Sargent-Fletcher type. (Author's collection)

BuNo 157350, the penultimate RF-4B, with VMCJ-2. Two AN/ALQ-81 ECM pods are carried on the inboard pylons. (via C. Moggeridge)

One of the design's most innovative proposals allowed the RF-4C to retain radar in its much-reduced radome area. Texas Instruments' AN/APQ-99 multi-mode radar could provide radar-mapping and terrain following. The latter required the pilot to follow in his "E-scope" radar display a "ride line" indicator set at a predetermined altitude above the terrain as seen by the forward-looking radar. Its scanner saw a five degree "slice" ahead of the aircraft horizontally and +10 degrees/-15 degrees vertically. It was therefore possible to maintain a fixed height above the ground, avoiding hills and other threatening terrain features.

The U.S. Marines' interest in the RF-4 concept paralleled that of the Air Force, though it did not result in a contract with McDonnell until 21 February 1963. The initial deal for twelve F4H-1P (later RF-4B, Model 98DH) was for the USMC only, the Navy having had to acknowledge that its order for 144 RF-8A Crusaders (completed in 1958) and 140 RA-5C Vigilantes would supply its carrier-borne reconnaissance needs. RF-4B BuNo 151975 made the type's first flight on 12 March 1965, and the order was later extended to 36 aircraft. Essentially, a "disarmed" F-4B with an RF-4C nose, the

RF-4B included the majority of the Air Force's sophisticated sensor fit, retaining the narrow 30 x 7.7 inch wheels and J79-GE-8B/C engines of the F-4B. A small follow-on batch of twelve aircraft ordered to meet Vietnam needs had the later "thick" wing-root and main gear wheels of the F-4J and J79-GE-10 engines. The last three in that batch (BuNo 15734 to 157351) also had the less "boxy" under-nose camera bulge of late production RF-4Cs.

The RF-4B's 60 ft 10.9 inches long fuselage housed a jettisonable cassette system for ejecting urgently-needed film by parachute over a ground collection point, flight controls in the front cockpit only (unlike the RF-4C), and a standard probe refueling system. RF-4B Phantoms also had a set of telescopic boarding steps in the front fuselage, another feature absent from the RF-4C. Marine Phantoms had to be self-supporting and operate with minimal ground equipment. It also had the slotted stabilator and fixed inboard slats which Phantoms received from Block 26 onwards. The RF-4C's AN/ALQ-102 sideways-looking radar (SLR) mapping set and AN/AAS-18 infra-red detection set were retained. The infra-red equipment produced a high-resolution film map of the land-

RF-4B 157346 shows the three-tank, two ECM pod configuration which was occasionally used. The *Playboys* flew the photo Phantom for ten years from 1965. (via C. Moggeridge)

Marty Lachow blasts off in an RF-4B at Patuxent River in 1965. (Marty Lachow)

RF-4B BuNo 151978 on pre-service trials with NATC. Marty Lachow was in the rear seat as Project NFO on the RF-4B (with LT Joe Acord). Project pilots were Jack Batzler, USN (later an Admiral) and LT COL Ken Curry. (Marty Lachow)

scape below the aircraft, monitored on a five-inch screen in the rear cockpit and imprinted on up to 350 ft of film. Radar mapping SLR was capable of producing a high-resolution (with the option of moving-target indication) radar map of the terrain to each side of the aircraft on data film at a scale of 1 to 400,000. In later RF-4Bs the system was replaced by the UPD-8 SLR.

Despite its radically different nasal contours the RF-4B handled similarly to other naval F-4 variants, though Fritz Klumpp, who test-flew most of the Phantom family at St Louis, reckoned it didn't seem quite as stable as the F-4B in high-speed flight and was comparatively "rather awkward" since the longer nose had inevitable slight effects on center of gravity.

The first RF-4Bs went to Marine Composite Squadrons VMCJ-2 at Cherry Point and VMCJ-3 at El Toro in 1965. The Marines' RF-8A Crusaders were passed back to the USN to be refurbished for another two decades of service. Within a year it was clear that the rapidly expanding demands of the Vietnam War would require the two units to supply aircraft to VMCJ-1, based at Iwakuni.

Marty Lachow was Project NFO at NATC evaluating the RF-4B prior to its acceptance by the Marine Corps, and he soon became involved in the aircraft's transfer to the Far East. He described some of his early assessment tasks with the new bird:

"I don't believe we changed much in the back cockpit from what the USAF had. Two things stand out vividly: the first was that on a bright day the green 'ON' lights appeared 'OFF'—not lighted at all. That was written on the gripe sheet and noted as unacceptable—just something for one to be aware of. The other was to check the night photo flare system. We were in the area between Patuxent River and Washington D.C., and at the briefed time popped all the seventy-two flares simultaneously. By the time we landed the base had been inundated with calls from everywhere up and down the coast. Believe me, if you were in New York you could see clear down to Florida.

I departed the Test Center in August 1965 and joined VMCJ-3 at El Toro. There we received RF-4Bs, beginning in October 1965, and began preparing for Vietnam. During that period of training all the systems worked well, and the most difficult part was getting the pilots (primarily) to trust the [terrain avoidance] system. That is, to fly hands off, 'nap of the earth.' You set the altitude at which you wanted to clear the ground, locked in the forward-looking radar, and if you had a good system you could cruise over mountains and valleys 100 ft off the ground. However, there was much stained underwear until they came to trust the system and the 'guy in back.' In August and early September 1966 we completed practice carrier

Marty Lachow's "100 Missions" patch. (Marty Lachow)

VMCJ-2's later "black tail" scheme. (Author's collection)

landings and in-flight refueling exercises and were ready to depart for Vietnam."

At Cherry Point Colonel Edgar J. Love, who had enlisted in 1950 and flew in every reconnaissance and electronic warfare squadron in the Marine Corps, participated in VMCJ-2's transition to the RF-4B.

"The squadron had helped to initiate the operational requirements for the aircraft as a replacement for the RF-8A Crusader and later contributed to the mock-up evaluation of the pre-production model. One somewhat humorous item in the mock-up was the fact that the brightness control for the viewfinder had to be rotated counterclockwise to increase the brightness. McDonnell's solution was to change the name to 'Dimness Control' so as to comply with the standard of having clockwise rotation to increase and counter-clockwise to decrease whatever was being controlled!

A startling incident occurred at Cherry Point a few months later. When a maintenance crew was sent to the salvage yard to find parts for the aging EF-10B 'Willy the Whale' they discovered new parts for the RF-4B had been discarded in the dump. Investigation showed

that the parts had been on the shelf for more than six months and had shown no usage. Therefore, in accordance with policy, they were discarded. The problem with the policy was that the aircraft had not been delivered and the parts were being thrown away. Needless to say, once the 2nd MAW staff were made aware of the problem it was quickly corrected by Navy Supply."

Shortly afterwards, in 1964 Colonel Ed "Alligator" Love was transferred to HQ USMC as Aviation Electronics Officer in DCS Air, where one of the primary projects within his purview was the RF-4B.

"Soon after the first delivery of RF-4Bs McDonnell was in HQ Marine Corps trying to sell us an upgrade consisting of improved infra-red detector, improved scale on the SLR, better radar scope, larger KVa generator, drooped ailerons, and slotted stabilator. The estimated cost of the proposed modifications was almost as much per aircraft as we had paid for them. I did recommend that the project should go ahead, but the squadron was due to deploy to Da Nang, so one of my primary duties was to ensure that they could be ready. One of our major problems was that the USAF had ordered 300

A VMFP-3 aircraft returns with a souvenir of a less-than-successful air-to-air refueling. (Bud Brown via Steven Albright)

VMCJ-3, the first composite unit to operate the RF-4B, received its aircraft at El Toro from May 1965. BuNo 153098 is seen there in 1971. (C. Moggeridge)

The original "chisel nose" camera bay, seen here, was replaced by a more rounded configuration on the last three aircraft. (Marty Lachow)

RF-4Cs and Marine requests had taken a back seat. However, McDonnell came through in the end and provided us excellent support in the last months before deployment. In fact, much of what we had perceived as a lack of support turned out to be a problem with Navy Supply at El Toro. The McDonnell reps visited Navy Supply with VMCJ-1 personnel and identified truckloads of parts and supplies that were promptly hauled to the squadron area.

The airframe and engines performed beautifully. I'm sorry I couldn't say the same about the camera system and radar altimeter. In late Summer 1966 VMCJ-1 deployed to MCAS Yuma for inten-

sive training. The cameras performed so miserably that we had to revert to the old Kodak cameras, and the aircraft had to be retrofitted for the Vietnam deployment. One thing is for certain, though, the pilots loved the RF-4B! The cockpits fit like a glove, and all the controls and switches were where you expected them to be. You added throttle, and there was an immediate burst of power. Responses from the other controls were equally satisfying."

Jerry O'Brien was at El Toro in 1963 with electronic warfare (EW) EF-10Bs and photo-recon RF-8As. He explained the functioning of the Marine Corps' unique Composite Reconnaissance

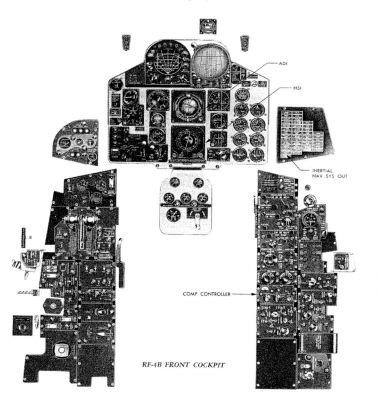

RF-4B FRONT COCKPIT

RF-4B front cockpit instrument layout. (via Marty Lachow)

VMCJ-1 devised an unmistakable tail logo for their Phantoms in 1975 shortly before their amalgamation into VMFP-3. This scheme was worn for their 1975 detachment on USS *Midway*, using three-digit side numbers, as shown here. (Simon Watson)

Another VMCJ-3 aircraft with the addition of stars on its rudder on a wet day at El Toro. (via C. Moggeridge)

squadrons, which also helps to explain the Corps' willingness to put its aircrew into a sophisticated multi-sensor aircraft such as the RF-4B:

"After the Korean War the Corps mingled the photo recon people with the electronic warfare folk. This was so that the limited assets of each mission could enjoy a common pilot and maintenance base. Except for the pilots assigned, the squadrons were something of a career backwater. Nobody outside of a few higher Staff Officers in Washington understood the electronic warfare mission." Partly, this was due to the high degree of classification which surrounded everything to do with electronic warfare. Even the term 'ELINT' was classified. "As a result of this stultifying secrecy there were few officers outside of the squadrons who even had a glimmering of what was going on. This aura of secrecy even extended to the photographic missions. For example, our VMCJ-2 RF-8As flew the first photo missions over Cuba. Contrary to popular history, this was long before Air Force U-2s were employed. Shortly

after the Bay of Pigs the USMC conducted its own private electronic investigation into the activities of the Cubans. The photo missions were flown as a result of that surveillance.

One anomaly about the squadron and these missions is that they were usually flown as single aircraft, and as a result most of our pilots were second-tour aviators. Contrast this with a fighter squadron where a young lieutenant would fly on a senior pilot's wing for a couple of years until he gained a reputation for good headwork and reliability. In the 'J' squadrons, as we referred to them, the pilots were frequently given the responsibility of single-aircraft flights.

Aerial photography was the glamour mission: the RF-8As were faster, more dynamic, and much more exciting to fly than the old EF-10B which, because it was slow and limited to 3g, was sheer drudgery for a pilot. After the USMC ELINT activities came to light the roles changed. It was still great to fly the RF-8A, but the EF-10B was where the action would be, or might be. Thus, while

VMFP-3 gave three of their aircraft bicentennial schemes: BuNos 153101 and 153107, as well as 153098, *Spirit of America*. (via C. Moggeridge)

After experimenting with a red horizontal stabilizer and upper fuselage color scheme in 1975, VMFP-3 used this black design in 1976. (James E. Rotramel)

The ubiquitous "spook" dominates the tail of this all-gray VMFP-3 aircraft. (Simon Watson)

the RF-4B might be the exciting machine to fly, EW was still the bread and butter of the squadron.

In Spring 1965 the squadron was notified that it would receive the RF-4B, but there were no back-seaters in the Corps who had any photographic experience. One has to take into consideration that photography was pretty much a daytime activity. Night missions were a rarity because lighting up the night sky in a combat zone was considered tantamount to suicide. On the other hand, EW thrived at night, and in the worst possible weather. Because of its dual role the squadron demanded that the back-seater in the RF-4B double as right-seater in the EW aircraft. Thus, the squadron found itself in something of a quandary. Most of the EW operators were Staff NCOs or Warrant Officers. While the F-4B fighter squadrons had assimilated most of the old EF-10B radar operators into the officer ranks, there were still quite a few of these sergeants flying right-seat in the 'J' squadrons. Thus, it was that Warrant Officers Petty and Cooke, Technical Sergeant Barkhardt, and Staff Sergeant O'Brien were sent to St Louis for RF-4B school. It was a fun couple of weeks. Most of it was old hat. Aerodynamics and navigation

were mentioned, hydraulic systems were analyzed, camera systems were explained, including the new and exciting IR and SLR systems. Then they got into the inertial navigation system. It is extremely difficult to understand the idiosyncrasies of the INS when one is hung over."

It was the INS which really enhanced the capability of the RF-4B as a photographic platform. This was particularly true in Vietnam where miles and miles of endless jungle made map-reading a chore. We would crank in the co-ordinates of the next target and could plan on being within a few miles of the proper location. Then it was a matter of picking out the right place. Most of our crews were quite good at map reading, and an experienced crew usually brought back the photos required. It was also the INS that kept us off the carriers. The system required an 'umbilical cord' to marry the aircraft into the ship's INS [SINS]. Foresight being what it is, no-one ordered the cords, or if they did we never received them."

Jerry found that his early experience of the "black arts" of EW put him in some interesting situations among those of more senior rank:

A SURE-modified RF-4B about to launch from USS Midway. (U.S. Navy)

"Lo viz" paint schemes in the mid-1980s left only a dark gray fin flash to identify VMFP-3's RF-4Bs. (Norman Taylor)

"On 9 June 1966 a conference was held at McDonnell for the re-configuration of the rear cockpit to accommodate the new SAM warning equipment. I was considered the most experienced ECM operator engaged in flying the F-4, and perhaps this is the reason I was chosen to represent the Marine Corps. When we arrived at Lambert Field we had a leaking nose-gear servo. A mechanic examined the leak and said that he and I could probably fix the leak by the time Major Jim Pierce (my pilot) came back from the conference. Jim explained that he would help the mechanic and that Staff Sergeant O'Brien would be the one to attend the meeting. I'll never forget the look on the 'mech's' face. I attended the meeting and listened respectfully to the Navy officers, most of whom were F-4 back-seaters, as they discussed the proper location for the device.

In my opinion all their solutions were wrong for the RF-4B. They wanted to move some of the camera counters, etc. I objected strenuously and wanted the device placed at the left of the counters, the only available space. The Navy men objected because that was their slot to look over the pilot's shoulder to view the ship's landing signals. They didn't want it on the right side either because that's where they watched the in-flight refueling probe. We all agreed to disagree, and they had to think that I was one presumptuous SOB of a sergeant to argue with all the Navy officers, but I felt that as I was the sole representative of Marine aviation I wasn't about to

back down. Subsequently, the SAM warning was installed where I suggested in the RF-4B" [and elsewhere in the F-4B].

The RF-4B had three camera bays in the nose, including a forward-looking camera (KA-87, Station 1) that could be rotated to the vertical position. There was a low-altitude horizon-to-horizon scanning camera (KA-56, Station 2) that could operate at 250 ft and 600 kts."

The third camera bay contained a KA-55 vertical or two KS-87 split vertical cameras, while Station 2 could hold either a left and right oblique pair of KS-87s looking through the side windows or a tri-camera array of KS-87s. The large, slab SLR antennas were in two cheek fairings behind the camera bay, and the AN/AAS-18 IR set was mounted in the lower fuselage between the intake splitter plates.

"The terrain following mode of the radar required a minimum speed of 0.93 Mach. If the pilot ignored this provision there was a likelihood that the aircraft would impact a hilltop. Obviously, the speed was a function of the attitude of the aircraft. One had to be slightly nose-down in trim for the gear to function properly. The forward radar beam in this mode of operation was also very narrow. Thus, you could fly down a very narrow valley and be feeling perfectly safe, never knowing that the hillsides were crowding in on you. Another problem with this mode was that the direction of this narrow beam was slewed to the INS, which provided information concerning the present relative wind. This meant that if the INS wasn't sensing absolutely correctly the forward-looking radar beam could be directed off to one side or the other. The slightest error in this regard would be hazardous to one's health if one was foolish enough to enter a narrow valley 'on instruments' [IFR]. Theoretically, you could plug the terrain-following into the autopilot mode, but I never knew anyone dumb enough to try it. I have often been surprised by random inputs into the autopilot at low altitude that could cause the aircraft to dive abruptly. I recall one such experience when we pulled about 2.5 negative g rapidly followed by 7g positive. I don't doubt that someone may have tested that autopilot provision and lived to tell about it, but they didn't do that in any aircraft that I was crewing. The mapping radar was quite good, and by tweaking the set properly a backseater could talk the pilot down through a self-contained approach—more of a trick than

VMFP-3 detached RF-4Bs to USS *Midway* from 1975 to 1980. (via Angelo Romano)

Another variation on VMFP-3's color scheme appeared on this SURE-modified RF-4B. (via Angelo Romano)

a useful exercise. Still, a competent RO [radar/reconnaissance systems operator] could get his pilot safely back to base in some pretty bad weather conditions."

A second group of pilots was assigned to the squadron, including the new CO, LT COL Jerry Fink. In Jerry O'Brien's opinion;

"Colonel Fink was an extraordinary man. He was probably the finest CO that I ever served with. He had been captured during the Korean War and spent quite a time as a POW. Fink was about 5ft 8inches and a bull of a man. When he first arrived at El Toro the squadron was situated in an old hangar that was far too small to accommodate the RF-4Bs, so we spilled over into a hangar assigned to the C-130 squadron. Candidly, the place was a mess. There were partitions separating the offices, and security was non-existent. As an EW section we were extremely sensitive to who might be listening to the subjects we discussed. In the process of investigating the rooms assigned we found that there were communications wires going every which way. Obviously, this was unacceptable to security-minded people, and one of our sergeants was tracing the wires that ran across the top of the partitions. Suddenly, he stopped and motioned excitedly for several of us to join him. Clambering on the desk tops we found that we were looking down into the Colonel's Office. Before his desk was a young private who was so frightened that his knees were shaking. Colonel Fink was reading from the charge sheet which listed the young Marine's many transgressions. We listened quietly to the charges. The Colonel finished reading and laid the paperwork on his desk. Then he looked up at the young man. As he did so he noticed the four or five sergeants leaning over the top of the partition. His eyes went wide as we offered our opinions by gesturing 'thumbs down.' A smile flickered across his face, and he forced his eyes down to the desk top. His voice dropped an octave as he announced his decision in a low, menacing growl. We all immediately dropped from view and found other things that needed to be done. By the time Colonel Fink rounded the corner our office was empty, but for a young corporal answering the telephone. "Where did those bastards go?" the Colonel bellowed. The kid dropped the phone and sprang to attention. "Which bastards, sir?" Frank smiled and went back to his office. The word was passed

from the Adjutant that we would never again be so presumptuous as to offer unsolicited advice to the Commanding Officer.

I first flew with Fink on a local familiarization flight. By this time I was really 'deeply experienced' with all of seventy hours in the RF-4B. (*None* of us had any depth of experience of the aircraft). As we were cleared onto the runway to take a position and hold, the Colonel asked if I was ready to go. I assumed he was talking about me personally because we had been told to hold our position. I replied that I was ready, and he lit the burners. I shouted to the tower that we were rolling and noticed that an R5D [C-54 transport plane] had just landed on the intercepting runway. Fortunately, we had no wing tanks and it was a cold day. We passed over the R5D at the runway intersection. The tower operator came up and said, 'That is a flight violation, Colonel.' Fink replied with a gruff, 'Rog.' To his credit, he later asked what occurred, and I explained. He never said another word about the situation, but our cockpit communication became much more precise.

On another occasion Colonel Fink and I were taxiing out at night and the right wheel hub failed. The aircraft literally fell down onto the right undercarriage strut. The crash crew came running out, and they tried to signal us to shut down. Fink wouldn't hear of it: he was set to go flying! I had to climb out, slide down the wing and off the wing tank to examine the problem. He was still vexed when I signaled to chop the engines, and he shouted down from the cockpit, 'This had better be good, O'Brien!' I replied, 'Your right wheel fell off.' This had never happened before, and he just couldn't believe it. He stood there looking at the white-hot wheel and asked, 'Think we can get another aircraft?'"

Colonel Fink also encountered an RF-4B spares situation even more bizarre than the one which had occurred at Cherry Point:

"The aircraft was new, and many parts were different from other F-4 parts. When Navy Supply was given various bits of equipment which they didn't recognize they would often send them back or sell them for scrap. This came to a head when a sergeant from our photo section went over to Navy supply to scrounge some salvage parts. He found that the high-altitude cameras and special gyro-stabilized mounts, for which we had been waiting months, were being sold for scrap. This was allegedly because no-one knew what they were or where they were slated to go.

BuNo 157350 was one of the final batch of three RF-4Bs that had the rounded camera door beneath the nose. (via Angelo Romano)

RF-4B BuNo 157345 at sea. (via Angelo Romano)

The sergeant returned and told Colonel Fink, who promptly ordered a couple of large trucks. Fink and the sergeant went back to Supply. Taking the money out of his own pocket, Fink personally purchased hundreds of thousands of dollars worth of equipment at the going scrap rate of 10 cents per pound. A meeting was established between the General, Fink, and the Colonel in charge of the Base Supply, who Fink openly accused of sabotaging the squadron. Fink told them about the material being sold for scrap, and they laughed at him. Then he opened the door to the conference room and had his troops cart in the boxes, all of which were clearly marked as having been sold for scrap."

Sadly, Col. Fink's effort to fight for his squadron did not go down well, and he was relieved of his command for "lack of judgment." Colonel Fleming took over VMCJ-2.

During the squadron's pre-deployment training at MCAS Yuma their Instructor Pilot, Captain Doxsee, checked out the new pilots:

"One of the procedures that Doxsee liked to demonstrate was a no-flaps take-off. I think this was designed to anticipate the possibility that the aircrew might neglect to use the check-list. The pro-

cedure for take-off in the F-4 was to hold the stick full back in the pilot's lap. Then, as the stabilator became effective the nose would start coming up and the pilot would move the stick forward, catching the desired flight attitude. If one was to forget the flaps in an aircraft equipped with drop tanks the nose would come up a bit more rapidly than with flaps, but it was still controllable as long as one was paying attention. On a clean bird the nose would literally shoot up, and you had better be quick or the engines would literally push the aircraft into the air at a very unusual angle of attack. That day, Captain Doxsee had another pilot, Captain Carpenter, in the back seat. For some reason he did not catch the nose in time and the aircraft rocketed into the air and promptly stalled. The right wing dropped, and at that point Carpenter ejected. These were not zero-zero seats, and his 'chute didn't fully open. He experienced severe injuries and spent a long time in hospital. In the meantime, the RF-4B, gear up, slammed back down onto the deck and began sliding towards a flight-line of A-4 Skyhawks. Doxsee elected to eject. He was still in the seat when it struck the ground. The Phantom came to a stop, intact but for the seats, about a hundred feet from the flight-line. The crash-crew was right on top of it and extinguished the fire immediately. Had Doxsee chosen to ride it out he might have survived. As it was, he died shortly thereafter."

After a brief period of familiarization with their new aircraft the "J" squadron headed off to war. Marty Lachow, who flew over 100 of his 326 combat missions in the RF-4B, was in the lead:

"We departed El Toro on 20 September 1966 with eighteen aircraft, including six RF-4Bs and LTCOL Bill Fleming, with me in the back, leading the pack. We 'Transpaced' to Da Nang, landing there on 28 September. Bill Fleming and I flew the first RF-4B mission on 31 September (a night flight), and we became VMCJ-1, 1st MAW. The most dangerous sorties requested of us were the 'mosaic' type; that is, flying a number of parallel lines, building up a map of an area. These were done by day or night using either cameras, or the infra-red at night. We were asked for a mosaic of the DMZ separating the Marines and the NVA. In order to avoid

Nose art on bicentennial bird BuNo 153101. (Angelo Romano)

only one aircraft doing that (sure death) I set up a plan using five aircraft in V-formation. We had each Radar Operator set up a radar position on the other aircraft with Fleming and myself in the lead. The formation took up position some 60 miles out to sea at 40,000 ft and heading in, descending, 'balls out.' Fleming and I centered on the DMZ. Suffice it to say that one pass was successful, and in all probability both our 'grunts' and the NVA all lost their hearing from the sonic boom of five Phantoms passing overhead at 200 ft."

Jerry O'Brien elaborated on the "mosaic mission":

"There was a continuous mission that consisted of approximately fifty-seven lines that lay immediately north of the so-called DMZ. Normally, each aircraft would be assigned two or three lines of this mosaic, which were flown at about 250 ft and 600 kts. This policy limited the exposure time in a very heavily defended area. Usually, we would go in from the sea and shoot one line, come off target over the hills, and drop down for another run to the sea."

The presence of RF-4Bs in the war zone massively increased the amount and quality of data available to ground commanders, and the aircraft's speed and agility protected it from many of the threats. However, its missions were far from safe, as Marty Lachow found later in his tour:

"Near the end of my tour Major Ed Love [who took over from Colonel Fleming] and I were tasked to recon/photo the A Shau Valley that our A-6s had bombed the night before. We were low and slow, and I commented on this to Ed, but about this time there was a flash off a hill to our left and BAM!, half our port wing disappeared. Home to Da Nang we went!"

Colonel Love recalled that one RF-4B was lost in September 1967 on an early morning mission. The pilot, Major Richard W. Hawthorn, and his RIO, Captain Richard King, were listed MIA. Another crew, airborne at the same time on the mission, reported seeing a bright flash in the highest terrain in the area. "We suspected that the pilot may have turned into the mountain in a 30 degree bank, and the terrain was out of the field of view of the radar until it was too late for the pilot to react." Jerry O'Brien added:

"I had argued with King earlier that he was briefing the mission wrong. It was his intent to use the terrain following mode to keep the aircraft at a more or less constant altitude. I pointed out that one had to stay at 0.93 Mach and that you could not use this mode in 'cross-compartment missions.' That is to say, we could go up a narrow valley at a fixed altitude, but we couldn't expect to go over ridgelines, drop down into valleys and then go up over the next ridge repeatedly. The reason for this is that the radar, not seeing the bottom of the next valley, would give you a 'dive' command as you topped the first ridge. Then it would sense the valley floor and give you a 'climb' command. The narrower the valley the more violent the commands. This could start off pilot-induced oscillation (PIO), a vertical oscillation that grows increasingly violent until the wing structures failed due to g-loads. To fly these missions at any speed below 0.93 Mach would cause the aircraft to clip the tops of hills. I don't think Hawthorn knew this."

Reflecting on the AAA risk, Ed Love said:

"The flight limits of the aircraft were constantly exceeded, especially in the areas of Vietnam where AAA or SAMs were encountered. Soon after checking into VMCJ-3, Jack Daily (who later became Assistant Commandant of the USMC) was on a mission in the DMZ and was hit by AAA. When he returned to the base and taxied into the flightline there was a hole in his right wing near the fuselage big enough for a man to stand in. We had nose radomes blown off, external fuel tanks blown apart, and took numerous hits from small-arms fire without loss of lives or aircraft. The RF-4Bs were a dream come true, and we flew the hell out of them.

I don't think we had one aircraft that was not hit by small arms at one time or another. However, we attracted a lot of AAA in Route Pack 1 and the DMZ. One special task involved flying numerous sorties at 5,000 ft and photographing the guns while they were firing at the aircraft. One aircraft straight and level at 5,000 ft didn't get much attention, but let another aircraft roll into a simulated bombing run at the same time from 'bout 10,000 ft and all hell broke loose. It appeared as though the ground erupted with the

BuNo 157342 from the extra batch of twelve war-attrition RF-4Bs with F-4J pattern wings and main landing gear. (Angelo Romano)

launching of grapefruit-sized rounds, which disappeared under the nose of the aircraft or turned into gray puffs before your eyes. You would wonder how you got through all that flak without getting hit. It was not unusual to count 15 to 30 muzzle flashes in a single 5.5 inch (1:5000) negative. Needless to say there were no second passes. One pilot, Captain Fred Carolan, flew so many of these missions that he was awarded a well-deserved DFC."

Elsewhere, repeat runs over the "target" were necessary. Colonel Dayton Robinson, Jr, recounted his tactics against the defenses in the A Shau Valley:

"There was AAA around the area, but they didn't usually shoot at a single photo aircraft, perhaps because they didn't want to give away their own positions. If you drew enemy fire there it was usually a short burst from behind when they thought you couldn't see them. You often knew you had been shot at only when you turned 'round and saw the smoke from their guns. Since the area was lightly defended you might make three or four runs for more coverage, but even here you played a cat and mouse game of seeming to retire, delaying, and then sneaking in another run with surprise on your side."

Even in Vietnam the *Cottonpickers'* (VMCJ-1 callsign) spares problems were not over, particularly where the radar altimeter was concerned. Ed Love:

"It was the heart of the RF-4B's sensor systems for providing altitude information. Without the radar altimeter the altitude and airspeed had to be entered manually, diverting the pilot's attention from the viewfinder and from the events outside the aircraft. It also required the crew to know the elevation of the terrain—a difficult task in mountainous areas. The parts problem was solved in the usual USMC manner. We placed SSGT 'Pappy' Britt on emergency leave to St Louis with a list of parts and an empty suitcase. McDonnell was understanding and helpful; SSGT Britt returned with sufficient parts to keep the RF-4B recce systems operating."

On the ground it took a while for the recce support facilities to catch up with the RF-4B's new technology:

"The III MAF Marines were not ready for the RF-4B. We had sent photographic interpreters to school for infra-red imagery and had them in place when the aircraft arrived in Vietnam. However, no plans had been made or tactics developed for use of the RF-4B other than for the normal black and white photography. I made several trips to III MAF in an attempt to generate some interest, all to no avail. The personnel were always poring over high-altitude panoramic photos of North Vietnam or Laos. Without exaggeration, one RF-4B could generate more imagery in one flight than the III MAF could interpret in a year. On about July 16, 1967, the attitude changed. The Viet Cong unleashed a rocket attack on Da Nang. The major damage occurred as a result of a hit on the VNAF bomb storage alongside the taxi-way. The storage area caught fire, and the explosives in the bombs melted and pooled in the bottom. When the temperature reached a super-critical point there was one hell of an explosion, causing structural damage to several WV-2 Super Constellations a quarter mile away. Third Wing HQ a mile away had all the clay roof tiles shattered. While there were no direct hits to aircraft or facilities, it caused a lot of concern about what could have happened.

Intelligence (III MAF, G-2) had information that the rockets were transported from the North using many trails and modes of travel. However, it seemed that the enemy would stop at dusk, set up camp, and cook meals on numerous small fires. This procedure was repeated before daybreak. VMCJ-1 was tasked to use IR sensors to map an area 30x 50 miles every day starting at dusk and again an hour and a half before day break. The missions were flown at 500 ft AGL and 360 to 400 kts. G-2 began plotting fires on a map, and the plots seemed to produce a pattern as they started to merge in one or two areas." Artillery fire was directed into these concentrations when the VC had settled for the night. "The RF-4B infra-red imagery played a major part in locating and keeping the rockets off Da Nang."

Others had less time for the SLR and IR systems. Colonel Robinson stated that he never used the SLR:

"It can be useful against shipping and port facilities, but we were never tasked to use it. I had a low opinion of IR missions over Vietnam. IR can be effective, but it is practically useless over the heavily forested areas. This mission was flown at minimum altitude at night and over mountainous terrain. It was a very dangerous mission with predictable, poor results. Flying an IR mission at night, in clouds, using the terrain-following mode was enough to give anyone prematurely gray hair.

The photo missions were very productive. With the careful planning of approach and retirement, photo missions obtained excellent results with acceptable risk factors. To achieve that balance, however, the photo pilot needed to plan the mission over high risk areas carefully, using both terrain and other activity in the area to achieve all the deception available to him. Making your photo run while strike aircraft had the enemy busy was a good trick.

For a typical photo mission in South Vietnam one was usually tasked to photograph at least three areas. The DMZ was heavily defended, and one usually flew one pass there. At low altitudes it was necessary to fly slow enough not to exceed camera limitations. Photography results were only good if the aircraft was on speed, altitude, and held an exact heading. The lower speed required for a 2,000 ft run made the RF-4B an easy target for AAA. On the DMZ mission I usually approached the area from the sea at minimum

The *Rhinos* launch another RF-4B sortie from El Toro. (Author's collection)

altitude using a salient coastal feature for initial line-up. At the last minute I climbed to 2,000 ft, established heading, stabilized altitude, and watched speed to stay within the cameras' capability. One held the run as steady as possible, ignoring the AAA until violent evasive action was possible on completion of the run. If another run was necessary one retired to plan another approach perhaps thirty minutes later. The RF-4B with external tanks had remarkable endurance. If I needed to make a second run I would make it by shielding my approach behind mountains. One could make a run through the A Shau Valley, retire, proceed to a routine mission and come back for a later run through the A Shau, perhaps 40 minutes after the first run.

The RF-4B was such a solid, tough airplane that I usually felt very comfortable with it and its ability to take punishment. The only missions I really detested were the night IR missions at a very low altitude. These were extremely dangerous, difficult flights, and results in South Vietnam were so negligible as to be a useless high risk. My basic principles did not include risking my neck to check out a bad theory when I knew it was a useless endeavor. I was not forgiving about some of the poor directions we received and some unnecessary risks we were required to run by those whose expertise was all theoretical. As a relatively senior flying officer I firmly believed that it was my duty to stop idiotic experimentation flying where the possible results were out of balance with the risk factors. The seasoned old pilot's rule is, 'Expect to be a target but don't make yourself an easy target.'"

Jerry O'Brien also had memories of the A Shau Valley:

"This was the infamous Ho Chi Minh trail, and we had to cover the area with IR and 'photo' every day. One would line up at altitude and then run down the valley at minimum altitude with all sensors recording. Looking up at the hillside above us we could see the automatic weapons firing down at us. Fortunately, they weren't leading us enough to hit us. In almost all our photos we would find that the ground was alive with anti-aircraft fire, but we weren't lingering along. We had a few aircraft come back with hits, but nothing that would disable them. This isn't to say that we didn't lose aircraft. Major D. C. 'Eski' Escalera lost one because as he landed at Da Nang a cable slapped up and struck the bottom of the aircraft, causing the horizontal stabilizer to go to a nose-up position. The nose rocketed skyward, Escalera hit the afterburners, and the aircraft went straight up. The Major ejected and survived, but by the time the RO, 1LT Tom A. Grud, ejected the aircraft was already beginning to settle back to earth. His 'chute didn't open and he died. It was that close—a matter of a fraction of a second was the difference between life and death.

Another RF-4B was lost over the Philippines. A highly competent crew left Da Nang for Cubi Point about 800 miles away. It was well-known around that area that the Philippine air control people were something less than absolutely competent. The trick was to announce the penetration of Philippine airspace and then cancel instrument flight rules (IFR) and proceed VFR ('visual') to Subic Bay. If you announced that you were capable of doing your own navigation the controllers would happily let you take over. I relayed this information to the aircrew in question, and they nodded knowingly. The next night they had to leave their aircraft over

Camera doors hang open on 31 RF in November 1986. Its dull paint finish bears the mark of many a maintainer's boot. (via B. Pickering)

Mindoro because the Philippine air controllers had lost track of them. Mindoro is a couple of hundred miles away from Cubi Point. If you have a modern aircraft with tons of navigation equipment on board it is absurd to lose that aircraft because someone gives you the wrong vector.

Night missions could occasionally consist of photo work with flares. These were flown at a fixed altitude with the flares lighting up the world at frequent intervals. If you were smart you did not linger over the target area. I recall one night when Captain Ron Sciepko and I went on a flare mission over the west of the A Shau Valley. We flew a multi-line mission at about 8,000 ft using the INS to line up the mosaic. Ron was an excellent pilot, and we just sort of fell into a routine. It was going so well that we went back to base, refueled, loaded flares, and went back for the rest of it. Ron kept asking if I had seen any ground fire. What with flares exploding it was difficult if not impossible to see anything, so I had to reply that I hadn't seen any. By the time we got back into the photo briefing section the troops were looking at our first photos and regarded us with awe. In each photo it appeared that the entire ground was afire. We were above the altitude for effective ground fire, but how we avoided getting hit that night is a mystery.

The A-6A squadron was based at Da Nang and had come under considerable adverse comment as to their accuracy in night bombing. VMCJ-1 RF-4Bs were assigned to follow the A-6As onto their target and conduct immediate damage assessment photos. Picture this, if you will; the A-6As stir up the natives, and all hell is breaking loose, and now we come in and start popping flares! We didn't lose any aircraft: I think the Vietnamese were as dumbfounded by this lunacy as we were. There were other idiot missions, of course, including one which may have seemed strange but did have some thought behind it, though the mission itself was dangerous. We were to fly over Vinh at 8,500 ft at high speed to excite the AAA batteries and were supposedly limited to a single run. We flew up to the north of Vinh and then, on being assured that we had ECM support, started our run. The first run was north to south at 8,500 ft and Mach 1.4. When the run was complete the pilot, LT COL Arve Realson, asked if I had seen any flak. I responded in the negative, and so we made a second, unauthorized run from south to north. Since we had to go south to get to Da Nang it seemed reasonable to make one more pass. The cameras were all running as we swept by each time, and we had some excellent photos of the fully-exposed anti-aircraft batteries. We climbed to 30,000 ft off the coast and then made a long descending approach to Da Nang. We were definitely in a low fuel state when we touched down.

As a matter of fact I can only recall seeing a great deal of flak on one mission up near the DMZ. I was flying with a Major who had just joined the squadron, and we were loafing along at about 10,000 ft and checking maps when the pilot looked up and...asked what the hell were those things floating by the canopy. I surmised that they were tennis-ball sized AAA tracers and strongly suggested that we hit burner and get the hell out of there. As we turned away I looked below and could swear I could have walked on the wall of flak coming up. That is absolutely the worst I had ever seen it. Even with the machine guns around Khe Sanh and in the A Shau Valley shooting down at us I had never seen so much concentrated ground fire.

I flew one Khe Sanh mission with Jim Pierce. In the late afternoon I had just returned from a couple of missions and was a bit weary. The bar at the O-Club was open, and I had just wrapped myself around cool when the Major came into the bar looking for a RO as we had an urgent mission to fly. I nobly recommended a couple of my younger chums, but as I was the soberest of the group I was nominated. The Marines were going to attack one of the mountains north of the base and wanted photographs of the hillside from a low perspective. We went full burner to Khe Sanh to beat the fading sunlight, then we let down to low altitude. Once again, we were looking up at the top of the mountain and the machine gunners were pouring fire down at us. We made several passes using the low oblique camera and the scanning camera, then we returned to base and I went back to the bar."

The majority of the RF-4B's missions were photographic, as Colonel Love explained:

"Visual recce was not an operational concept for VMCJ pilots. If we saw it, we photographed it. We reported unusual happenings at our debriefings, such as aerial bursts, unusually heavy fire, or aircraft sightings. I don't believe the S-2/G-2 (Intelligence) ever acted upon any visual reports. On one occasion an aircraft called in that the VC were preparing to launch rockets from a sand spit near the DMZ. VMCJ-1 received a request to photograph the location. We launched a RF-4B, photographed the location, recovered, processed the film, and delivered the prints in an hour. Of course, the enemy was alerted, and by the time [strike] aircraft arrived the VC had vanished and only mounds of sand remained.

The RF-4B was not equipped with any external ECM pod during my tour. One thing the Marine Corps had at the time was the best tactical electronic warfare capability in the world, in the same squadron as the RF-4Bs. We had developed practical tactics against SAMs, AAA, and fighters, and the operators had years of experience. One of my most satisfying moments was when the squadron received a message from the CAG on USS *Coral Sea* when they were going off line, saying that he had never lost an aircraft over North Vietnam when VMCJ-1 was flying electronic warfare sup-

port. During 1967 we had the mission to photograph the DMZ on a weekly basis, and we always had an EF-10B or EA-6A on station in support of these missions north of the DMZ. The EW operator could monitor the appropriate frequencies, and if a threat appeared he had the equipment to neutralize it. During my time in Vietnam we didn't use fighter escort for recce flights, mainly because it was in conflict with the mission. Fighter pilots were always looking for a fight. Our mission was to get photos and get out with as little contact as possible. With the RF-4B you could clean it up and outrun anything in the air."

Former Crusader pilot CDR Peter Mersky assured the author that RF-8A pilots also preferred to fly unescorted. However, by 1969 the Marine F-4 squadrons were occasionally asked to provide fighter escort for RF-4 sorties. Jim Sherman flew a couple while with VMFA-334 *Falcons*; both memorable hops. His F-4J carried six Rockeye CBUs on the centerline and three pods of Zunis on wing stations 2 and 4. "Lots of drag. I am about half a mile behind and stepped up 1,000 ft. I am running at 710 kts so I can slide from side to side looking for ground fire or missiles. At the same time I must keep sight of the RF-4B. It is very easy to lose sight of him if we are popping in and out of clouds. Fortunately, he flies straight and level while on his recce run." However, the mission was worthwhile. On this 2 January 1969 sortie the photo interpreters picked out 700 trucks heading south on the trails network.

On another flight on 27 January Colonel Sherman made three high-speed runs across an area near Hue with the photo bird and then noted a bright red light in the upper right corner of his instrument panel. "It said the dreaded words LOW FUEL. I instinctively told my RIO what our problem was and said we were going to try and get as far into southern Laos as possible." He then warned the RF-4B pilot to keep clear, as he was 'cleaning up.' "With that, I looked on the rear of my left console for the button, raised the safety cover, and pushed down hard. We could feel the airplane jump. This button blew everything off the bottom of the aircraft. I just jettisoned about 4,000 lbs, but reduced the drag to practically nothing."

A very literal toning-down of the squadron's "green tail" scheme on BuNo 157347, which crashed in a spin off San Clemente. (Simon Watson)

A weather-worn tactical paint scheme on BuNo 157350 in June 1988. (via B. Pickering)

Climbing to 31,000 ft he set maximum endurance power and called Mayday for a tanker. He was initially offered incompatible boom system KC-135s, but then a Marine KC-130 with the required refueling system was located over south Vietnam. Red Crown finally brought Sherman's *Falcon 06*, with only 400 lbs of fuel remaining, to within 100 miles of the tanker, *Otis 12*.

"When I commenced the climb I had 2,300 lbs of fuel and was 375 miles from the tanker. Normally we planned to be in the landing pattern with 2,300 lbs. My RIO and I were saying a few 'Hail Marys,' and he recommended at 50 miles that we should throttle back and start our descent to the tanker's altitude. [*Falcon 6's* radio transmissions carried the story]:

FALCON 6 Otis, we hold you on our nose at 50 miles and we have started to descend to your altitude. At 25 miles I will ask you to start a standard-rate turn to reverse course. I will call you when I have a visual and you can stream your hoses. Fuel state is now 250 lbs.

FALCON 6 Otis, now at 30 miles, 29, 28, 27, 26, 25. We hold you visually.

OTIS 12 Roger.

FALCON 6 Falcon 6 now has 200 lbs, 8 miles to go, closure 150 kts.

FALCON 6 Fuel state is about 150 lbs. I don't want to touch the throttles if I can help it....Now two miles out, have your hoses in sight. Overtake 75 kts....Otis, stop your turn, roll wings level. My probe is out and overtake is under control. I hope your response is set properly. Stand by for a jab.

(We hit the basket, the hose started forward, and I had to add power to hold my position. As I did, I saw a green light come on the

tanker pod which meant fuel was flowing. My RIO was sure he felt the left engine surge when I added power.)

FALCON 6 I'm shaking so bad I can hardly talk. Otis, give me 6,000 lbs.

(I have never before or since shaken so *hard*! I would try to stop, but couldn't. Thanks for the autopilot! I never flew another photo escort.)"

CWO Jerry O'Brien returned to the States and VMCJ-3 where he trained new RF-4B pilots, revisiting Da Nang in 1968 and remaining with the RF-4B and EF-10B until the latter was retired in 1971. The *Cottonpickers* held out at Da Nang until July 1970, when they re-deployed to Iwakuni for five years until their assets were divided, the EA-6As going to VMAQ units and the photo-Phantoms to VMFP-3, which absorbed the surviving RF-4Bs from all the "J" squadrons.

In November 1968 LT COL E. B. Parker took over VMCJ-1 at a time when the unit was mounting over 200 missions a month, many of them with RF-4Bs, and maintained that level up to 1970. With such a high degree of utilization the amount of data generated was prodigious. In June 1970 alone over 310,000 ft of film were passed to the photo lab, and IR imagery was produced showing 6,735 nm of terrain in 235 sorties. The next CO, LT COL P. A. Manning, flew the squadron's 25,000th combat sortie on 29 June 1970. These totals were achieved with an average strength of eight aircraft, of which four or five would be available each day. Throughout the conflict only three RF-4Bs were lost to enemy action. The "Recce Phantom," in Jerry O'Brien's words was, "a magnificent airplane; tough and responsive—but it took a good pilot to handle it."

6

The Second Round

Superior Specter

Progressive development of the Phantom for the USAF's F-4C and F-4D versions resulted in evolutionary improvements to the aircraft, some of which passed over to a new Navy version, the F-4J. The F-4C's heavier landing gear and wider wheels improved taxiing, shore-based ground handling, and tire life. Updates to the F-4C's attack performance led to the AN/APQ-109 radar with an air-to-ground ranging mode, and the linking of the AJB-7 bombing system to an AN/ASQ-91 weapons release computer which enabled the aircraft to deliver a much wider range of weaponry more accurately. Meanwhile, General Electric proceeded with more powerful J79 models for what was seen as the definitive USAF Phantom, the F-4E. The Navy also wanted improved weapons delivery (having largely jury-rigged its F-4Bs for the job), and a better air-to-air radar which could "see" targets against ground clutter.

Another parallel Phantom project, the F-4K/M program for the UK, led to examination of the aircraft's carrier landing and take-off performance to see whether it would be possible to operate the F-4 from British carriers with their limited deck area. Collaborative work with Hawker Siddeley designers in the UK led to the slotted stabilator leading edge, which countered the aircraft's tendency towards nose-heaviness on approach and deck-launch, reducing landing speed. A firm development contract for the F-4K was signed on 30 September 1964. Two months later the USN's YF-4J flew, and by the end of that year the definitive F-4J (Model 98 EV) configuration was authorized by NAVAIR. It used the slotted stabilator and USAF-type landing gear, and also an extra fuel cell, the 95 gallon Number 7 cell, located above the jet pipes. At the same time the Number 1 fuel cell behind the cockpit was reduced in size to 231 gallons in order to provide a trough-shaped space for new avionics. These included the AN/AWG-10 fire-control system, linked to an AN/APG-59 radar using the standard 32 inch dish. A carry-over from the limited F-4G experiment was the AN/ASW-25A data-link, though the Fleet was still not ready to utilize this gear when the aircraft entered service in 1967. William D. Knutson, CO of VF-33, pointed out that, "neither the E-2A Hawkeye nor the carrier was really up to speed on using it for vectors or control. It was not used in combat."

Factory-fresh from the second F-4J production batch BuNo 153790 was on loan from VF-121 to the Air Force Flight Test Center at Edwards AFB in May 1967. (Warren Bodie via Norman Taylor)

Although it has a temporary "180" number for a Paris air show, this VF-33 F-4B was normally assigned to LT Gene Tucker, who later destroyed a MiG-21J in a *Tarsiers* F-4J. The squadron transitioned to the F-4J in 1967. (via C. Moggeridge)

The *Tarsiers* (later *Starfighters*) changed their tail design several times in the 1970s, arriving at this version in 1976. (Norman Taylor)

A further color scheme modification for one of the squadron's late-1970s cruises on USS *Independence* took the nose "anti-dazzle" over the radome, too. VF-33 flew the F-4 longer than any other USN squadron. (via Norman Taylor)

General Electric J79-GE-10 engines each producing 17,900 lbs of thrust and fitted with the longer F-4E type afterburners were installed. The AAA-4 IR housing under the radome was removed, since its original contents had long since been made redundant by improvements in Sidewinder performance and it was no longer needed for the new AN/APR-32 RHAW system, either. The auxiliary IR sensor had received very limited use in any case. For Roger Carlquist, who was Initial Project Test Pilot on the F4H BIS trials, "The AAA-4 was purely a designation as far as I was concerned. I never laid eyes on one other than in mock-up from the day I first flew the aircraft to the day I relinquished command of a Phantom squadron six years later."

McDonnell tested the innovations in an F-4B BuNo 151473, re-designated YF-4J, which first flew on 4 November 1964. The AWG-10 was test-flown in F-4B BuNo 151497. As a further aid to improving deck landing performance the F-4J also introduced drooped ailerons. With landing gear and flaps lowered, the ailerons also deflected down 16.5 degrees. Together with the slotted stabilator this knocked 12 kts off the landing speed, which would otherwise have reached 137 Kts. This enhancement of carrier landing performance enabled the Phantom to earn an even better safety record than its predecessors. The F-8 in particular had a fearsome reputation for ramp strikes (hitting the ship's stern on low approaches) on the smaller *Essex*-class carriers.

On 27 May 1966 the first F-4J, BuNo 153072, lifted off, celebrating the eighth anniversary of Bob Little's first F4H flight and the third anniversary of the F-4C's debut hop. F-4J BIS trials began in December 1966, and the first production aircraft reached VF-101 at Key West on the 21st of that month. It achieved Initial Fleet Operating Capability (IOC) and made its first deployment on 18 April. VF-84 and VF-41 were the first deployable squadrons to receive the F-4J, in February. Compared with the F-4B, Bill Knutson

A VF-103 *Sluggers* F-4J makes ready to fly. The Sanders AN/ALQ-126 DECM installation is visible above the national insignia. (Simon Watson)

One of the F-4Js assigned to VF-121 at Miramar, where the squadron continued to develop effective ACM strategies for the F-4. (Norman Taylor)

found the F-4J, "heavier, but with the new stabilator, wing roots, and more responsive engines it was a dream. Over-rotation [off the catapult] was still possible with the new stabilator, but it pretty well eliminated that problem." His squadron, VF-33 *Tarsiers*, was the third F-4J unit (with VF-102 close behind), and it received its first aircraft, BuNo 153838, on 1 October 1967. However, to their disappointment, it was a "lead nose" aircraft with ballast replacing the AWG-10 radar, as it did in the first fifty F-4Js while radar development problems were tackled. Fred Staudenmayer, also in VF-33, recalled, "We got wonderful, reliable new airframes, but took on terrific problems with the AWG-10 Doppler radar, which posed great challenges until some experience was gained. Some of these radars would operate for forty to fifty consecutive flight hours without a hiccup, while others never worked at all!"

Combat Debut
VF-33 and VF-102 became the first F-4J squadrons in combat when it was decided in December 1967 that USS *America* would become the next East Coast carrier to join the war effort in S. E. Asia. The *Tarsiers* received a dozen new, radar-capable F-4Js to replace their "153" series machines. These were from the BuNo 1555 batch and had their final engine and electronic upgrades installed at Oceana prior to deployment. Captain Knutson found the new Phantom to be, "a great aircraft, more stable than the F-4B, but it had a faster landing speed. The extra engine thrust was great for ACM, and when the radar worked it provided a great increase in combat capability with its look-down ability. Best of all, the F-4J was super to bring aboard the carrier, and the large tires greatly reduced blow-outs on landing."

RIO Steve Rudloff found the AWG-10, "a major improvement in our ability to track aircraft, especially against low-flying aircraft." With the APQ-72, "invariably a MiG would come up from a lower altitude, which meant that for the most part until the AWG-10 came along, you weren't going to pick him up except visually. Of course, MiG-17s and -19s were rather small and difficult to pick up visu-

ally. One of the things that we did when we went 'over the beach' was radar mapping, and that is probably the most use of the radar that I got."

VF-33 and VF-102 departed Norfolk, Virginia, aboard USS *America* on 10 April 1968 for an eventful cruise in the West Pacific. Bill Knutson flew the F-4J's combat debut (in BuNo 155551, *Rootbeer* 201) on 31 May 1968.

"The first combat mission in the F-4J was an air-to-ground mission over the North. The normal procedure was to start a new Air Wing with air-to-ground or other low-risk missions in relatively lightly defended areas to let the crews get their feet wet before Alpha strikes into the heart and heavily defended areas of Vietnam. Most of the missions were air-to-ground or flak suppression."

In preparing his squadron for combat, Bill Knutson took as his model the doctrines of his former VF-84 CO, Jack Waits.

"All COs must be the leader of the squadron. They have to set the example, take the hard missions, and lead the big strikes. It

This Crusader pilots' motto sums up the philosophy of F-8 drivers who, at first, resented the move to the heavier, gunless, two-seat Phantom. (via Jerry B. Houston)

F-4Js from VF-92 and VF-96, flying from USS *Constellation*. (USN via CDR James Carlton)

doesn't always work that way, which has a direct effect on morale and battle efficiency. Also, if the aircrews knew their aircraft and systems capabilities and limitations, knew the rules of engagement and didn't violate them, they would have the best chance of surviving and inflicting maximum damage to the enemy. The rest would take care of itself. I felt my job was in front of the troops. I wanted to be at every briefing and debrief if I wasn't flying. I wanted to be on the hangar deck and flight deck ensuring the enlisted men knew how critical their work was to the success of the mission. VF-33 won the Battle E and Safety Awards every year during my time with the unit. On safety, I felt that most accidents occur when the pilot is faced with multiple failures or problems. For the maintenance troops the aim was to give the pilot an aircraft with zero problems; not just an 'up' aircraft, but one with all systems working to peak efficiency."

Some of the maintenance problems were outside the "phixers" remit. One unexpected and alarming fault was found in the canopy ejection process. It occurred initially during the *Tarsiers*' first line period when LT Eric Brice and LTJG Bill Simmons in BuNo155554 were hit by 37 mm AAA on a strike near Vinh. With their Phantom on fire, they managed to go "feet wet," and Simmons ejected at 400 kts, sustaining injuries. Brice was unable to get his canopy off and then fire his seat. As his F-4J headed down towards the ocean he was seen to be desperately pushing at the plexiglas to free the canopy, but without success. "Tex" Elliott, VF-33's XO, had a similar experience in BuNo 155551, *Rootbeer* 203, over a month later on 24 July 1968:

"I went onto a secondary target after finding no resistance at the primary target as flak suppressers. On the pull-out we were hit by AAA, which immediately knocked out my primary hydraulic (PC-1) and utility hydraulic systems. I got a fire warning light very shortly afterwards. We were positioned about 10 to 15 miles northwest of Vinh. I headed out towards the coast, climbing, and very soon found that my throttles were stuck at full military power and my PC-2 hydraulics system was beginning to fluctuate, meaning we would shortly have no hydraulic pressure at all. Normally in a Phantom, if you lost hydraulic pressure on the PC-1 and PC-2 the horizontal stabilizer went leading-edge down, which threw the airplane into a nose-high, loop maneuver. Duke Hernandez had this happen to him a couple of years earlier over Haiphong, but still had the utility pressure so he came out in a kind of corkscrewing ma-

neuver until they could get 'feet wet.' McDonnell engineers said that if you were able to neutralize the selector valves as the hydraulic pressure faded to zero you could lock the stabilator. In reading the F-4 'Field Service Digest' (which I had devoured as the F-4 NATOPS evaluator) I discovered this information, and it was at the back of my mind when this emergency occurred. I had the stick in the neutral position, and this locked the 'stab.' Very shortly afterwards I was driving a jet-powered rocket. The stick was frozen in concrete, but this was OK because we were climbing at the standard climb rate of 10 degrees and were interested in reaching the water before we had to eject. My wingman (Lieutenant, later RADM Fred Lewis) was giving me 'sitreps,' and as we got feet wet the overheat light came on, which told me the fuselage was burning. I told my RIO, LTJG Andris Dambekaln, to eject, and I used the alternative ejection handle to put us both out [command ejection, initiated by the pilot using his seat-pan mounted handle]. Unfortunately, with the speed we were at, once the RIO's canopy came off, the front canopy wouldn't release. I knew that once I rotated the black canopy lock handle aft the canopy was unlocked and just sitting there, but I couldn't do anything about it. I turned the engine masters off to shut them down, and figured I had better tell someone why I didn't get out. I keyed the mike and said, 'I can't get...,' but then the generators went off-line and I lost the radio. So I went

Returning safely to USS *Ranger*, NE 106 of VF-21 completes another mission over Vietnam in 1970. (USN/R. L. Lawson collection via Norman Taylor)

A VF-143 *Pukin' Dogs* F-4J awaits its crew at Miramar in August 1969. (D. Kasulka via Norman Taylor)

back to pushing on the canopy and pretty soon I had it off. Everything after that was normal. I was picked up by a helo, taken back to the carrier, and was back in combat two or three days later.

I wrote up a report, and they made a service change so that when you pulled the ejection handle, little thrusters kicked the front end of the canopy upwards to release it. They had figured out that with the rear canopy gone there was a venturi effect at 400 to 500 kts which held the front canopy down. I could have dumped the cabin pressure and opened the vents if I had thought about it."

This was VF-33's third and final loss of the war. The second had involved *Rootbeer* 210 (BuNo 155546) on 18 June. LCDR James Holtzclaw and his RIO, LCDR James Burns, took a SAM during an armed recce. In "Tex" Elliott's words, "They managed to duck the first two, but by the third one they had used up a lot of energy. It hit them in the right wing root and blew them out of the sky." Both men ejected with injuries, and their recovery became one of the most famous of wartime rescues. They landed in a paddy field and managed to crawl some distance to the comparative secu-

rity of a forested karst slope. Tex Elliott: "We listened on the radio back at the ship, and I was afraid it would be a tragedy. The rescue effort went on for four hours." LTJG Clive Lassen and LTJG Clarence Cook of HC-7 flew their HH-2C helicopter sixty miles, sighted the wrecked Phantom, and searched in the darkness for a signal from its crew. As they descended from 5,000 ft they narrowly missed a SAM aimed at them, but spotted pistol flares from the F-4J crew men. Under constant small-arms fire they made a fifty foot hover between trees which were about 150 ft apart, in the dark. When flares dropped by the other RESCAP aircraft went out Lassen was unable to avoid hitting a tree, damaging his rotor blades. With only thirty minutes fuel remaining and intense fire coming at them from both sides Lassen turned on his landing lights and moved in to pick up the two F-4 aviators. He returned to the destroyer, minus his right-side door, with five minutes fuel remaining. LTJG Cook received the Navy Cross, and Clive Lassen the Medal of Honor, the second such award to a Naval aviator for a Vietnam combat rescue.

On the same cruise VF-33 *Tarsiers* scored their only MiG kill, which was also the first for the F-4J, by LT Roy "Outlaw" Cash and J. Edward Kain (in *Rootbeer* 212 BuNo 155553). Bill Knutson commented; "MiG activity was very light, and although we always flew CAP there was very little action. Roy and 'Killer' Kain were fortunate to engage a MiG." It was also the first aerial victory for an Atlantic Coast F-4 squadron and the only kill for an aircraft from USS *America*. Roy Cash (whose famous uncle was about to release his seminal *Johnny Cash at Folsom Prison* recording at the time, and whose daughter, Kellye, became 'Miss America' in 1987) takes up the story:

"We had been on Yankee Station since the end of May, with one visit to Cubi Pt. in late June, so we were back on station after a 'July 4th' break at the Cubi 'O' Club. On 10 July 1968 I was scheduled for a MiGCAP as wingman to Major Charlie Wilson (*Rootbeer* 202, as I recall), a USAF exchange pilot who had been in the squadron for about a year and had joined us on our Mediterranean cruise the year before. I made the whole Med cruise, so I was fairly expe-

VF-121 instructors fly an exemplary formation over the home base. (USN via CDR James Carlton)

Ghostriders F-4Js line up at Miramar in 1972. The nearest aircraft (BuNo 155894) completed combat cruises with VF-154 in 1970-71, VF-142 (1972-73), and then served with VMFA-212 and VMFAT-101. (C.Moggeridge)

Another view of F-4J BuNo 155894, which was later sold to the UK and flew with 74 Squadron, RAF as ZE 364/Z until 1991. (C. Moggeridge)

rienced in the aircraft and squadron. The aircraft were brand-new F-4Js; our Med cruise F-4Bs had been traded in for them. Charlie's RIO was LTJG Bill Williams.

Charlie and I launched mid-afternoon around 1500 and were assigned MiGCAP station about 15 miles off Vinh, clear of the beach but close enough to 'Buster' feet dry if needed. We determined soon after launch that Charlie's radar was marginal to non-existent, so it was agreed that if we took a vector for bandits I would assume the lead. We established CAP station quickly, then about 45 minutes to one hour into the flight our controller, 'Raider' (USS *Horne*) called us over to 'Cipher' frequency to alert us to impending MiG activity. Basically, the information boiled down to the fact that MiGs were about to launch and sortie down to attack the A-7s on their strike missions just below the 'no-bomb' line just north of Vinh. 'Raider' kept us apprised of the increasing activity and MiG communications (our ECM and 'spy' planes had picked up good info on the MiGs, apparently), switching us back and forth from clear to cipher frequencies. We told Raider that, in the event we were vectored, we wanted to fly a specific attack profile, and they concurred. That profile was as follows: we would vector west at high speed and low altitude to gain a position south-west of the approaching MiGs so as to be able to vector north-west with the afternoon sun over our left shoulders. That might provide surprise and put us in a position so the bandits could not see us well—coming out of the sun. The MiGs' tactics at this point in the war were to dash in over the 'no-bomb' line, shoot the A-7s and retreat north before fighters could be vectored for them.

The MiGs finally launched and started south. 'Raider' vectored us west, we jettisoned our centerline tanks, armed missiles, and hit the deck. We went down to 1,500 ft and got to the karst ridgeline just as the MiGs headed south and crossed the line. We were vectored north-east, turned and pointed to the area they were coming from, and immediately got a PD [pulse-Doppler] radar contact; at 32 miles, as I recall. We were still low, and the MiGs were at around 5,000 ft. On cipher we were told that they were two 'blue bandits,' which identified them as MiG-21s [MiG-17s were 'red'], and there were no other known bandits in the area. Also, we were told that the MiGs' communications were being jammed by our

EA-3 ECM bird, sitting just off the coast. That meant they probably would not know we were coming. Great sport!

We continued at low level and around 550 kts with smoke off (the anti-smoke device on the F-4J diminished the amount of smoke emitted by the J79s), in combat spread formation with Charlie at my 3 o'clock position, so he could look through me at 'bad guy' country. He still had no radar. Since the MiGs had been 'positively' ID'ed I asked for 'clearance,' meaning clearance to fire. To my utter amazement, 'Raider' responded, 'Roger, contacts are two blue bandits; you are cleared to fire!' Ed and I were ecstatic, since it was normal to have to gain VID [visual ID]. I checked switches were armed and ready and made sure missiles indicated good. We were loaded with two AIM-7E Sparrows and four AIM-9G Sidewinders. I reviewed in my mind procedures for switching from 'radar' to 'heat,' and we kept on tracking.

We maintained radar contact continuously, down to twenty miles, and we checked everything again, keeping Charlie up to speed

By 31 May 1968 the 3,000th F-4 was ready for its first flight. Issued to VF-92 in September 1968, BuNo 155772 also flew with VMFA-212. (Boeing/McDonnell Douglas)

VF-33 F-4J BuNo 155554 is catapulted from USS *America* on 26 May 1968 in a haze of afterburner. A few days later it was hit by AAA over North Vietnam, and its pilot, LT Eric Brice, was unable to eject when his canopy would not open. (U.S. Navy)

The *Black Lions* traded up to the F-4J in August 1969, receiving the first aircraft with the automatic carrier landing system (ACLS) installed. (via C. Moggeridge)

on the situation. He was to maintain visual lookout for other bandits who might be hiding in the weeds. At twelve miles I reconfirmed 'clear to fire' with 'Raider' and began looking intently for any sign of bogies. At eight miles I called, 'tally ho two, on the nose.' What I really saw was two glints from the bright sun behind us on the silver fuselages of the MiGs, not the aircraft themselves, but from eight miles in I never lost sight. Locked on, dot in the center, MiGs head-on, it looked good for Sparrow shots down the throat. At five to six miles the missile launch circle began to expand, indicating max range, expanding to mid or optimum range. At four miles the circle reached its largest diameter, indicating optimum firing parameters had been met. I fired off two Sparrows and called, 'Fox one. Fox one.' The Sparrows appeared to guide, heading for what looked like imminent kill. The range on radar suddenly appeared to freeze at 3-4 miles, and I watched as the MiGs, now fully in sight and looking like planes and not sun glints, began a lazy left turn away from us....and the missiles! Guess what the Sparrows did? They saw the decreasing Doppler, and by the time they got to the MiGs the missiles were looking at a belly-up, beam aspect. They exploded harmlessly at the wingman's 2 o'clock position, about 100 yards away. Until the Sparrows exploded the MiGs did not know we were there. The wingman, apparently startled by the Sparrows, broke into the explosions, but then turned back left to stay with his leader. He then apparently realized I was quickly approaching a good 6-7 o'clock firing position, and again broke hard into me, by this time rapidly closing to a firing position. The MiGs were only flying at about 350 kts, so the wingman quickly came into me and was just as quickly inside minimum range. I had switched to heat and fired off a Sidewinder, but the aspect was almost 90 degrees off at less than 1,000 ft so the Sidewinder missed. However, it scared him so badly he continued his descending right break, hit the deck and headed north out of the fight. Meantime, I was performing a high-g left barrel roll to get in behind the leader who, by this time had figured out the program and was breaking right into me. My wingman broke left over the top of me and spotted two more MiG-21s down in the weeds, about three miles away.

Simultaneously, Red Crown (USS *Long Beach*) broadcast, 'Heads up, Rootbeers. You got two more bandits west.' I was too busy to respond, and Charlie was telling me he saw them too so I continued my turning and with my energy, combined with the MiG leader's bad position and slow speed, I quickly attained the six o'clock at about 1,500 to 1,800 yards and fired an AIM-9G. I watched it guide and impact the tail area of the MiG, blowing the empennage completely off. The pilot obviously knew he was had because almost simultaneously with the impact I saw his chute. It appeared he had ejected either just before impact or as it occurred.

Meantime, Charlie had called out something to me about breaking. I didn't hear it, but what he said was, 'Outlaw, break left...I mean RIGHT!!' (I got to hear it on the tape, later). One of the MiGs hiding in the weeds had fired off an Atoll, well out of range, and I was vaguely aware of its smoke trail corkscrewing lazily across the sky, well away from me. I broke back to where the other MiGs were coming from and saw them hit the deck about 2-3 miles away. They turned tail and ran. As soon as they were tail-on they disappeared, vanished—couldn't find them, visually or on radar, so I

In one of the world's most hazardous working environments, deck crew check that all is in place for another successful F-4 cat shot. (U.S. Navy)

called to Charlie to 'Unload, unload. Bug out, bug out!' and we headed for the water. We called 'feet wet,' and Raider called and confirmed, 'Splash one blue bandit, Rootbeers.' I responded with something like, 'You betcha, Raider. I got that son of a gun!'

We hit the tanker, took on enough gas to get to the ship, and I performed the best rendition of a victory roll I could imagine. The ship and Air Wing crews swarmed me after landing in much the way depicted by the movie 'Top Gun' when Maverick and Iceman return to the ship after shooting down the bad guys. It was a neat feeling to be a hero for the day; in fact, hero for the cruise. I gave up smoking as a result of that kill. I had told some guys jokingly, 'If I shoot down a MiG today I'm going to quit smoking.' I suppose God said, 'Oh yeah? Let's see if you really mean it.' I haven't smoked since."

"Outlaw" and Ed Kain received Silver Stars, but there were no other shoot-downs on that cruise, though CDR Wilbur and his wingman, Emory Brown, shot at a retreating MiG. Gene Tucker was in a favorable firing position on another occasion, but his Sparrows failed to arm, and in another dogfight CDR Dave Shepherd, Bill Knutson's successor as skipper of the Tarsiers, got mixed up with some F-8s and was nearly fired on by one of them. VF-33's

The VF-142 F-4J flown by Jerry Beaulier and Steve Barkley as *Dakota 201* on 28 March 1970 when they shot down a MiG-21. (USN via Angelo Romano)

A white-nosed F-4J of VF-41 recovers aboard USS *Franklin D. Roosevelt* in July 1972. (USN via CDR James Carlton)

sister squadron, VF-102 *Diamondbacks*, took one of the rare hits from an Atoll missile on 16 June during a two-on-one engagement in which the Phantoms fired four AIM-7s unsuccessfully. The stricken aircraft had radio failure and was hit from behind, unaware of the wingman's warning. On this occasion the pilot was VF-102's CO, CDR Gene Wilbur, who became a POW, but his RIO, LTJG B. F. Rupinski, was killed.

Although the F-4J had marked up its first MiG, it would be over eighteen months before Navy Phantoms scored again. As Tex observed, "It was kind of ironic; when the carrier was on line there wasn't much MiG activity, but when the F-4s were pulled back and the '27 Charlie' carriers with F-8s were on line there were quite a few engagements." Crusaders knocked down five MiGs in the second part of 1968.

"We were operating below the 'parallel,' primarily for interdiction. We flew 'targets of opportunity' flights at night looking for truck lights. One night near Vinh my wingman and I were on such a mission. LT Fred Lewis spotted some lights in the trees, and we rolled in with six Mk 82s to bomb the target. I decided to bomb singly rather than salvo the bombs. You could count the flashes, 1 - 2 - 3 - 4. When the fourth flash went off it just grew and grew. We could see we had hit an ammo or petrol store on the north side of the river. We concentrated on that for the next few sorties and wiped up a big complex. The North Vietnamese had come down and cached their material on the edge of the river, to be taken south by barge. They were very innovative. You could knock out a bridge and they would just build by-passes on either side; three or four each side. Sometimes they would have them under water so they were difficult to see, or they'd have floating bridges in segments that they'd uncouple before daylight and hang them along the bank under the trees. A lot of times we were going after less-than-desirable targets. One time I remember putting six Mk 82s on a road and railroad. One bomb put a crater right in the middle of the road, but I thought it would take half a dozen workers thirty minutes to make it serviceable again."

Patrolling over the clouds near Hawaii, F-4J BuNo 158362 displays VF-154's tasteful color scheme. (Jan Jacobs)

Falcon Phantoms

In February 1968 Jim Sherman became XO of the Marines' first F-4J squadron, VMFA-334 *Falcons*, at El Toro. At the time they had radar-less, "lead nose" BuNo 153-series aircraft while Westinghouse and Naval Air Systems Command sorted out the AWG-10's problems, but they were promised the first fifteen "full-systems" F-4Js off the line. Meanwhile, RIOs were promised twelve intercept flights in Miramar RAG F-4Js and extra time in their simulator, while pilots had to settle for three flights. There was considerable urgency about this, as the *Falcons* were scheduled to deploy to Da Nang on 1 September 1968 as soon as they received combat-capable F-4Js. Jim Sherman: "As the first F-4J squadron to deploy there was no supply support in the Far East (only fifteen per cent of spares in the F-4B were interchangeable with the F-4J), and none for the AWG-10 except for the factory. As a result, we received factory supply support for our first six months and deployed with seven civilian tech reps." The Transpac flight commenced on 20 August with refueling support by VMGR-352 KC-130s. On arrival at Da Nang after a flight in which the stages were limited mainly by the F-4's 4-5 hour oxygen supply, Colonel Sherman found the base much improved compared with its state in 1965: "Instead of six hangars there were now twenty (MAG-11s were far superior to what we had at El Toro), and a second, parallel runway." Also on the base was VMFA-542, flying "very old and very tired F-4Bs." The *Falcons* took up residence almost opposite them. "Amazingly, when the Falcons arrived from Cubi Point they immediately knew where home was. It was the hangar with the 50-foot red falcon painted on the roof." Protective revetments had been built from twelve-foot corrugated steel sections in double walls with a filling of earth between them. One row (revetments 9-16) had a back wall, and parking an F-4 in them meant taxiing it slightly past the revetment, folding the wings, and shutting down the engines. A towbar was then attached to the nosewheel, and the aircraft was backed in by a tow tractor. Use of these vehicles was kept to a minimum: "A fully-loaded F-4 weighed 30 tons; the tow tractor weighed twenty tons

and suffered numerous hydraulic leaks, causing the wheels to seize up and make the vehicle a monument in the middle of the flight line until it could be moved by a crane that could lift twenty tons." Maintenance crews worked 13-14 hour shifts, with check crews working at nights, when it was cooler. Life was eased a little by having the "biggest beer-cooler available in the Far East, which held over 50 cases of soda-pop cans." Among the other residents were Navy EC-121K/M Warning Stars (Super Constellations). "No-one was allowed in their areas; no-one knew what or where they did their thing. If one of them was ready to taxi out as you taxied out for take-off you kept your fingers crossed, for they had been known to fire the F-4's jettison cartridges [by electronic interference] and dump your fuel tanks on the runway."

Another innovation at Da Nang was the arresting gear:

"The old runway had huge puddles of water after it rained. When they built the new runway they made a 'crown' of about twelve inches so the water would run off. Unfortunately, so did the landing aircraft. If you were not straddling the runway centerline you would drift towards the side of the runway and into the mud. As a result, when it was raining our SOP was to land on the new runway with touchdown at about 150 kts, with no drag chute and enough engine power to hit the arresting gear at about 120 kts. At that speed you could 'fly' the aircraft using rudder and aileron to keep the nosewheel on the centerline. There were distance markers on each side of the runway and big arrows marking the arresting gear at 4,000 ft. You dropped your tailhook about 1,000 ft short to ensure it was fully down when you hit the cables. Power was still at 80 per cent, and when you engaged the cable it gradually increased drag and stopped you in about 500 ft. You could then pull the throttles to idle, and the arresting gear automatically pulled the Phantom backwards after you stopped so that you could raise the tailhook and taxi off at the 6,000 ft marker."

Prior to deployment the squadron had no time to study the new DECM gear in their brand-new BuNo 155-series F-4Js. This was done in a series of five one-hour lectures at Da Nang. For operations, the aircraft always carried two wing-tanks on stations 1 and 5 (outboard wing pylons) and two AIM-7s. Ninety per cent of the

A split second before hook grabs arresting wire, *Showtime* 103 is seemingly suspended above USS *Constellation's* deck. (USN via CDR James Carlton)

missions were CAS, DAS, or BARCAP. On the latter, eight missiles were hauled along, while CAS required "nape and snake," gunpod, and 2.75 inch rockets. DAS loads were slick 500 lbs bombs, Rockeyes, and Zunis for interdiction of troops not in contact with "friendlies" via delivery from a safer stand-off range. Initially, about half of the eight hours turnaround time between missions was for ordnance loading. This was reduced when the squadron introduced a supervisor on the flight line equipped with, "a VHF radio, telephone, hand-held radio, PA system, and hand-held bull-horn" to coordinate the effort better. The use of pre-loaded MERS and TERs and a SATS loader also helped. Refueling times were often cut by "hot refueling," with engines running, in a special pit which operated 'round the clock. On entering the pit, Phantoms had safety pins put in any external stores and extended their refueling probes to de-pressurize their fuel systems. Refueling took less than eight minutes.

There were sixteen revetments (eight of them "drive-through") in the VMFA-334 area, the first four of which were covered and used to park aircraft which were "up and ready to go again." F-4s with minor glitches went to revetments 5-8, while the more serious maintenance cases went to revetments 9-16, which had a back wall.

On the Trails

On 1 October 1968, the day Colonel Sherman took over the *Falcons* command, the JCS began its massive campaign against the flow of war material south along the Ho Chi Minh Trails. "When we received Mk 20 Rockeyes and no-one else had them our popularity soared, and we went after the truck convoys in earnest. In December 1968 we flew 664 combat missions. Half of these were on the convoys, and that month we got confirmed kills on over 600 trucks. About 1 November 1968 Rockeye CBU arrived at Da Nang. The DoD had rushed the shipment of the bomb before it had been fully tested. One major drawback was that the bomblets in each canister could not be 'safetied.' Accordingly, the Navy would not load them on carriers, and they all came to Da Nang."

Vulnerability of the F-4's hydraulics to the ever-present AAA on the trails missions continued to be a hazard. However, the F-4J had an improved system, whereby the PC-1 system operated flight controls on the left of the aircraft and the PC-2 powered those on

VF-92 *Silver Kings* pose in front of *Silver Kite* 201 on their 1970 cruise. (Curt Dosé)

"Boards" (airbrakes) are extended on F-4J BuNo 148359 as it makes a section VFR high-speed approach to a break for landing at Miramar in 1973. (Jan Jacobs)

the right. On the F-4B, a direct hit took out both systems, but the F-4J arrangement (later retro-fitted to some F-4Bs) kept one half of the controls in use.

"We had two instances in which F-4Js with battle damage lost one hydraulic system, returned to Da Nang, and made a successful landing. Both blew down the landing gear pneumatically and landed with no flaps. With the power back to the 80 per cent range the slightest stick actuation caused the flight controls to seize momentarily. Both aircraft touched down at 250 kts plus. They had shut down both engines and hit the mid-field MOREST gear at around 200 kts. The cable played out, and when it hit the stops the cable broke and they continued on down the runway, catching the overrun chain gear, and went off the end of the runway about 1,200 ft. Both were repaired and flew again. We had two others that didn't make it all the way home. One lost control over the Gulf and ejected. The crew was picked up by an Australian destroyer, and when the USAF rescue helo landed on the destroyer he radioed that they would not give him our aircrew, as they were holding them for ransom. I asked them what the ransom was and the reply was, 'Fifty gallons of ice-cream.' I had no problem getting a helo to make the delivery, but finding 50 gallons of ice-cream was quite a chore."

Rockeye Mission

Colonel Sherman re-creates a typical truck hunt, such as he flew in November or December 1968:

"Normally when you were scheduled to go to Laos to hit truck convoys they were back-to-back missions, because you knew where you were going and knew what the threat was. If the weather was clear and the moon was up it was bad luck, because the guns could see you and would try their best to nail you. Our flight meets in the Falcon ready-room, and at the scheduled briefing time we walk across the street to the MAG-11 briefing room. The main thing we want to know is the en-route weather, the weather in the target area including ambient light conditions, the number and location of guns in the target area (which we have reported to them), and the forecast weather at Da Nang when we return.

Showtime 100 (BuNo 155800), made famous by Randy Cunningham and Willy Driscoll in 1972, flies formation on NG 102 on return from a strike mission. (USN via CDR James Carlton)

When we return to the ready room I inform my wingman that he will brief and lead the flight, and I will be his wingman. We were scheduled to take off at 2030 for a 2100 time on target (TOT). Our aircraft are not assigned yet, but lead will be loaded with 12 Rockeyes, and I will have six Rockeyes on the centerline and three pods of Zuni, armed with VT fuzes set to explode ten feet in the air, on each inboard station. We covered call-signs, radio frequencies. Our take-off would be single-aircraft with a left turn to head up the coast. The wingman will make a running rendezvous, and when closed we will turn left to 330 degrees and level at 21,000 ft. Shortly after entering Laos we will switch to a clandestine TACAN and switch our radios to the USAF control center to contact a FAC (or a given frequency) who will tell us to proceed to our target area and hold high while he sees what he can find. If the weather is good and the ambient light is bright, lead will drop all his ordnance on the first run. I will be behind him by about 3-5,000 ft and offset to the left side about 30 degrees from his run-in. When the guns open up I will be in the dive and able to fire rockets at the gun positions. I

will make a second run and, depending upon the outcome of lead's run, I can play 'clean up' on the trucks, or I can use the Rockeye on the gun positions while firing out my remaining rockets. Coming off target I will call 'clear,' lead will give me his heading and altitude, and we will head for home. The FAC will give us BDA, and we will switch to 'squadron common,' turn on our external lights, and check each other's aircraft for hung ordnance, fluid leaks, or damage. At fifty miles from Da Nang we will switch to Da Nang approach control and commence our let-down. Since the weather at Da Nang is IFR I will make one orbit after lead descends to get three minutes separation and then follow him down. I will get my own radar vector and GCA pick-up, and GCA will inform me when lead is clear of the MOREST before he clears me to land. I will then make a no-chute landing and trap in the MOREST. We will hot-refuel in the pits and turn in as directed by the maintenance coordinator. After shut-down we will meet in the flight-line office. This concludes the brief.

At 1950 the duty Officer was informed that our aircraft were ready. I had *Falcon* 6 (BuNo155735), and the Lieutenant leading the flight had *Falcon* 4 (BuNo 155563). We proceeded to the flight line and filled out the yellow sheet on our aircraft. As we came out of the office the maintenance coordinator walked over and told us which revetments our aircraft were parked in. I told the coordinator that *Falcon* 4 would taxi out first. Our plane captain and an assistant greeted us as we approached our aircraft. My RIO and I had flown together for three months, and we could go through our 'walk-thru' inspection of the aircraft and strap in without conversation. There is some light in the revetment, but you need your flashlight to check for details. The first thing you always do is put on your helmet. There are many things under and around the aircraft that will split your head wide open (most embarrassing). Besides normal integrity of the aircraft you also double-check your ordnance load. Everything looks good, and I tighten my straps and zip up my survival vest. I am now wearing about 45 lbs of harness and survival gear. As I sit down in the cockpit the plane captain assists my hooking up the shoulder harness, and when I have fully strapped in he pulls the safety pin on the firing mechanism for the ejection seat and shows me the pin.

A pair of *Aardvarks* F-4Js take it down over some spectacular terrain. (Harry Gann/McDonnell Douglas)

VF-161 *Chargers* tail logo on F-4B BuNo 149415, which also has both AN/APR-30 fin-cap antennas. (via J. T. Thompson)

VF-161's NF201 thunders away from USS *Midway* in 1973. The squadron scored four MiG-17s and two MiG-19s during the war years, all but one of them in 1972-73. (U.S. Navy via Norman Taylor)

The external power is applied to the aircraft as I plug in my radios, oxygen mask, and suit. I can hear my RIO breathing, so I check in with him on the ICS. He reports everything looks good so far. I finish my sweep through the cockpit to make sure all the switches are to my liking. I double-check that the wing-fold handle is in the 'fold' position so that the wings don't spread when the engines start (another potential screw-up). When I am ready to start I give the plane captain the 'crank' signal, select the starboard engine, and hear it start to turn. I check the rpm, and when it reaches 35 per cent I bring the throttle up to idle and at the same time hit the igniter button. When the engine reaches idle rpm I switch the air to the port engine and repeat the procedure, meanwhile turning on the radios, TACAN, and other equipment. The RIO is busy cranking things up back in his department. The plane captain has unhooked the starter unit and checked everything is clear of the aircraft. I signal to him to pull the chocks from the main wheels, and we are ready to taxi out and point towards the taxiway, where we will spread the wings and lock them, and the plane captain will check for free movement of the control surfaces. I report to my leader that I am ready to taxi. He acknowledges, and we switch to ground control. In the vicinity of the fuel pits we halt while a final checker pulls the pins on our external fuel tanks, MERs, and TERs. He checks that the rocket pods are plugged in, and with his hand holding up a bunch of red flags we continue our taxi.

Approaching the take-off runway, leader switches to tower frequency, and we are cleared for take-off. We pull onto the runway and apply full power to each engine separately, checking that the afterburners light properly. If you did both engines at the same time you would skid the tires. Lead transmits to the tower, 'Falcon 4 rolling,' and his afterburner lights as he accelerates down the runway. I watch and see him rotate the nose up to a take-off position, and the afterburners are bouncing flames off the concrete. As his aircraft breaks contact I release the brakes and apply full power, then select afterburner. I transmit 'Falcon six rolling' as I see leader disengage his afterburner and disappear into the overcast. I rotate

at 150 kts, the aircraft skips once, and we are airborne. Gear up...180 kts, flaps up...240 kts, out of afterburner and commencing left turn to find my leader. 'Bump, bump,' as we get jolted around by the turbulence in the rainshowers that are moving in. I switch to the assigned frequency and broadcast, 'Falcon 6 up.' Lead responds, 'Roger, Six, passing the Da Nang 060 at 16 passing 12,000.' By this time the RIO has the radar working and has our leader dead ahead at 8 miles. We lock up our leader, see that we have a 50 kts closure rate on him, and unlock so that we can look out for other aircraft. As we are about to join our lead while climbing in the clouds our DECM gear is indicating we are being swept by a search radar, bearing 020 degrees. My RIO lowers his radar antenna so we can see the ground, and we look at the 020 bearing and confirm what we expected. It was the USMC TPQ radar at Dong Ha.

Soon we popped out of the clouds and leveled off at 21,000 ft. We were on schedule to meet our 2100 TOT, so we throttled back to conserve fuel. About 40 miles out we contacted USAF control, 'Falcon 04, flight of two on the 140 at 40 miles 21,000 TOT 2100.' 'Roger, Falcon 04, stand by.' [Contact was made with the USAF FAC]. 'Hi, Falcon 04, you're the sh*t hot Marines from Da Nang. Meet me at the usual place and hold high while I call control to see what's happening.' Several minutes later: 'Falcon 04, this is FAC. You're in luck tonight. We got movers about twelve minutes out, what is your position?' 'Roger, FAC, we are over the target area at 21,000, lights out.' 'Okay, Falcons, I am going to diddy bop up the road to pick these guys up and escort them down. Have you picked up the 'lights' west of the road? As you see, there are six lights, and they are about 150 ft apart.' Each night at sundown the USAF dropped 'lights,' 'sticks,' or 'logs,' which were all the same, along the west side of the highway. These devices looked like a flashlight that is used to taxi aircraft. They buried themselves in the earth, and the light could only be seen from directly above. Airborne controllers used these as references to tell us where the convoys were in relation to the 'lights.' 'Falcon 04 has all the lights.' 'Falcon 06 has them.'

VF-103 *Sluggers* **F-4Js with various combinations of external tanks and TERs. The squadron did a single war cruise in 1972 aboard USS** *Saratoga.* **(via Norman Taylor)**

The sky was clear of clouds, although there was some haze. We were in a left hand orbit of 7-19 miles in circumference. I had dropped back to 500 ft in trail of the lead. I couldn't make out the road, but I knew where it must be in relation to the lights. 'OK, Falcons, we are three or four minutes out at this time. We have eight customers for you. As usual I will call the position of the lead vehicle in relation to the lights and also where the last vehicle is. Try to time your run to drop as the first mover is abeam the first light.'

'Falcon 04 Roger, we will make a couple more orbits. I will roll in north to south, offset on the lights, and release my bombs in pairs starting on [truck] number eight and ending at number one. Falcon six will be to my left and above with rockets to suppress the guns.' 'Sounds good to me, Falcons. Lead vehicle is 300 ft above the number six light.' 'Falcon 04 descending, switches on,' 'Falcon 06 roger, switches on.' As we approached our roll-in positions Falcon 04 was at 14,000 ft and 350 kts, and I was a quarter mile back at 16,000 ft. 'Falcon 04, this is FAC. Lead truck is abeam light 3 and last truck is approaching light six. Vehicles are evenly spaced, 50 ft apart. You are cleared hot. Good luck.' 'Falcon 04 is in hot, do you copy 06?' 'OK, roger, in right behind you.'

I pulled my nose down, pointed at where the road hit the river, and throttled back to achieve correct TOT. I was probably about 13,000 ft, and my lead must have been around 10,000 ft when the 37 mm guns opened up on him from two positions, one on the east side of the road and the other across the river and to the west. I yanked the nose hard left about 25 degrees to get two pairs of rockets off on the easterly guns, but disregarded the guns across the river. If I had pulled my nose way down there I would not have been able to get back down on the ZSUs [guns] when they opened fire. Meanwhile, I saw the flash of the 37 mm shells self-detonating behind and above us in my rear-view mirror. At the same time one ZSU opened fire, followed immediately by the other. I yanked the nose over to the first gun and fired two pairs of rockets. Before they reached the ground I had the nose coming over on the other gun. I

could tell by the erratic movement of the rounds coming up that the second ZSU could see Falcon 04, but also that 04 was jinking as he pulled off target. Meanwhile, the first ZSU ceased firing, and the second ceased before my rockets got to his position.

How I loved those 5 inch Zuni rockets. They came off my aircraft in a flash of fire and continued to burn most of the way to the gun position. When they burned out they would be doing Mach 4 plus my launch speed. To make matters worse they detonate ten feet above the ground, spewing shrapnel all over the target. The folks on the ground had high regard for this, for they would quit shooting and start digging as soon as they saw the rockets coming. Meanwhile, the FAC started chattering when the ZSUs opened fire. 'Falcons, watch the ZSUs. Lead, they are on you. Good show, the bombs are walking right down the convoy. Must not be carrying ordnance. Jink, leader, jink. Beautiful rockets. Gunners are old-timers. They quit shooting as soon as the rockets are fired.' I had fired

Ghostriders **F-4J BuNo 155764 bears the Fleet Excellence and Safety Award letters on its splitter plate. Compared with some of the earlier naval jets, the F-4 had an outstanding safety record, and a number of squadrons completed long cruises without losses. (via R. F. Smith)**

my final pair of rockets at about 400 kts and immediately went to full throttle to get some speed to jink and climb back to altitude. Never go into afterburner over a target and give away your position. If someone down below has a hand-held, heat-seeker missile you are certain to make him a hero.

We had gone back to 16,000 ft orbiting the target area while the FAC sorted things out on the ground. 'FAC, this is Falcon 06. I still have six bombs if you need them on the convoy. Otherwise I will dump them on the easterly gun positions, over.' 'Roger, Falcon six. I don't think there is any need for them on the road. Go ahead.' With that, I rolled over, throttled back, and pulled my nose down to the closest gun position, and at 10,000 ft began pickling off my bombs as my gunsight ran across the area. At 8,000 ft my last bomb was off, and I added full power and began pulling the nose up with almost 6gs on the aircraft. As the nose passed the horizon I began jinking left and right, and I brought the aircraft into a 45 degree climb. 'No one fired at you on that drop, six. I wonder why? I can't give you any BDA except for one small secondary explosion. Falcon four, you got all light movers. I estimate they were five-tonners and were carrying 'stuff.'' 'Roger, FAC. We are bugging out for the homeplate, but will be back at 2400. Falcon 04 out.' 'Nice working with you, Falcons. FAC out.'

Back at Da Nang we clear the runway after a MOREST landing. I switch to ground control and say, 'Falcon six clear of runway 18 right back taxi to my line.' 'Falcon six cleared to your line. Good night.' As we approach the fuel pits I turn the landing lights off, on and off again to alert them, fold my wings, extend the probe, and switch to squadron common frequency. I leave my canopy closed, as it is raining cats and dogs. (Flying in heavy rain, you notice the noise of the rain hitting the windscreen. If you have driven a car in a thunder storm and been astounded by the noise of the rain, imagine what it is like at eight times that speed.) Two very wet Marines drag out the fuel hose to the aircraft. One has a handful of pins attached to red flags that he inserts into the five ordnance stations, and he unplugs the electrical power to the rocket pods. Normally, my RIO would get out here and walk in to the debrief, but he doesn't like to get wet and hopes that we will get a covered revetment to park in. The maintenance coordinator comes up on his radio to ask the status of our aircraft. I inform him we are up, but we have twelve rockets unexpended and we will take the same aircraft out if possible. He acknowledges that he can make that happen and directs us to (covered) revetment 2. I also inform him that Falcon 04 took shrapnel down both engines, and they are fine but we should inspect the turbine blades.

Refueling complete, I retract the probe, check visually that the wings are folded, turn my internal lights on bright, and taxi down the alley to revetment 2, which is nice and dry, and shut down. As soon as the chocks are in place, on come the floodlights, and there is an ordnance crew with two SATS loaders and the new ordnance. Rather than drop pylons two and four to hang new rocket pods they just refill the empty tubes on the existing pods. That takes five minutes, and dropping the centerline pylon to hang the replacement with six Rockeyes already attached takes five to ten minutes. We got so good at this evolution that some non-aviator suggested we hot re-arm the aircraft, thereby saving thirty minutes. It worked,

but after most aircrew had done it we polled them and they unanimously agreed not to do that. They would rather get out of the aircraft, remove their helmets, scratch their head and their ass (not necessarily in that order), and chain-smoke several cigarettes. Back then we all smoked."

When Falcon 4 taxied in, the engine shop had a supervisor and two young Marines with short ladders and flashlights standing by to check the engines for damaged turbine blades. The engines had just quit windmilling when they headed down the intakes literally head first, flashlights in hand. As expected, they found nothing damaged, so we sent the F-4 back out. It was 2230, and our night was only half over. We had to do this all over again."

On his second mission of the night Colonel Sherman encountered a FAC who had to order them to drop on a "suspected truck park" in a patch of jungle north of their previous target. Objecting that he was not prepared to waste ordnance "shooting enemy trees," Jim Sherman got the FAC O-2A "Mixmaster" to assign him their original target area where the FAC's "starlight" night-vision scope showed a grader and truck repairing the road from the previous strike. Once again, the AAA opposition was effective, but Colonel Sherman first ruined their night vision (and aim) by pumping out flares, and then presented them with a load of Rockeyes and Zuni. Meanwhile, his wingman undid the road repairs.

BARCAP Blow-back
Like other Marine F-4 units, the *Falcons* were periodically called on to fly BARCAP for TF 77. As compensation, the tedium was relieved only by "shooting the breeze" with the Air Controller on the "Crown" destroyer, who would sometimes play music for the bored BARCAPpers. Usually, Colonel Sherman found the experience "relaxing," but there were exceptions:

"There is an old-timer, like myself, who shows up at happy hour every Friday night at the 'O' Club at Kep airbase, north of Hanoi. About 8 p.m. when he's well-lubricated he tells the story of the day he thought he shot down an F-4. He had it in his gunsight and was firing as fast as he could when the F-4 turned into a ball of fire and disappeared. He had heard many pilots talk about last-ditch maneuvers, but this F-4 had one like he couldn't believe. Here is what really happened.

Aboard USS *Ranger* in the South China Sea, F-4J BuNo 158346 is just about to drop the arresting wire, which will snake back to the correct tension for the next trap, and taxi clear. The *Black Knights* made four war cruises with the F-4J, following two with the F-4B and one with F-8D Crusaders. (Jan Jacobs)

My wingman and I had been on the BARCAP station about an hour when I went to the tanker and topped off. When I returned my wingman headed for the tanker. A few minutes later on guard channel came, 'This is Crown on guard. Two MiG-21s airborne Kep, climbing west.' Our Air Controller at Crown sat beside the controller on guard, and he said, 'Falcon six. You copy?'"

A few minutes later the warning was repeated, with the additional news that the MiGs had turned east at 20,000 ft. Colonel Sherman was told to arm his missiles while the controller ran through the current Rules of Engagement with him: 1) the MiGs have to be beyond the twelve mile limit before they could be engaged 2) you must make a visual ID before you can shoot 3) they should make some attempt to engage you before you shoot. The F-4J crew were then vectored on the two MiGs from 48 miles.

"On the intercom I told my RIO we would intercept like Top Gun taught, by descending well below the MiGs and getting the afterburner lit to get rid of our smoke trail, then pull up so as to pass them close abeam, climbing vertically and supersonic. The chance of them seeing us under their nose was nil. As we climbed vertically we looked back to see if the MiGs had broken formation to engage us. To our shock all we saw was orange 'tennis balls' coming up at us and passing close by on all sides, with a third MiG-21 firing them. I told my RIO, 'Not to worry, he can't continue firing that big cannon going straight up. He will drop off any moment, and when he does we will follow him down and put a Sidewinder up his ass. I reached over and switched to 'heat.'"

My RIO and I both kept our eyes locked on the MiG, and soon he shuddered and 'fell off.' This was just what we were waiting for, except that he fell off in such a way that he ended up behind us and out of sight. In order to see him I had to turn my Phantom around as I got the nose of the aircraft pointed down. To do so I kicked full left rudder, which was followed by one huge explosion. It threw us forward in our harness, and we were one large fireball. My first thought was that we had been hit by an SA-2 missile, because we were flipping end over end and I had no control over it. Soon the flames disappeared, but we were still toppling end over end. The airplane seemed to be intact, but we had no airspeed and both engines were dead. I don't remember talking to my RIO, but I imagine I said something like, 'We are OK, don't eject yet. We have plenty of altitude left.'

Soon the aircraft resumed forward flight, and the airspeed built up rapidly while the engines began windmilling. As they passed the idle rpm figure I lit off both engines, and they began running normally. We were pointed nearly straight down and almost supersonic, passing 20,000 ft. I pulled back on the stick, and the aircraft responded normally as we leveled off at 9,000 ft. I added power to climb as I turned south and told my RIO, 'Let's go home and come back another day.' Crown called and asked, 'What the hell happened?' I replied, 'You don't really want to know.' I called my wingman and told him to join on me or follow in trail.

What had happened? No, the MiG had not hit us. No, we were not hit by an SA-2. When I kicked the rudder to turn the aircraft around we were not still climbing, although we were at 36,000 ft in full afterburner. In fact, we had run out of airspeed, and we were falling backwards at about 60 kts. My rudder deflection interrupted

the airflow through the engines, which caused both engines to 'blow back.' Air was coming up the tailpipe, and the fire and thrust was coming out of the intakes. Is it any wonder that there was a hell of an explosion, the aircraft was covered in flame, and we stopped flying instantaneously? It is a real tribute to General Electric that those engines could stand catastrophic forces like that with no damage. As we headed for Da Nang I regained my composure enough to give Crown a piece of my mind. He had broadcast at least four or five times *two* MiGs. Why hadn't they seen the third? What kind of maneuver had he made to get behind me in firing range when I had been supersonic and climbing vertically? We made two major mistakes, also. When we started our intercept we should have jettisoned the two external fuel tanks, which were probably half full. The loss of that drag and 3,000 lbs weight would have helped our performance greatly. Secondly, we should have broken lock on the two MiGs from time to time during our ID run to look for other aircraft."

Colonel Sherman's last task with the Falcons before relinquishing command was to organize their move to Chu Lai from 1 January 1969 in order to create room for an A-6 squadron at Da Nang. This was accomplished without missing a single combat sortie, and despite the fact that the squadron had to organize their own dedicated space at the new base. They continued a frantic rate of combat flying, concentrating on CAS. During one day soon after their arrival they generated 46 sorties with eight aircraft between 6 am and 8 pm, an average of six sorties per Phantom. During May 1969 they flew their 5,000th combat sortie, and on 1 August they departed for Iwakuni as part of the general force reduction, returning to the USA on 1 September.

Tackling MiGs

Although ACM had been a component of Navy F-4 training from the outset, the dominant Fleet philosophy was still centered on air defense tactics and a belief that "dogfighting" was essentially redundant in an era of BVR missiles. Phantom pilots were often unused and in some cases unwilling to fly their aircraft hard and explore its performance limits. Although the F-4 had more than held its own against the nimbler MiG-17 and MiG-21 in Vietnam, with fifteen kills against five losses to MiGs by the end of 1968, it was realized that there could be distinct improvements in this 3:1 ratio in a number of ways. First, the Rules of Engagement required a

A late production F-4J of VF-154 in June 1972 at NAS Willow Grove. (R. Harrison via Norman Taylor)

visual identification of the enemy, which often denied the Phantom the use of its long-range Sparrow main armament. When forced to engage at closer range the Phantom was unable to match the maneuverability of the MiG fighters, which had been designed for more basic, close-in fighting. Navy pilots were often unable to bring their more sophisticated weapons to bear on the elusive enemy, and were at the same time in severe danger of finding MiGs on their own tails.

There were many additional factors which were peculiar to the Vietnam scenario. MiGs operated under the control of a very effective Russian GCI network and were always close to their bases, whereas USN fighters were more fuel-limited and reliant on more distant radar controllers. However, Phantoms were usually vectored to their "bogies" very effectively, and a much greater source of concern was the failure of the F-4's weapons to destroy those targets once battle was joined. In 1966 the Navy began to investigate all these problems in an attempt to improve the Phantom's success rate against its much simpler Soviet adversaries. A CNO study called *Project Plan* looked at the Navy's use of fighters and missiles, while VX-4 flew the F-4B against a variety of contemporary fighters, including the F-8, F-105, and F-104. CNO also had the advantage of access to the Air Force's data on MiGs, gleaned from samples provided by Iraq and Israel. The USAF Project *Have Drill* had evaluated a captured Syrian MiG-17F, while *Have Donut* dissected an Iraqi Mig-21's performance. Earlier information came from the Air Force's *Feather Duster* program, which pitched ANG F-86H MiG simulators against F-105s and F-4Cs. Its findings had suggested that U.S. pilots would be at a real disadvantage, particularly if they were sucked into a turning fight below Mach 0.9. *Project Plan* rated the F-4 more highly as a dogfighter and recommended that pilots should keep their speed up and attempt hit and run tactics to avoid heavy losses. At this early stage the advisability of using the F-4's superior climb performance to give vertical separation from the adversary and the chance to turn back down into the MiG for a missile shot was questioned.

These tactics were later supported by hands-on experience with real MiGs, and U.S. pilots were impressed by the capability of their covertly-operated Russian fighters. The MiG-17, essentially an improved version of the Korean war MiG-15, had half the F-4's wing loading and weighed 75 per cent less. It could turn very quickly and get behind a Phantom at lower speeds, while its simple con-

Showtime 105 traps aboard USS *America*. VF-96 made a single cruise on the carrier from April to December 1970 with no combat losses to any of the CVW-9 squadrons aboard ship. (USN via CDR James Carlton)

struction and lack of hydraulics made it hard to damage lethally. There were handicaps, too. Lack of powered flying controls meant that the aircraft was limited to 575 kts and hard to maneuver above 450 kts. At this speed it also lacked stability, making it a poor gun platform, and its afterburner could only be used for about three minutes continuously. The three slow-firing 37 mm and 23 mm cannon were hard-hitting, but only had ammunition for five seconds firing.

The MiG-21F, in use by the VPAF in 1967-68, had the considerable advantage of relatively small size which, together with its smokeless engine, made it hard to detect, particularly head-on. It could also turn with the F-4 (which was twice its weight), though its smaller delta wing gave it less of an edge than the MiG-17 in a turning fight. Like the MiG-17 it required heavy control forces in its upper and lower speed ranges, and its R-11F-300 engine was slower to accelerate than the Phantom's J79s. Armament was limited to a single 23 mm cannon with five seconds' ammunition and two K-13A Atoll IR missiles, copied from the AIM-9B Sidewinder. A very simple radar ranging device was available for targets within a four-mile range, but pilots relied mainly on their rudimentary gunsight. Visibility from the cockpit was restricted by a thick windshield, heavy cockpit framing, and a 50 degree blind zone to the rear. The F-4 and F-105 could out-run it below 15,000 ft by accelerating above 550 kts, at which point the MiG began to suffer from severe buffeting.

These revelations reinforced the *Feather Duster* conclusions, and tactics evolved which required Phantom crews to keep up their speed and acceleration and attempt to climb or dive way from MiG-17s, avoiding the temptation to engage in a turning fight. Against the less numerous MiG-21 the rule was to stay low and fast and to try and get into the MiG's blind spot. The F-4 was found to have a small turning advantage over the MiG-21 at low altitude if USN pilots were prepared to push their aircraft into buffet. Against either fighter the Phantom was best in the vertical dimension, where it could out-accelerate the foe and gain separation in order to use its missiles from a favorable distance.

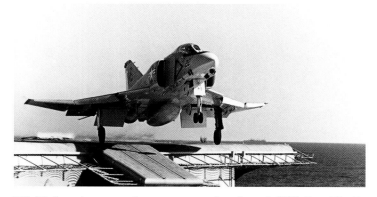

VF-32 *Swordsmen* made one war cruise aboard USS *Franklin D. Roosevelt* in 1966-67 with F-4Bs. (USN via Norman Taylor)

Faced with the fact that neither USAF nor USN had shown a sufficiently convincing superiority over the enemy, both forces put aside their rivalries, and a small number of USN crews were allowed to experience the Air Force's highly classified MiG asset. The CO of VF-121 *Pacemakers*, "Doc" Townsend, was among the first to visit the USAF's secret MiG base at Site 51 on the outskirts of Nellis AFB, and he was so impressed that he suggested that the Navy needed its own training program to pass on the tactics that were being developed there. He began to build up the ACM training, which had been a comparatively small part of the RAG's syllabus since 1962. From 1965, under Scott Lamoreaux, F-8 Crusaders had been used occasionally as MiG-21 simulators, and in 1967-68 "Doc" added agile TA-4 Skyhawks seconded from VA-126. Also, the RAG began to teach "loose deuce," a battle formation in which two F-4s flew abeam roughly one mile apart instead of behind each other in "trail," with three miles separation, as they had done previously. In this way either aircraft was able to become the "shooter" depending on the way the target maneuvered, while the other Phantom watched the shooter's tail. Air Force Phantoms continued to use the four-ship "fluid four" formation, in which one aircraft (usually the senior officer) was nominated shooter and the others attempted to cover him. Although this "welded wing" (as the critics called it) formation had been declared unsuitable for combat with MiGs by the *Feather Duster* survey, it continued in use by USAF Fighter Wings throughout the war and probably decreased their success rate against the MiGs, while at the same time increasing losses to unsuspecting members of the quartet who were too intent upon formation keeping.

The revision of Navy tactics was accelerated by a wide-ranging report instigated after the Summer of 1968 with its succession of unsuccessful missile launches and two F-4B losses to MiGs. Prior to the shoot-down of the VF-102 CO's *Milkvine* 101 in June 1968, VF-92 F-4B BuNo 151485 from USS *Enterprise* had been destroyed by a MiG-21's Atoll. It happened during a dogfight in which the five participating F-4s only managed to get off two AIM-7s against the pair of MiG-21s which they engaged, without a hit. On this occasion the EKA-3B "Queer Whale," which normally jammed the MiGs' GCI and radar control links, was unavailable, and the MiGs were able to take full advantage of their excellent ground control and the hazy atmospheric conditions to intercept the Phantoms successfully. The VPAF's success rate had risen from 3 per cent of U.S. aircraft combat losses in 1966 to 22 per cent in the first quarter of 1968.

As the Captain of the USS *Coral Sea* in 1966-67, Captain Frank Ault had plenty of experience of the air war and its cost. In its first three war cruises his Air Wing sustained fifty-two combat losses (the rough equivalent of four squadrons) in exchange for one air combat victory. Three were lost to MiGs. Following his suggestion for improving AIM-7 performance he was commissioned in May 1968 to investigate 140 aspects of the role of fighters in the air war. During the period of time in which the investigations were conducted over thirty AIM-7s were fired by USN F-4B aircrews unsuccessfully. In many cases it could be demonstrated that they had been correctly fired within their range and maneuvering parameters. Several F-4s fired two or three Sparrows, and all failed to hit.

There was no shortage of targets. Following the pause in bombing the North, from the end of March 1968, MiGs were more daring, and they often ventured south to try and catch USN attack packages working in Route Packs II and III. It was usual to clear strike aircraft from the area if MiGs were detected, giving the fighters a free fire zone. Missile performance was therefore high on Captain "Whip" Ault's list of priorities. During his inquiries his team visited carriers on line, including USS *America*, and interviewed many flyers, including "Outlaw" Cash, but also maintenance crews.

Of Missiles and Men
As a background, he could observe, from existing reports, that USAF and USN Phantoms had fired over 330 AIM-7Ds and AIM-7Es from 1965 to the end of *Rolling Thunder* in 1968. Of these, only 27 had scored kills. Overall, U.S. missiles had shown a one in ten probability of achieving a kill, far less than pre-war training results in ideal conditions had indicated. Navy and Air Force results were comparable in this respect. Launching an AIM-7 was a relatively complex procedure. The missile was originally meant to be launched from around twelve miles at large, non-maneuvering targets at high altitude in an environment where there would be few external distractions, such as AAA, SAMs, or fighters. Captain Ault noted the complex nature of the launching process, requiring carefully coordinated switch setting by pilot and RIO, an understanding that missile parameters varied with altitude and temperature, and some intense in-cockpit scope-gazing to keep the narrow radar guidance beam locked on target. While this could be done effectively in a low-threat, BVR situation, it was too much to ask of crews in a multiple-threat dogfight. Success in that scenario with either type of missile required extremely thorough training (rather than three or four ACM sorties, which was the norm) and considerable skill. Rather than the all-aspect AIM-7 intercept mode for which crews were trained, in Vietnam the traditional tail-chase was favored as a means of achieving a tracking solution.

Although it was possible to effect a quick boresight lock-on and launch, this gave the AIM-7 a minimal chance of success. The correct "full systems" lock-up required three seconds between initial radar lock-on for the missile to be programmed successfully, and over five seconds for the entire launch procedure. This often seemed an age in combat against a maneuvering target, and pilots

A VF-21 F-4J in landing configuration for an instrument approach. The *Freelancers* made seven war cruises, five of them aboard USS *Ranger* with CVW-2. BuNo 158363 made its final appearance with VMFA-112, the last F-4 unit within the U.S. Sea Service force. (Jan Jacobs)

sometimes impatiently launched a second missile, believing the first had failed. Sometimes this would confuse the fire control system, causing both weapons to miss. Having started the AIM-7 on its trajectory the F-4 crew had to keep a radar lock-on to the target for the missile's guidance system to "ride" to its destination. If either the target or the launch aircraft engaged in maneuvers of 3g or more during this time the lock could be lost and the missile would go ballistic. Obviously, no pilot would wish to fly a predictable course in this way when under threat from fighters or ground defenses. The additional possibility of radar failure, which was common, meant the F-4's main armament was negated, and AIM-9 Sidewinder then became the only option.

Navy Crusaders had pioneered the use of the AIM-9B/C Sidewinder in combat, and it remained in service with F-4B squadrons until 1967. The AIM-9B was comparatively simple, cheap, and fairly reliable, though only about a half of those used in *Rolling Thunder* launched successfully, and only 15 per cent scored kills. A launch could be initiated once the pilot got a "growl" tone in his headphones, showing that the missile's lead sulfide nose dome had picked up the target's heat source. It could then be left to find its prey. However, the launch still had to be made within the correct parameters of range: the minimum was 3,000 yards, as for the AIM-7. It could not be launched accurately if the Phantom was turning at more than 2g, or against a target which was closing in on the launcher. Also, the pilot had to make sure it was not "growling" at another heat source, such as the sun (known as sun-capture). From 1966 onwards the USN introduced the AIM-9D, with a more sensitive, liquid nitrogen cooled seeker head, with an eleven mile range and doubled warhead size (to 22.4 lbs). Coolant was stored in the missile's LAU-7 launcher rail.

Ault's report also focused on missile maintenance. Rather than up-loading fresh, newly-checked Sparrows for each mission and downloading unused missiles afterwards, the hard-pressed Air Wing armament teams often kept the delicate AIM-7s on the aircraft for months. They were subjected to the jarring impact of deck launches and recovery, strike missions, and the effects of moisture, salt air, and frequent temperature changes. Air Force Sparrows were at least bench-checked after around ten sorties, whereas the pressures of Alpha strikes and cyclic carrier operations often meant that Navy missiles could go fifty sorties before a check. Quite minor changes in maintenance procedures could bring substantial improvements. CDR William Greer was CO of VF-11 *Red Rippers*, and recalled that one of Ault's discoveries was that, "Sparrow missiles suffered a large number of launches where the motor did not fire. Among the corrective actions was that the wafer switch making the electrical connection between the aircraft and the missile was discarded and replaced each time a missile was downloaded. This caused a marked decrease in the number of 'no motor fire' launches. Despite bitter complaints on my part, USS *Forrestal* [with the Sixth Fleet] continued to re-use the wafer switch many times for economic reasons, and our real air defense capability was probably much less than was thought."

Sidewinder was a much simpler device, better able to stand the rigors of carrier life. Initially, it had to be mounted on an inboard pylon by itself, but VF-21 maintainers re-arranged the wiring so

that TERs with ordnance could be carried, too. Better versions of the missile became available from 1967. The USN received over 2,000 AIM-9Gs with improved off-boresight target acquisition, known as Sidewinder expanded acquisition mode (SEAM), and better maneuverability. Further development led to the AIM-9J with electronic improvements and "double delta" fins. Ault concentrated more on improving the Sparrow, and the result was the AIM-7E-2 "Dogfight Sparrow," which became available in 1969. Minimum firing range was reduced, and there were improvements to maneuverability, but problems continued with fuzing: some exploded too soon after launch. Others still failed to ignite, and became costly "dumb" ordnance on release. Many of Ault's far-reaching proposals had to wait for a later generation of pilots: head-up displays for armament control, an ACM maneuvering range, and, in due course, a new Navy fighter.

Top Missile

As a priority Ault scrutinized the ACM training of Phantom aircrews in a Navy where the basic philosophy was still very much focused on the attack concept, with air-to-air skills accepting a fairly low status. He noted that the Fleet Air Gunner Unit (FAGU) had been phased out in 1960 with the advent of all-missile armament. FAGU's approach had been to train a small number of pilots to a very high level in air gunnery and then send them back to their squadrons to disseminate their expertise. The Ault Report suggested a new unit, eventually to be called the Naval Fighter Weapons School (NFWS), which would be manned by experienced air fighters, particularly those from F-8 units. It would bring together all the ACM work done by VX-4, the RAG, and operational squadrons, and focus it within a purposeful syllabus.

Among the initial cadre of pilots at NFWS (soon to be called Top Gun, though it taught missile shooting) was John Nash. He had recently returned to VF-121 following an exchange tour at the USAF Fighter Weapons School after the curtailment of his training on the GD F-111A. He was to have been CO of the first USN F-111B squadron prior to the cancellation of the project. Until July 1970 when he returned to combat with VF-142, he flew as an instructor with both VF-121 and NFWS teaching the principles of ACM.

"The guys who had the most trouble adapting from the air defense mission were some of the older ones who had flown F3H Demons. They had never really done ACM, whereas the F-8 people were well versed in it. The F-4 was a superior ACM airplane, flown by the right guy. Some of the F-8 pilots were a little apprehensive about the Phantom because it would do maneuvers at slow speed that you'd never consider in a F-8. A Crusader would depart and spin end-over-end, whereas you could fly the F-4 through such maneuvers very comfortably. Doing ACM in the F-4 was more of an art than a science, and some of the older guys weren't happy with flying the F-4 to the edge of the envelope."

One of the reasons for their hesitation was the Phantom's known problem of recovery after a departure from controlled flight at low altitude. This was popularly known as the "stall-spin condition," but was more accurately described as adverse yaw. John Nash:

"Probably there were few spins in the Phantom. It would let you know when it didn't want to fly any more, but it would also let you know when it didn't like the conditions it was in. You could fly the airplane forever at high angles of attack, but there was a point where it wasn't going to fly any more and it would depart, but most of the losses were low-altitude departures rather than spins. In the slick-wing [un-slatted] Phantom, if you departed the airplane below 10,000 ft; certainly below 5,000 ft, the odds on you recovering were almost zero. You had to 'unload' the airplane, reduce the angle of attack, fly it straight down and then very gingerly milk the nose up to the horizon. It was touch and go whether it was possible to do that, even with a skilled aviator in it. There were many folks who were afraid of the Phantom in the post-stall departure condition. At altitude, it was just a fun ride, but at low-level you were in big trouble because you didn't have the g available to get the nose up. You'd just wind up in wing rock and heavy airframe buffet. It was possible to fly the airplane, but if you looked at your vertical speed indicator you were going down at about 4,000 ft per minute. I don't think I've ever spun a Phantom, though I've had it in various conditions of out-of-control flight. Maybe if it had departed earlier people would have recovered earlier.

"During the recovery procedure you didn't touch the ailerons, since you'd be out of control again. You reduced the angle of attack and 'ruddered' the airplane to the horizon you wanted at 5 units of angle of attack (equivalent to 1g). The stall speed at that g level is around 80 kts. There was enough power to maintain 100-120 kts in almost any condition—but you'd be going straight down."

Mike Shaw emphasized the importance of rudder in controlling the F-4: "All F-4s except those with leading-edge slats would buffet when an angle of attack of 15 units was exceeded. At that point rudder became the primary roll control. Harsh use of the ailerons and spoilers would result in rapid departure. The advice was, 'When in **B**uffet use your **B**oots.'"

In VF-121/NFWS John Nash introduced pilots to the Phantom's recovery characteristics early in the program:

"We used to make the old guys nervous. On our first flight over to Yuma (for air-to-ground before the ACM syllabus) we'd put the pilot in the back seat, go straight up until we reached zero airspeed, and then say, 'OK, I'm putting my hands on the console'

and let the airplane fall out of the air. After about two or three twists and flops the thing would recover, going almost straight down with airspeed accelerating. Then we'd do the same thing again and say, 'Here's full left (or right) rudder.' The rudder didn't basically affect the flying qualities of the airplane, but it was the only control you had at that point. The guy in the back seat would stop breathing for a while. Old guys would say, 'You couldn't do this in an F-8, you'd be tumbling end over end!' I've had F-8 commanders say, 'I've never been this slow in an airplane before.' It was a great confidence builder, but a lot of old guys were never comfortable. They'd do the maneuver when forced, but they were terrified. They would never fly the airplane well or hard, and they were the first choice to lose an airplane if they got out of control because they panicked."

When Top Gun was being established new syllabuses had to be written using the combined wisdom of the pilots concerned and also reviving the lessons of earlier generations of fighter tacticians. "Nothing changed from WWII," John Nash observed, "apart from the distances and speeds. Most of the maneuvers were very similar and accomplished the same thing—barrel rolls, high and low yo-yos, lag pursuit, defensive maneuvers. You would go back and research techniques recorded from WWII (for example, in Heinz Knoke's 'I Flew for the Fuehrer'), and it was all the same." Top Gun became a department within VF-121 at Miramar, initially commanded by VF-121's Ops Officer, Dan Pedersen. Former VF-21 pilot Jim Ruliffson wrote the syllabus on maximizing missile potential, and Mel Holmes from VF-143 focused on aerodynamics and tactics. They were soon joined by others from the cream of VF-121's instructors.

Because John Nash had done the USAF air-to-ground course in 1967 he became the "air-to-ground guy" at Top Gun, writing the syllabus on that equally problematic area of Phantom operations.

"There were few people in the Navy, and certainly in the fighter community, who knew why a bomb hit the ground other than via gravity. The USAF had gone into it mathematically to the point where it was ridiculous—'What happens if your airspeed is 2 kts off?', etc. VX-5 did a study in the mid-1960s and found out that the average first-drop CEP was 300 ft on an unfamiliar target in the pop-up and roll-in to the target, using the 'iron sight' and with no AAA. An experienced bomber could do better, but there were always the winds in the target area which have a significant effect on a free-fall. We weren't very good. It would take 300 bombs to do what they're doing with one bomb today."

MiG Awareness

In June 1968 John Nash left Top Gun and flew with VF-142 *Ghostriders* on a Westpac cruise as Ops Officer, eventually taking command of VF-161 *Chargers* in 1976. In between, he had his first of two assignments to VX-4 *Evaluators* at Point Mugu, where he had first-hand experience of the USAF's *Red Hat* MiG squadron.

"The only thing the Phantom didn't do well was turn. I think the only airplane in the inventory we could out-turn was the F-104! In our evaluation against the MiG-21 the slick-wing Phantom didn't stand a chance if you tried to turn. The best you could do was to keep your energy up, use the vertical, and try to make him turn. The MiG turned well, but not many times. I flew the MiG-17 and MiG-

The *Sundowners* made good use of the F-4B's predatory looks and adapted a traditional design for their fearsome nose-art. (via C. Moggeridge)

21 at VX-4 during evaluations against all the Navy airplanes. The USAF said we couldn't do training, only 'operational evaluation' on the MiGs, so we tried to do it with as many Navy guys as possible. Shortly afterwards we got MiG-23s.

The MiG-21 was a tremendous airplane. It could out-accelerate the Phantom. On my first flight in it I had an RF-4C chasing me. We started at Mach 0.9 at about 40,000 ft. I was at Mach 2 when the Phantom was at Mach 1.6, and I was four miles ahead. It was a 10g airplane; you couldn't turn it more than once, but it was an eye-watering turn. It had a real high stick, so you could pull a lot of g on it. There were hydraulic ailerons, but a manual rudder, so when you went straight up and reached zero airspeed, in a tail slide the rudder would kick your foot over. Visibility was not good, but not a lot worse than the F-4 and not nearly as bad as people said. The MiG-21 was highly unstable, which made it a very good ACM airplane. USAF test pilots were all wrapped around the axle about instability and inability to do accurate gunsight tracking, but much better to get there and not track well than never to get there. I wish they had put a tail hook on the MiG-21: if I had been a MiG-21 pilot in Vietnam I would have had a lot of kills. You could kill guys before they even knew you were there. I did an evaluation against the F-14A Tomcat in VX-4 which, well-flown in a good close-in fight, was a match for the MiG-21—but it wasn't really superior to it."

On his second VX-4 tour in the mid-1970s John Nash became Director of Project *Have Idea*, which matched the MiGs against a variety of Navy aircraft and acquainted many more pilots with its characteristics. He was responsible for developing tactics for all USN tactical jets in combat with the Soviet fighters. "*Have Idea* exposed to the average Navy pilot how good the airplanes really were and what their weak points were, such as the MiG-17's terrible turning ability at high speeds. When a guy recognized that, for example, 'Mugs' McKeown who shot down a MiG when he saw it trying to turn at high speed, he recognized what the MiG's problem was, and so he was able to attack it with a lot of confidence. In Project *Have Drill* I flew against a MiG-17, and that was an eye-watering experience. That airplane, at 300 kts, could do a 6g, 360 degree turn and not lose more than 5 kts, whereas the Phantom would have done that and been down to landing speed. People were quick to criticize the Russians for the quality of their equipment, but, flying those airplanes, I was impressed. Guys made fun of all the rivets sticking out on the MiG-21; I don't know what it might *not* have achieved if the rivets weren't sticking out. Also, we pulled the engines out and sent them to Westinghouse for overhaul. They were amazed at the compression ratios and the capabilities of those engines."

Another aim of the Top Gun tactical courses was to develop the all-important cockpit teamwork in the F-4, and between members of a flight of Phantoms. Although the NATOPS F-4 "Dash 1" Flight Manual contained a highly detailed chapter called Flight Crew Co-ordination, which specified pilot/RIO responsibilities for all situations, its guidance on ACM in that section was limited to nine lines. Such skills were refined in practice, but much could be taught in the hard school of Top Gun training. John Nash:

"At first, the single-seat pilots hated RIOs and thought they could do it all by themselves. When we were doing the Top Gun routine the first requirement for thorough and accurate communication came when RIOs were making the radar intercept. Since I had flown the F3H Demon I could run my own intercepts, so I knew what was going on, but quite frequently your wingman or your entire flight might not have had contact with the target. There was a requirement for the RIO to talk over the radio, telling the rest of the flight what he was seeing and doing; the orientation, what the attack was going to look like, and who was going to take each enemy airplane. If you are a flight of two attacking several enemy airplanes and you both end up attacking the same plane, you've left a lot of guys unaccounted for and you're probably going to get shot.

Your wingman may have been carrying Sidewinders only, simply because his radar wasn't working, so you were trying to describe the attack verbally so that he knew what was going on. Seeing airplanes at four or five miles at high speeds in the heat of battle is really difficult. The chances of picking out another Phantom-sized airplane at that distance are not very good if you are not looking in the right direction, or excited. It is very easy to miss seeing a part of a flight you are attacking, so the first requirement came for the RIOs to talk in the cockpit and tell the pilot what he was going to see. It's always nice to know from what aspect you are going to attack the enemy, how many bogies there are, and what sort of formation they are in. For a while, it became complicated to tell everybody what the enemy formation looked like. If you said they were riding in echelon, to me that might mean the wingman is flying to the right side of the echelon, but to a lot of RIOs, who were looking at a mirror image because they were looking at a radar scope, in fact the formation was in left-hand echelon. It ruins your whole day if you attack a guy and think his wingman is on the right side but it turns out he's on the left.

The RIO may have asked the pilot, 'Do you want to take them down the middle, left or right?', and what was going out over the air is the RIO's comments. Once you had a 'tally ho,' RIOs were talking to the pilots within the airplane, and pilots were going to be talking to each other externally. When we took students up, for example, in a 'two versus two' fight, I had to know where the student was all the time and where the bogies were. If I had a RIO I could depend on I could have him keep an eye on the bogies and let me know if they got to be a threat. Otherwise, the pilot had to do all this and also direct the student.

Typically, when you were in a turning fight and you were head-on with a guy at five miles and you saw him turn you were going to have a tail shot in a few seconds. Usually, the student would be off on the perch or on the right, and when he saw the bogies turning he'd say, 'OK, I'm going to have a tail shot,' so he'd roll in. By that time, though, he would probably be looking at a head-on shot again. As an instructor you were using radio a lot just to talk the students into taking firing shots or getting them primed to the point where they knew when to roll in on a bogie. It's hard to practice that, and we didn't have ACM simulators in those days. RIO communications were really important, and I know a lot of guys who had their RIOs save them in combat. Just going out and finding the bogies on radar intercepts wasn't that difficult. Their real test in ACM was whether they could keep track of the fight and keep their pilot in-

A bombed up and ready to go F-4J, BuNo 157268, begins another mission for VF- 96. (USN via Angelo Romano)

formed of what was going on around them. Jim Laing and J. C. Smith [among the first cadre of Top Gun RIO Instructors] could listen to a pilot's commentary on the target, get a radar lock-on in an ACM environment, and get a missile off, rather than have the pilot go boresight mode, point his nose at the target, and then lock the target up. The ideal way to acquire a target is in lag pursuit or in off-boresight mode, not pointing at the bogie. Once you're pointing at him you're using up the separation distance in which the missile can perform properly. A reasonable visual range is five to six miles at most. You're not going to fight a guy beyond about four miles, and typically you need two or three miles in a turning fight. You need at least that to get a radar missile locked up, timed-out, and fired."

The first Top Gun class arrived on 3 March 1969, and John Nash and his colleagues set about imparting their combined wisdom and experience in a grueling schedule. The demands of the RAG on instructors had already been great, not so much in flying hours as in the number of sorties required.

"I had more sorties than flying hours; 350 a year, approximately. All were ACM or air-to-ground, plus night proficiency. I flew the A-4 Skyhawk as an adversary pilot. Neither the A-4 nor the F-4 was air-conditioned on the ground, and you couldn't taxi round with the canopies open all the time. A lot of the fatigue was just the heat. I went to Yuma once a month for ten days with ten

airplanes, students, and instructors. We would have to quit flying by noon because it was about 115 degrees F out there in summer. My first launch was at first light. Just the heat and the energy expended in an ACM sortie really took it out of you. Many times we would be out at the Top Gun trailer briefing the first ACM flight for a dawn take-off. You'd do up to three flights a day, and if you were in the Phantom you'd be in a pretty hot cockpit environment with a lot of pulling on the stick. If you were flying the A-4 you'd be fighting two sections of F-4s, typically. You would have one guy sitting in 'warm up' with one F-4 section, and the other section airborne. When the first section got down to about 5,000 lbs of fuel the second would launch, and so you'd have a pretty constant forty minutes of ACM in the A-4, which was a pretty good work-out.

Top Gun was a controlled environment. Guys were going to be there for a month and fly a set number of sorties, which meant that an instructor could fly once or twice a day. When I was in Top Gun and the RAG I might have flown a Top Gun sortie, then turned around and flown something for the RAG the same day. It's hard flying an ACM sortie with a student who is out left field somewhere when the last thing you want to do is get whipped yourself by your buddy flying the bogie aircraft. It was tough to fly your own ACM and also control the student and fly his mission for him. The good ones required some attention, and the bad ones...well, you would rather be out there by yourself!"

7

Linebacking

Making it Work

The good students rapidly absorbed the characteristics of their aircraft, which Top Gun encouraged them to explore directly. Matt Connelly, later to score two MiGs in one engagement, explained the basic philosophy: "The F-4 was actually ill-suited to close-in dogfighting due to its limited g available below 420 kts and the lack of a 'gatling gun.' The MiG-17 and MiG-21 had one full g advantage below 420 kts. Top Gun taught us how to fly the F-4 to its advantage and capitalize on the weakness of the enemy aircraft."

Among the students in that first Top Gun class was Jerry Beaulier from VF-142 *Ghostriders*. Paired with RIO Steve Barkley early in 1969, they were both first-tour members of the squadron, though Jerry had made a previous Westpac cruise.

"Since we were both formerly enlisted," recounted Steve, who had previously been in the USAF, "we found ourselves several years older than most of the first-tour people. That didn't make much difference, except that our sense of humor was at a higher level. Not to say that all was always on the 'up and up' in the cockpit. Beaulier and I often joke about several 'divorces' during sorties.

But we never could stay mad at each other (divorced) for an entire flight. Being a bit older and perhaps more focused on quality and survival, our missions tended to be fairly well planned and 'by our book.' By this, I mean we had a set of procedures, checks, and counter checks in which we (privately) took a great deal of pride: look-out doctrine (never miss a bogie), radar search doctrine (never miss a target), equipment checks, navigation, and airmanship. We spent a great deal of time together apart from flying, and, I suppose, we discussed every aspect of our various missions. We wanted to have a plan. For example, we discussed whether it was prudent to drop our tanks at the slightest provocation. Our external tanks, given the length of our missions, were certainly necessary. Our concern was, why drop the tanks if we probably won't engage? At that point in the war (1969) almost never did anyone lay eyes on a MiG, much less get within Sparrow range of a bandit. We concluded it was more prudent to keep the tanks until the first turn of the engagement. With that kind of drag it wouldn't be tough to get below proper jettison speed for the tanks. The problem was, of course, to remain aware of your speed.

A pair of VF-92 F-4Js taking on fuel from a VAQ-130 EKA-3B in March 1972. (Norman Taylor collection)

After their single war cruise on USS *America* in 1968 the Diamondbacks returned to USS *Independence*. BuNo 153886 is seen at NAS Miramar on 2 August 1969. (Duane Kasulka via Norman Taylor)

Garry Weigand (left) and Bill Freckleton after their MiG kill. Bill recalled, "We had been surrounded by an entire flight deck of men who had swarmed up to congratulate us. We made our way to the VF-111 ready room, where the ship's photographer told us he wanted one more shot. He said "Smile and give the victory sign." One of us complied, the other was giving the peace sign!" (U.S. Navy via William C. Freckleton)

Beaulier was a recent graduate of Top Gun. His class was the first one at the school which spawned the likes of Tom Cruise(!). He attended Top Gun while they were still in the middle hangar of Hangar One at Miramar. That was early in the genesis of Top Gun, and the philosophy at that time was a mix of simply making good pilots better and to provide Fleet squadrons with, perhaps, people prepared with a recent infusion of tactics to be training officers. Although the squadrons more or less supported the idea of Top Gun, it wasn't apparent at the time that they were ready to embrace everything that the returning warriors had to say. Beaulier was awarded 'NATOPS Officer' duties on his return from the course. So much for immediate assignment to Training Officer. But he had frequent opportunities to hold forth in the ready room, so some of what was absorbed at Top Gun made its way to a mix of willing ears. So, in a way, Top Gun was working even then, but not as some would suggest. It took a few years for Top Gun to evolve a compelling mission statement, and even then it was a matter of personality and personal ability to make a difference at the squadron level (silk purses and sows ears!).

Certainly it was clear to everyone that speed was life in the McDonnell Phantom. The need for speed was never greater than it was for an engaged Phantom, and management of energy was paramount. That message made it from Top Gun and VF-121 instructors to the squadron loud and clear, and the Crusader pilots were always willing to emphasize that point to the F-4 crews.

My class at VF-121 was the first F-4J class for the RAG. Our class enjoyed several distinctions: flights in the F-4A (really!), the F-4B, and the newly introduced F-4J. Our first flights were in the 'lead nose' F-4J—without a radar. More important, however, our class was the first to enjoy a period of intense tactics training. Before our class, believe it or not, replacement pilots and RIOs would

fly ACM now and then, but never more than a couple of flights in a row. ACM flights happened whenever 'things came together.' It wasn't thought prudent to pack all those flights together and get some focus on what it was all about. We were given three aircraft and four instructors and enjoyed a three week period of only ACM flights, interspersed with periods of classroom discussion and ACM 'homework,' such as mathematically describing maneuvering 'eggs' of various 'g,' and graphic illustrations of comparative maneuvering by the F-4 and MiG variants. Who knows if it worked, but it did make a lot more sense than occasional ACM flights accomplishing who knows what? Thereafter, the RAG followed this approach, except the 'middle hangar' at Miramar was out: Top Gun grabbed it after our class finished.

By March 1970 Jerry Beaulier and I were well into the routine of a Westpac cruise aboard the USS *Constellation*. The ship would sortie out of Subic Bay to Vietnam for about thirty days, followed by five or six days in the Philippines, Hong Kong, or Japan. We had fifteen crews flying 20 to 22 sorties a day. This translated into one or two sorties a day each, plus standing a 'spare' for one sortie. The spare was meant to ensure the squadron launched the requisite number of aircraft, usually two, for each mission. A spare aircrew briefed, manned the aircraft, started engines, and then watched their buddies launch on a terrific mission. Spare aircrews were launched when one of the main aircraft went 'down.' This only happened when the mission was either boring (such as BARCAP, approximately 10,000 miles from the nearest threat), or when it was a moonless night and bad weather. Combine a couple of missions with a spare, plus one or two five-minute Alert spells, and you have a full day. Our flying 'period' was typically twelve hours, but Alert 5s were 24 hours a day. This meant that your time for Alert always came in the middle of what would appear to be prime sleeping time.

Our sorties were mainly into Laos and South Vietnam for bombing missions, a lot of BARCAPs, and occasionally a *Blue Tree* (RA-5C escort) into North Vietnam. *Blue Trees* were relatively short and invariably a non-event, as we would sweep through the North at very high airspeeds, spending only a few short minutes overland. There was never any defense by the North Vietnamese apart from an occasional 'paint' by a fire-control radar. We never saw a missile fired. Sorties into Laos or South Vietnam were more eventful, as they were usually at night and flak (usually 37 mm) was forthcoming, especially in Laos near the North Vietnam border. It could easily be seen at night, and there was enough to be interesting, but

BuNo 153019, the F-4B used by Garry Weigand and Bill Freckleton for their 6 March 1972 MiG kill. In 1966 LT "Barrel" McCrea and ENS David Nichols had destroyed an An-2 using the same aircraft. (via C. Moggeridge)

don't for a minute think about 'thirty minutes over Tokyo or Berlin.' I wasn't in that war, but I'm sure our sorties were pastoral in comparison. Occasionally someone would be hit or shot down, so you didn't want to duel with what you saw.

We were about two or three weeks into a line period when a USAF rescue helicopter was shot down in Laos by a MiG-21 staging out of the southernmost airfield in North Vietnam. This was quite unexpected, as the North Vietnamese hadn't done anything like this in some time. We were a couple of years into the 'no bombing' strategy, so encounters such as these were very rare. They were probably enjoying the lack of attention, and we found it interesting that they should seemingly thumb their noses at us in this manner.

On 27 March a two-aircraft section from our sister squadron, VF-143 *Pukin' Dogs*, was vectored towards two MiGs. Both F-4Js dumped all tanks, and one aircraft shot a couple of missiles to no avail. I'm not sure anyone even saw an aircraft, but they were in radar range. That got everyone pumped up. Now we were in a real shooting war, we thought. That the encounter was the first in over a year and that the probability of another was very unlikely was not to be thought about. Incidentally, the encounter left the ship short of drop tanks. One section of aircraft dropped tanks, and the ship was short—good thing it wasn't a real war! We dispatched a couple of F-4Js to Cubi to pick up some more. That night Beau and I revisited our plan for tanks and decided to go with our conviction to drop tanks in a fight when we got slow.

That night too I was up late writing the flight schedule for 28 March. As Schedule Officer I was responsible for assigning flights to everyone, including Alerts. My job was to make sure everyone got a fair share at the good flights, that pilots kept night-proficient, and Alerts were evenly distributed. The skipper usually reviewed the flight schedule each night. His big deal was that pilots were absolutely even in everything: he didn't keep much track of the RIOs. This meant that a few of us fire-breathing RIOs were a little more even than others. The schedule was posted for the 28th, and there we were, standing Alert 5 during a mid-day launch. It didn't help that 'scuttlebutt' had it that the ship would be making a fighter sweep over the North during that launch. So, Beaulier and I were

Sundowners and *Screaming Eagles* F-4Bs release Mk 82 LDGPs over cloud cover, risking unseen SAMs. (U.S. Navy)

VF-111's CAG aircraft comes aboard USS *Coral Sea* like a bird of prey. (USN via CDR James Carlton)

destined to sit on deck and watch our buddies launch into glory. And it was all my fault!

We manned up for the Alert 5 [in *Dakota* 201] while the rest of our buddies were manning their aircraft for the scheduled launch. We didn't always start the aircraft at the change of Alert crews, but Beaulier decided he wanted to start up and await the launch with engines running. The Plane Captain cranked the 'huffer' (external starter), and we started up and began preflght checks. Meanwhile, VF-142 Skipper, Ruel Gardner, and Ed Scudder were spotted on the waist cat and were having trouble getting started. At this point I recall hearing indications on the radio that something was up. We had switched to the CAP frequency and were hearing chatter that led us to believe that, impossibly, the scuttlebutt was right. MiGs were airborne and were being tracked! We quickly switched back to the launch frequency and scanned the deck. Gardner and Scudder hadn't started yet, and the rest were beginning pre-flight checks. Things looked reasonable for our launch, as we were started and ready to go. The next thing we heard over the flight deck loudspeaker was, 'Launch the Alert 5.'

Our attention went up several notches. Gardner was on the catapult, but something wasn't going right. The flight deck crew immediately moved the aircraft off the waist catapult and signaled us to begin to taxi to catapult 3 on the waist. With the precision borne of a thousand launches, our aircraft was quickly moved into position and made ready for an immediate launch. The nose kicked up, and the Catapult Officer gave Beaulier the signal for full power, then afterburners. Everything looked good, and Beaulier gave me assurance he was signaling for a launch. With a snappy salute from the Cat Officer the aircraft moved from zero to 140 kts in less than two seconds and we were airborne. We left the ship with two wing tanks, a centerline, full internal fuel plus two Sparrows and three AIM-9Ds.

After the usual VFR departure from *Constellation* we began an outward track toward Red Crown which, on that day, was USS *Horne*, about 50 miles northwest of the carrier. As other aircraft were being launched we were instructed to circle as they caught up and radar status was checked. Our radar was just fine, and the mis-

For their final war cruise in 1973, VF-111 reduced their sunburst tail design but retained the sharkmouth. The squadron returned to the Westpac area with F-4Ns in 1974-75 under CDR John "Hot Dog" Brickner to cover the evacuation of South Vietnam. (Tom Patterson via Norm Taylor)

siles were OK except for the Number 1 Sidewinder, which was approximately 30 mils off centerline. Two F-4Js showed up, crewed by CDR Paul Speer (the CAG) with LTJG John Carter, and LCDR Gary Hakanson with LTJG Dave Van Asdlen. Dave reported his radar was down, and John said his was OK. Hakanson and Van Asdlen were instructed to maintain an orbit in reserve. We joined CAG's aircraft to form the section, which was immediately given a vector: 'Vector 273 for bandits at 43 miles.' CAG Speer (whose fighter experience went back to Korea and three F-8 Crusader tours of Vietnam, including a MiG-17 kill with VF-211) took the lead, and we began to close on the coast. Our controller (Petty Officer White) kept us apprised of what was going on: 'Bandits at 25 miles, you're cleared to fire.' '21 miles, they've dropped their tanks.' (It's worth wondering how he knew this, but it would be a pretty good guess to conclude that someone was monitoring their frequency. Either that, or White had some really good binoculars!) All aircraft were accelerating, and the closure was probably greater than 1,000 nm per hour, so information was useless almost immediately, but White was really doing his stuff. He sounded cool, calm, and organized. I was desperately looking for a radar contact as we crossed the beach—then the radar died! I fought it for about five miles and then gave up. This was no place to run Built-In Tests! I was really disappointed; the perfect set-up for a head-on shot, and no radar! I went heads-up, as we were about to be in a fight and the radar problem could be sorted out later.

By this time our section was really 'hauling the mail.' We were in the neighborhood of 550 kts and had moved down to about 12,000 ft, as White had estimated the MiGs' altitude at about 20-22,000 ft. CAG was north of us about a mile, and we were stepped down about 2-3,000 ft from them. We still had our tanks, since significant fuel remained, plus we were being 'cheap' about our tanks. Beaulier got first sight of the MiGs [and so had led the engagement] as they 'flew' out of the canopy bow at one o'clock. He called the sightings and a turn into the MiGs. We started the fight at 12,000 ft, with the MiGs at 2 o'clock and 22,000 ft, beginning a right-hand climbing turn into the bandits, who apparently didn't see us. Someone in our

aircraft commented, 'They don't see us!' Neither Beaulier nor myself will admit to making that call, but somebody did and, as you might expect, that's all it took for both MiGs to see us and begin aggressive maneuvering. The fight began with our nose on the MiGs as they crossed in front of us, left to right. CAG was 'sucked' [behind us] and trailing us about two miles back. The MiGs turned right into us and began a vertical maneuver down into us as they split into singles from the 'fighting wing' formation they were maintaining. We turned right into the nearest MiG and kept a visual on the second aircraft, beginning our turn at about 18,000 ft and pulling 5g as we chased the lead MiG through a 360 degree turn. The F-4J rapidly lost energy through one of these maneuvers, especially with the three tanks that we were still dragging. As we struggled through the last 90 degrees of turn our airspeed was about 400 kts and appropriate for you know what! So, Beaulier's trusty RIO called 'Tanks' when the speed was right, and off went the wing tanks as planned. We kept the centerline, as we were still transferring fuel. I've often wondered what the MiGs thought when they saw an aircraft calmly jettison tanks in the middle of a life or death fight. Did they think we were organized, or just nuts?

Anyway, we were now fighting for airspeed to keep up with the MiGs, who were now joining up as a fighting wing again (can you believe it?). As they joined up somehow they were head-on with CAG Speer who, somehow, was out of phase with us and the MiGs and coming from the opposite direction. The lead MiG launched an Atoll at CAG. Remember, this was not CAG's first real fight, but it was certainly John Carter's baptism of fire and John was focused. He saw the Atoll launch and called CAG's attention to it, probably in a near-falsetto voice. CAG saw the launch (a big puff of smoke) and the missile leave the MiG. His response to Carter's frantic call was, 'No chance.' He never varied his course one degree, and the missile passed harmlessly. That's experience! The MiGs then split up again with one aircraft going left and the other right. We followed the right-turning one and kept a visual on the second MiG, who passed from right to left at our six o'clock as we turned right, chasing the lead aircraft. Beaulier was onto him, and my responsibility was to track the remainder of the fight. I

Crewmen from USS *America's* catapults One and Two muscle *Showtime 107* into the correct position over the shuttle for an October 1970 cat shot. (USN via CDR James Carlton)

Rolling out on Miramar's sun-baked runway, another F-4J sortie begins. (David Daniels)

Silver Kite 106 **(BuNo 155569). Curt Dosé recalled that the F-4 would accelerate through Mach 1 with ease in a dive, even in idle power, and with speed brakes fully extended. (Curt Dosé)**

followed the second MiG through our six, and then it turned west and disappeared with CAG in close pursuit. They chased it nearly to Hanoi and then turned back [Cdr Speer was a little late in noticing that the MiG had lit afterburner and dived for home. Alone over enemy territory and overcast, he heard several SAM warnings and broke off the engagement]. Carter, who thought he had a good radar, could never get contact with the MiG. After they were 'feet wet' and tanking he discovered the radar had an effective range of one mile! What a great radar we had in the F-4J!

Beaulier and I were lagging our descending MiG about 40 degrees when he apparently lost sight of us and reversed his course. Jerry quickly corrected his turn and put our Phantom at the MiG's dead six o'clock with Sidewinder selected. Of course, the first Sidewinder to come up was the one with the 30 mils offset from centerline. He had checked this earlier and anticipated the correction, which he made. With the Sidewinder furiously buzzing we fired the first missile at the MiG's six o'clock at approximately half a mile. It hit the MiG's tailpipe and exploded. With the second Sidewinder now howling I suggested maybe a second shot might be in order, so we shot it and there was another explosion on the MiG, which was now engulfed in a fireball from the leading edge of the wing backwards. The nose and canopy were visible in front of the fireball. We pulled up at the MiG's four o'clock and took a look as the doomed aircraft descended into the cloud below us. We never saw an ejection.

One point I always make to younger fighter crews is that looking at a stricken aircraft is probably the worst thing you can do, as well as the most difficult to ignore. The first kill only happens once, and it's nearly impossible not to want to sneak a peek. My point is that unless someone has given absolute assurance that there is only one bandit airborne (an unlikely assurance), your self-congratulatory stare at your own handiwork is just asking for it. We weren't that smart, so we looked and looked some more. Then it was time to get out of town, and where was CAG? We decided we weren't going to circle Than Hoa looking for him, so I gave Beaulier a vector and we hauled ass. Almost immediately we were feet wet and were vectored to a tanker for refueling—the same KA-3B that CAG and Carter eventually joined. Beaulier and I were jubilant

about our recent good fortune and looking forward to getting back to the ship to begin the fun of telling what happened. We, of course, felt that heroes such as ourselves would get the King's treatment with a red carpet extending fifty miles behind the ship. That wasn't the case, however. The *Constellation* was still fighting a war, and our MiG didn't warrant any change in the flight schedules. We were told to 'delta' for an hour and ten minutes until the next scheduled recovery. Can you imagine the effrontery? Telling two red-blooded heroes to just wait? Of course we waited, and recovered with Beaulier taking credit for an underlined three wire trap—best you can get! And it was the best day we ever had.

The F-4J with all its foibles was a great machine which, although not as nimble as the MiG-21, could be properly fought, as many MiG drivers learned to their misfortune. The aircraft has now gone, but Beaulier and I can proudly remember the day we grabbed the brass ring over Than Hoa in the Phantom."

Jerry and Steve had a near-textbook kill which reflected credit on Top Gun in that Jerry felt the training experience gave him the confidence and clarity of thought to tackle the situation when it arose. However, political expediency meant that the Navy had to play down the kill for a while, and the crew of *Dakota* 201 (BuNo 155875) could not be named. The 1970 rules of engagement allowed USN fighters to respond to MiGs only if attacked during an escort or BARCAP sortie, and the JCS didn't want to encourage those who believed that there was a conscious attempt to draw the MiGs out into combat. There were no more MiG kills for nearly two years, though on their 1971 cruise the *Ghostriders* XO (then CO), "Smoke" Wilson, came close to a victory. At that time missions over the North were still confined to *Blue Tree* escorts, though there was increasing concern about the MiG pilots' obvious desire to "come down south" and bag a B-52 on one of the incessant *Arc Light* bombing missions in Northern Laos. "Smoke" Wilson described his tactics for dealing with the threat:

"We had developed a tactic for intercepting the night flyer. It was pretty simple. We established CAP stations off the coast, and when the MiG passed by we would fall in behind him and run him down, or if ever he turned around just shoot him head-on (the F-4J's weapons system was well suited for this shot). Several nights

A VF-51 trio led by the CAG ship with its multi-colored *Screaming Eagle* artwork. (USN via CDR James Carlton)

went by with no MiG activity. As luck would have it I was on the northern CAP (4 December 1971) when a MiG finally came down. The controller said, 'Go covered' (secure voice radio channel) and then said, 'Vector 265 for bandit. You are cleared to fire—repeat, cleared to fire.' Needless to say, my heart jumped to my throat. We (Jim Holds, my RIO, and I) had a 4.0 weapons system (both pulse and pulse Doppler modes working well), two AIM-7Es and two AIM-9s. As soon as I came onto 265 degrees I immediately locked onto the MiG, who was crossing from north to south. Because of the geometry we fell into a tail chase with him. I was doing almost 600 kts at 2,000 ft in the hills of North Vietnam. I was in and out of a thin overcast, and the flames from the afterburner made it seem like sitting on the inside of a frosted light bulb.

We were closing on him, but slowly. At that speed and low altitude the maximum missile range (shooter to target) becomes quite short, although the missile goes a long way over the ground. The MiG turned west, went to even lower altitude, and at about two miles we lost radar contact in the ground clutter. About the same time the controller called, 'Skip it,' and vectored us for a tanker. So close and yet so far! After landing back on USS *Enterprise* my Duty Officer grabbed me and said, 'The Admiral wants to see you.' The Admiral was as disappointed as I was, and asked what happened. After explaining the geometry and low-altitude speed vs range he accepted my explanation. In parting he said, 'Well, Smoke, if it's any consolation he knew you were after him. He diverted to Quan Lang airfield in west central North Vietnam.' When he landed they turned out the runway lights, apparently in fear that I would shoot him on the ground, and he ran off the runway. Later, we learned that the plane was destroyed, hence my 'lame duck' kill that was never officially recognized."

In fact, VF-142's next kill came over a year later when LTJGs Scott H. Davis and Geoff Ulrich got a MiG-21 in *Dakota* 214 (BuNo 155846) on 28 December 1972, but by then the whole nature of the air war had changed.

Linebacker

It wasn't until the beginning of 1972 that the air war began to gather pace again, and the first successful MiG encounter for twenty-two months occurred on 19 January. At the helm of F-4J BuNo 157267 was LT Randy Cunningham, with his RIO, LTJG Willy Driscoll. The previous occupant of Cunningham's back seat, on the relatively uneventful 1970 USS *America* cruise, was Lynn R. Batterman, who left VF-96 in 1971. Their missions in 1970 were the usual mixture of BARCAPs, recce escort, and strikes in the *Steel Tiger* area. "Most of the bombing missions were pre-planned, with a FAC on a site to pinpoint the target. Usually there were four Phantoms, and the RIO's job was to read off airspeed, altitude, dive angle, and keep an eye out for MiGs." Both Cunningham and Batterman were "nuggets," learning from their mistakes. Their first combat cruise included a near ramp-strike on their first night carqual, and a break and approach (briefly) to the wrong boat on return from a mission. However, it was very clear to Lynn Batterman that his pilot was completely single-minded about scoring a MiG kill. "He worked harder than average and was better than average because of it. He and I were the only crew to consistently check out and re-read the secret manuals we had on MiGs (we even had some MiG repair/'NATOPS' manuals) and the *Have Drill/Have Donut* manuals, which were controlled by the skipper, CDR Al Newman, but any crew could check them out. There were also the *Red Baron* reports and U.S. pilots' accounts of MiG encounters." Randy Cunningham made sure he studied the projected location of all MiGs before every sortie. "We would 'bait' the MiGs by flying fifty feet or so above the sea and approaching [the shore] as close as possible, trolling up and down the coast."

The *Black Falcons* had also received a thorough, practical preparation for the forthcoming increase in air combat activity. Matt Connelly, later to score two MiGs himself, was the squadron's Weapons Training Officer, and he commented: "On VF-96's turn-around between the 1970 and 1971 cruises the whole squadron was given the Top Gun academic syllabus. In addition, we flew tactics hops against Top Gun, as well as other dissimilar aircraft. We even flew against Air Force F-106s at McChord AFB. During our turn-

For its 1973 *Coral Sea* cruise VF-51 abandoned its "supersonic can-opener" design in favor of the fashionable "black tail" look. Ironically, the assigned RIO on BuNo 150406 (NL 112), LT John Letter, was largely responsible for the original audacious eagle design. This F-4 was lost in May 1973 when it rolled off the deck. (Fred Roos via Norman Taylor)

around missile shoot VF-96 expended the entire West Coast training allowance for missiles. This did not make us popular with the Westpac staff, but it later paid handsome dividends."

Cunningham's 19 January mission was a *Blue Tree* over Quan Lang airfield, from which MiGs had once again been threatening B-52 *Arc Light* operations rather more effectively than before. The RA-5C recce bird was heavily escorted by F-4s and a strike force, bringing the total to 35 aircraft, in case AAA provided the excuse for a "protective reaction" strike on the airfield defenses. Despite the official bombing halt, the opportunity to reduce the MiG risk by neutralizing an airfield or two was attractive. Cunningham's *Showtime* element took up station north of Quan Lang and found themselves over an active SAM site with two SA-2s lifting off. His wingman was LT Brian Grant, with his RIO, "Seacow" Sullivan (so-called because of his 'build, droopy mustache, and colorful demeanor,' according to his former pilot). Randy Cunningham watched one of the SAMs following Grant's F-4J and called for him to make a maximum-g turn to evade the SA-2 as the A-7 strikers moved in to attack the SAM site. The two Phantoms then came under attack from a site to the south of Quan Lang, and a total of eighteen missiles were fired, two of which narrowly missed the twisting F-4s and separated them. Brian Grant; "This mission was the first time I was opposed by SAMs. The first missile was a very close call, and the remaining two or three diverted my attention from Randy to the point where I lost sight and did not regain formation until after the kill."

The MiG kill was one of a pair of MiG-21Js which entered the arena and fell to Cunningham's second AIM-9, fired at very low altitude and catching the MiG just as it reversed out of a high-g turn. Randy Cunningham's dream was fulfilled, but Brian Grant drew his attention to the second MiG, which was leaving the scene at high speed. A low-level chase at 650 kts followed, broken off only when RIO Driscoll pointed out that their fuel state was critical. They attempted a last-ditch Sparrow shot before turning back, but the missile failed to activate.

Despite the best efforts of the Navy squadrons, MiG encounters remained a rarity in the early part of 1972. While life aboard ship was as busy as ever, there was still time in the off-duty hours for, "chess, booze, ready-room movies, and cribbage," the last of these being Brian's favored pastime. The Air Force bagged their

first on 21 February 1972 after a gap of nearly four years since their last *Rolling Thunder* kill. Task Force 77 missions continued to concentrate on strike and escort, as Brian Grant recounted:

"Strike and flak suppression employed the same 40 degree dive angle. Entry was at about 12,000 ft, and pullout was above 3,500 ft or so to avoid the predicted bomb fragmentation pattern. Mk 20 Rockeye CBU was used for flak suppression, loaded two on each bomb rack for a total of four. We delivered all of them in a ripple for a more even spread. The F-4J had an 'iron cross' fixed gunsight with 'mil' depression depending on the weapon carried or the dive angle. It was fairly accurate, but due to more limited training in bomb delivery I can take an educated guess that our average fighter pilot CEP was 150 ft; maybe 100 ft on a good day. The A-7 and A-6 squadrons probably attained 50 to 75 ft CEP."

Phantoms normally carried air-to-air missiles, whatever else was aboard their pylons. CDR Bill Freckleton, then a LTJG with VF-111 *Sundowners*, explained: "We were always configured with both types of missile [AIM-7 and AIM-9] whenever we were on an interdiction or air defense mission. We even carried both missiles when we were loaded with six Mk 82 bombs for a strike mission." Bill was teamed with LT Garry Weigand, who had come to VF-111 from a VQ squadron and F-8 Crusaders. The squadron had converted from the Vought fighter in 1971, picking up a batch of F-4Bs and deploying on USS *Coral Sea* in November, joining the similarly-equipped VF-51 on a cruise which yielded five MiG kills. Operation *Proud Deep* was well under way when the ship moved to Yankee Station, marking the resumption of major interdiction operations in North Vietnam. Their first strike, on 15 December, was in support of the Air Wing's "guest" USMC A-6 squadron, when twelve F-4s led by the two squadron COs bombed through clouds and lost a *Sundowner* F-4 (BuNo 150418) to SAMs. LCDR D.W. Hoffman and LTJG Norris Charles went to the Hanoi Hilton.

About half of CDR Bob Pearl's VF-111 pilots had come direct from the F-8C, in which LT Tony Nargi had scored a MiG-21 in September 1968. This ACM experience was very valuable on transition to the F-4B, as the basic RAG syllabus still included a relatively minor emphasis on air combat. Bill Freckleton:

"Fleet defense interception was not relevant to the situation in Vietnam, other than that it laid the basic ground-work for systems familiarity and crew coordination. The actual systematic procedures

As CO of the USS *Kitty Hawk* Captain "Tooter" Teague maintained his tradition of naming his aircraft *Bossier City Bearcat*. F-4J BuNo 153804 was the VF-213 example. (C. Moggeridge)

BuNo 157242 was the VF-114 *Bossier City Bearcat* for Captain Teague. (MAP)

A VPAF MiG-17, caught by the gun camera of a U.S. fighter. (National Archives)

of a '150 to 180 degrees to go' intercept with conversion on a straight and level bogie were not applicable. The greater preponderance of RAG training in the F-4B and F-4J at VF-121 was Fleet defensive. A very few (less than six) hops were dedicated to actual ACM, and even then one or two were sometimes waived to get guys into the Fleet." The former F-8 jockeys had plenty of dogfighting in their blood, but most made the transition to the two-man crew situation easily enough. In Bill's view this was, "because some of the more experienced RIOs really helped the new F-4 pilots to fully understand the aircraft and its weapons systems. I can think of one or two pilots who had flown the F-8 who were still hesitant to give up some of their responsibilities to the back cockpit."

Also in the squadron to facilitate the introduction of the F-4 were two of the architects of Top Gun, Jim Ruliffson and Jim Laing. Most of the Crusader veterans were philosophical about moving to a gunless fighter. (Although it had been dubbed "The Last of the Gunfighters" by its aficionados, only one of the F-8's twenty kills had been solely attributable to 20 mm gunfire, though guns played a part in three others.) Bill Freckleton's pilot was probably typical in this respect: "Having flown over 110 of my 117 missions with Garry Weigand I never once heard him, an ex-Crusader pilot, express a desire for 20 mm cannon."

Good cockpit teamwork was a big factor in the next MiG kill. Garry and "Farkle" in F-4B BuNo 153019, *Old Nick* 201, were due to fly a FORECAP (photo escort) over Quan Lang airfield, one of many MiG airfields which had been extensively re-stocked with fighters during the four-year "bombing halt." In this time the MiG force had increased from around 150 (mainly MiG-17s) at the end of *Rolling Thunder* to over 260, including 95 MiG-21s and 30 MiG-19s by mid-1970. The three main airfields used in *Rolling Thunder*; Kep, Gia Lam, and Phuc Yen, had been joined by many other newly-built bases. Some, like Quan Lang, were in southern North Vietnam, and the very effective Soviet-supplied GCI network extended much further south and west to control MiGs over Laos and the DMZ.

Old Nick's flight was to maintain a CAP offshore, only crossing the beach if the photo aircraft came under attack. The mission

began unpromisingly, though the word was that MiGs were likely to appear. Bill Freckleton described some of the back-seat duties and problems on this kind of mission:

"We were part of a *Blue Tree* strike with the object of drawing up some MiGs. The truth is that, since we were supposed to orbit the USS *Chicago* just off the coast I personally didn't expect to see anything. I guess that was because every time we were CAP we stayed 'feet wet.' I do recall this was a big deal strike, because the Admiral, RADM Ferris, said that anyone who got a MiG that day could go to Paris, France, for R and R. My radar was down prior to launch. I had no search or track capability, so I knew we were 'Sidewinder only.' In a combat situation you sure as hell tried everything to obtain a perfect radar. The AQs (avionics technicians) did a stupendous job of keeping these radar systems in an 'up' status. We used hand signals a lot in Vietnam to communicate between aircraft, and one of the most-used signals was the left-to-right sweeping motion of the hand followed by a thumbs down or thumbs up, or a rocking motion of the palm to indicate whether the radar was up, down or 'shaky' (degraded). After electrical power and engine start on deck the radar was fired up in the standby mode and the BIT checks (about seven) were performed. In the F-4B this included manually tuning the crystals to set radar operating frequencies and monitoring analog needle indicators to ensure everything was within the correct parameters. On the radar scope there was a series of displays that were checked, dealing mainly with the search and track capability of the system. The 'steering dot' was checked in relation to the azimuth steering error (ASE) circle, and other tests were performed. Transmitter power was critical to operation of the radar. We could tweak the pulse video and gain of the radar display to attain an optimum presentation. The Sparrow missiles were tuned on deck, then placed on standby until airborne.

Sometime after the tanking that always followed launching from the ship RIOs would run the airborne radar checks and lock up an aircraft to check the system with an actual lock. The pilot could select Sidewinder, get a tone, and see where the boresight was for that weapon (here you had to avoid shooting down a friendly!). By

A neat *Sundowners* formation traverses some inhospitable landscape in the South Western USA. The last F-4 in the group began life as a USAF loan "F-110A" and eventually became a QF-4N drone. (USN via CDR James Carlton)

locking an actual airplane we could see the relationship of the radar steering indications to a target that was on boresight, and we were then prepared to make the mental adjustments to the firing solution should we ever lock on to an enemy aircraft. The Sparrow missiles would be tuned, and we would get the relevant indications from the pilot. They would stay tuned and in standby mode until we armed up prior to crossing the beach.

There were a few times, crossing the beach, when aircraft didn't have a Sparrow capability because the single-target track capability was inoperative or degraded. Of course, when this happened we always hoped that somehow by re-cycling the system from 'on' to 'off' and back 'on' again, then re-running BIT checks, that the radar would magically cure itself. Sometimes this procedure worked. In a combat situation you would try anything and everything to get a full 'up' and operational radar. I remember many times having to tell my pilot I was shutting the radar down in an attempt to regain a capability. Sometimes, putting 'g' on the aircraft would magically bring a dead radar back to life. Occasionally, the wave-guides would have a leak and could not maintain the required pressure for radar operation. In those rare cases the transmitter would not work above a certain altitude due to lack of pressurization, which meant we either stayed low with an operable radar or went high with no radar.

As I recall it was not unusual at all in our Air Wing to launch with at least one aircraft in a section with a known, pre-launch radar discrepancy, such as search-only or limited antenna sweep. The AN/APR-30 RHAW gear worked extremely well, and it was also BIT-checked on deck prior to each flight and then again airborne. Several times I was able to sight SA-3 SAMs by correlating the strobes on the radar warning display to the outside world. This piece of gear wasn't essential to cross the beach, as long as one aircraft in each section had an operable one. The one piece of gear that *was* required was the AN/ALQ-100 deception/repeater. If this was down the aircraft couldn't go 'feet dry.'

Prior to coasting in we used the radar's mapping mode to look at the features along the coastline. Once we got about twenty miles from the beach we concentrated on the air search, operating the search radar in a scan pattern that ensued the best possible radar coverage in the direction of the potential airborne threat. We hardly

CDR Jerry Houston (right) and RIO Kevin Moore "debriefing the Admiral" after their MiG kill. ("Damn right, sir. That MiG driver was an excellent pilot" was Devil's wry suggestion for a caption line) (USN via Jerry B. Houston)

used the system for radar navigation or targeting, but used it extensively in the search and track modes.

The F-4B's Aero 1A system, a pulse-only radar, had a more cumbersome and archaic method of checking its operational status compared with the F-4J's AWG-10, which had pulse-Doppler and a more automated method of running the BIT checks. Unlike more modern fighters the F-4 had no special 'war reserve' mode, restricted only for use in combat."

On their 6 March 1972 mission Garry Weigand positioned their radar-less Phantom ready for deck launch and "hot refueled," awaiting a call to protect the recce bird.

"Garry and I were originally supposed to fly wing on Jim Ruliffson, but when he went down on deck with a hydraulic failure LT Jim 'Yosemite' Stillinger [a recent Top Gun graduate] became our flight lead. His RIO, LTJG Rick Olin, had a search-only pulse radar, which was atrocious in the clutter environment of an overland, look-down situation. So he, too, was without a useable radar-guided missile [for the forthcoming low-level engagement]."

As the recce pilot approached Quan Lang he sighted two "Blue" bandits (MiG-21s) and three "Reds" (MiG-17s). The two F-4 CAPs, one led by Jim Stillinger and the other by CDR Foster S. Teague of VF-51, orbiting over Laos, immediately closed in.

"Tooter" Teague, with Ronald "Mugs" McKeown, had been among the first of the VX-4 cadre to fly the *Have Drill* MiG-17s. He knew the aircraft intimately, and had been responsible for devising a substantial part of the Top Gun air-to-air syllabus. Top Gun policy had been to place a few graduates in each F-4 squadron to spread the word, but Tooter wanted to extend the awareness of the realities of combat in other ways, too. Jerry Houston: "A first, and Tooter's idea, was to let any squadron ground officer or CPO who wanted to do so fly a combat mission in the back seat while we were bombing south of the DMZ. This was a tremendous boost for morale, and most of the eligible people chose to do so. The only restriction was that they fly with Tooter, 'Blackjack' Finley, me, or Chuck Schroeder." (Tooter had taken over VF-51 on 21 June 1971, with CDR Finley as XO and "Devil" Houston as Ops Officer).

VF-51 hold tight formation for McDonnell Douglas photographer, Harry Gann. (McDonnell Douglas via CDR James Carlton)

VF-51's "hard core of ex-F-8 gunfighter jocks." (Standing, left to right): MiG killers Ken Cannon and Tooter Teague, Tom Tucker, Dave Palmer, Don Scott. (Kneeling, left to right): Rick Bradley, Chuck Schroeder, and MiG killer "Devil" Houston. The last of twenty MiG kills by F-8s occurred on 19 September 1968. Two fell to VF-51 pilots in Crusaders, and four more when they flew F-4s. (Jerry B. Houston)

As CDR Teague's flight closed in from the west his RIO, LT Ralph M. Howell, tracked a MiG-17 head-on, and Tooter swung 'round to get on its six o'clock, firing an AIM-9 when he was within range. His missile exploded close enough to the MiG to cause what Tooter felt was mortal damage, but he had no time to establish its fate, as three other MiGs were threatening their *Screaming Eagle* F-4B. He tried another Sidewinder on a second MiG, but his over-shoot speed didn't give the missile time to arm. Teague and Howell evaded the MiGs with a steep climb and returned to the boat with only a damaged MiG to their credit.

The offshore CAP was then vectored in by *Red Crown* Radarman Larry H. Nowell on USS *Chicago*. (By the end of May 1972 Larry Nowell had guided Navy fighters to MiG kills on thirteen occasions.) Bill Freckleton:

"During our vectors for the MiG I was totally visual [no radar]. Checking heading, altitude, and airspeed indications periodically, I was mostly looking left, right, and aft for any sign of the MiG we were being vectored on. Going through the weapons checks once again and loosening my straps in order to move around in the cockpit, I told Garry, 'You've done this before—you can do it again,' referring to our ACM training. Visibility in the F-4B was marginal from the back seat looking forward, but good to fair on the sides and towards the rear. My first sighting of the MiG was as we were in a starboard turn, descending through about 8,000 ft. He was on my port side at about ten o'clock and heading up towards us. From that point of view, I would say the visibility from the rear cockpit was spectacular!"

Jim Stillinger rolled down after the MiG, and Weigand pulled up to "cover" position at higher altitude. The MiG pilot kept Stillinger in a turning fight, reversing into him repeatedly and forcing him into a series of high yo-yo maneuvers after overshooting. Garry Weigand watched the fight, losing sight briefly at the start, but then tracking it closely as Stillinger's Phantom continued unsuccessfully to get on the MiG's tail. Meanwhile, Bill Freckleton had swiveled around in his seat, checking for other MiGs to their rear. Finally, Jim Stillinger realized that he wouldn't be able to gain an advantage on the MiG, and Weigand decided to enter the fight. As Stillinger rolled level he headed out, dragging the MiG behind him and giving Weigand and Freckleton the opportunity for the Sidewinder tail-shot which, without radar, was their only chance. In an attempt to catch Stillinger's Phantom the MiG-17 pilot, clearly unaware of the approach of *Old Nick* 201, lit his afterburner, thereby making his aircraft even more attractive to an AIM-9.

By now the fight had dropped closer to the tree-tops. "We were definitely low," Bill said, "The last glance at my barometric altimeter showed 500 ft. Garry and I were dead six [to the MiG], a half mile away at 500 kts when we shot." Thinking that their first AIM-9 might not have found its target, "Greyhound" Weigand decided to set up another shot just as the MiG's tail came away and it crashed into the forest. For Bill Freckleton it was a kill for a pilot and RIO working closely with their wingman, and for a RIO who, ironically, "never had the chance to search actively [on radar] for bogies or run an intercept in combat." It was the second kill for the USS *Coral Sea*, and the second too for the shark-mouthed BuNo 153019, which had also been used for VF-213's 1966 An-2 shootdown. The Admiral's offer of Paris didn't quite happen, though: "Garry and I ended up going to Bangkok instead."

Old, Bold Phantoms

Sister squadron to VF-111 aboard *Coral Sea* was VF-51 *Screaming Eagles*, commanded prior to June 1971 by CDR Tom Tucker. Jerry "Devil" Houston was with them when orders came to transition from the F-8H to F-4Bs.

"That was exciting news. What Tucker didn't know was that our F-4s would be coming from USMC rejects. The F-4Bs we were

USS *Coral Sea* CO, Captain Bill Harris, meets Devil Houston and Kevin Moore on the flight deck after their 6 May 1972 MiG kill. Devil claims that he, "kept his sunglasses on to hide tears of joy." (USN via Jerry B. Houston)

An *Aardvark* F-4J kicks up a plume of smoke off the USS *Kitty Hawk's* blast deflector as it strains at the hold-back bar. The squadron shot down five MiGs during the war. (USN via CDR James Carlton)

Fighting Falcons F-4J BuNo 155543 carries the squadron's MiG kill "scoreboard" on its splitter plate after its final deployment on USS *Constellation*. (James E. Rotramel)

assigned had been preserved in whatever state they had been in a couple of years earlier when the Corps had declared them unairworthy. Maintenance *Tiger Teams* from all the F-4 squadrons at Miramar were assigned the job of going to MCAS El Toro and performing overdue scheduled maintenance and getting those hulks capable of flight from El Toro to Miramar. LCDR Chuck Schroeder was assigned to VF-51 as our Maintenance Officer, and his accomplishment, getting those over-the-hill rustbuckets ready for deployment, was a Herculean undertaking. He did it on time, despite competing against my back-breaking training schedule."

Ken Baldry's VF-96 had some similar F-4Bs back from the Marines after hard use at Chu Lai when the squadron handed over their Block 26 F-4Bs (15000 BuNos and up) at the end of their 1966-67 USS *Enterprise* cruise. The battle-worn Spooks they got were, "the worst bunch of hangar queens I have ever seen! As pilots popped their canopies after reaching the tie-down spot, *Enterprise* was still into the wind, and each airplane had a miniature sandstorm coming out of the cockpit area!" The *Screaming Eagles* "Marine relics" suffered from faulty cockpit pressurization seals, stress cracks in the wings, and other peculiarities. Devil's aircraft would bank to port whenever its crew used the radio. Fatigue problems in another F-4B caused the tail section to drop off during straight and level flight.

Devil Houston explained the origins of the squadron's extravagant "Screaming Eagle" color scheme:

"Tucker sent ENS John Letter to a short corrosion control and aircraft painting course at NAS North Island to determine which areas made the most sense to paint and which ones couldn't be painted on the F-4. He found out which panels came off most during routine maintenance and then devised a squadron paint scheme to avoid them. Then there was a squadron competition at designing paint scheme proposals. Our supersonic eagle won hands down. The Miramar wags said all the design lacked were mud flaps and a long raccoon tail on some aerial. This was especially true since most outfits were looking for ways to hide their airplanes, blend with the clouds and terrain, etc. We didn't care. We were going

hunting and prayed to be found. Our goal was to be audacious; a magnet for the MiGs, we hoped. Detractors called our supersonic eagle a supersonic can-opener. We didn't mind. Intruders into our airspace (anything above sea level) didn't have to speculate about who just whipped their ass. To their credit, our new sister squadron VF-111 [replacing VF-53] had a similarly bold paint scheme. They would get one MiG during their deployment. We would get four.

There were problems of course. We had been single-seat F-8 pilots, and as such we'd ridiculed all multi-crewed aircraft, especially the 'Phantom phlyers.' Now we were becoming some of, well...*them*. Our hard-core philosophy going in was that there wasn't then, nor would there ever be, an NFO we'd rather have than an extra 500 lbs of fuel and the 20 mm cannons we were sacrificing by leaving the F-8s. Word got about how we felt, and it was with a great deal of reluctance that F-4 pilots and NFOs accepted orders to VF-51. However, Tom Tucker ordered us all to 'dummy up'—keep our mouths shut—during our transition training in VF-121. It was hard, but we did it. Having instructor NFOs flying in our back seats was about as much fun as having our mothers along on a date. But the 'muzzle' orders were serious, so we swallowed our bile during the day and tried to wash it down with an extra beer or two at night. Some of the pretty good F-4 ground instructors probably thought we weren't smart enough to understand their jokes. Truth be told, it would have been murder to have been the first Crusader jock to have acknowledged anything even slightly acceptable in the up-'til-then rejected brotherhood of interceptor pilots with their goofus guys in the back seats.

It didn't take us long, however, to become enraptured with the F-4's stronger rubber bands. Those two J79s could flat move you from 'a' to 'b,' and the F-4 didn't lose as much energy as quickly as the F-8 had under g-loading, especially at lower altitudes. We envisioned being a 'Super F-8' squadron with another set of eyeballs in each plane. This would allow both pilots in a section to continually search ahead for enemy planes, while the RIOs cleared the wingman's six o'clock position. It was a good concept, and we eventually got there. But it was only after bruised RIO egos were assuaged, and unreliable F-4B radar systems ran us out of options.

Every time we thought like 'Phantom phlyers' something bad happened. One time we got a plane shot down (by a MiG-21); once we (two second-tour F-4 pilots) fired two Sparrows at a cloud of chaff. As soon as we started killing MiGs even the RIOs endorsed the F-8 philosophy—with another set of eyeballs. In F-8s we had always assigned the most junior wingman to the CO. The thinking was that the CO's experience would keep the youngster out of trouble. It was a good practice, and it worked. As new F-4 drivers we adopted the same philosophy, but with a different slant. We assigned the newest pilots to the most experienced RIOs. The experienced RIOs went ballistic; they felt they had paid their dues by already flying with inexperienced pilots. This was another difference in the communities. F-4 senior pilots got senior RIOs. We didn't agree, and our senior RIOs wore resigned looks and rued the day they got orders to VF-51.

Soon after we completed RAG training and had accepted the first of our Marine-reject F-4Bs I came back to the squadron area from lunch to receive a double dose of bad news. Tom Tucker and Tooter Teague had divided the squadron into tactical divisions and assigned wingmen and RIOs. Normally, the Ops Officer would have been included in such decisions, but it didn't take me long to figure out why I was excluded from the planning. They had given me the least desirable wingman and the least competent RIO. Behind huge grins they explained it was a training challenge and, as Ops Officer, certainly due me. Both of my new charges were straight from the RAG, where they had been properly tagged as 'least likely to succeed.' They truly kept me from enjoying any benefits of our aircraft transition. And after all the misery we were allowed to leave them behind when we deployed for combat.

During work-up training at Fallon a couple of our ex-Crusader pilots asked for a cross-country flight back to San Diego. They may have been going to a wedding or something, but we had already decided we could spare one aircraft and they were going together. Their request was approved, and all hell broke loose. The RIOs believed the back seats belonged to them exclusively, and they were pissed off. They came to me and asked to talk about it, so I listened.

LCDR J. B. Souder was the senior RIO, and he made a good case. To my surprise and chagrin I agreed with them. It was my first inkling that there were people in the back seats much more capable and eloquent than the poor examples I'd been flying with. At last, after about a year of exasperation I started to see some talented RIOs. I'd sort of settled on Kevin Moore as my favorite. Then, after our 6 May MiG kill, I flew with him exclusively."

Before Devil and Kevin Moore faced their MiG there was one more opportunity for Tooter Teague to get his. Once again he came away frustrated and disappointed, while the encounter cost VF-51 an aircraft, with senior RIO J. B. Souder and his pilot LT Al Molinare aboard. The two crews were on a 27 April MiGCAP and received a vector on a MiG-21 from Noi Bai, heading south. Both crews were apparently intent upon their radars, trying for a head-on Sparrow attack, but a misunderstanding with Red Crown caused them to turn away from the MiG they were trying to follow since the controller incorrectly believed them to be actually ahead of the enemy aircraft. The MiG's own GCI then vectored its pilot (Hoang Quoc Dung of the 921st FR) behind Molinare's F-4, which was then downed with an Atoll missile before either Phantom crew detected the MiG two miles behind them. The crew of NL 102 (BuNo 153025) ejected from their blazing jet and became POWs. Jerry Houston, who was on the mission but was sent to refuel while the other two F-4Bs engaged, recalled that, "in less than a year they were landing back at Clark AFB. Fortunately, VF-51's carrier was in Cubi Point, so J. B. Souder and Al Molinare were met by some squadron mates."

By 6 May full Alpha packages were being sent to attack airfields, including Bai Thuong, when significant numbers of MiGs were sighted on them. "Devil" Houston had Chuck Schroeder and RIO Rick Webb on his wing, and two *Sundowners* F-4Bs completed the TARCAP. As they approached the target the section of four VMA(AW)-224 A-6A Intruders made a low-level Rockeye attack, and soon reported "MiGs everywhere." The strike had arrived in the middle of a busy day's training at the base. They hit Bai Thuong's defenses and pulled off target in single file, heading down a long karst valley at low altitude. Jim Ruliffson, leading the strike

Showtime 106, BuNo 155769, destroyed two MiG-17s on 10 May 1972. LT Matt Connelly and LT Tom Blonski became involved in the unprecedented fighter melée in which Randy Cunningham and Willy Driscoll shot another three MiGs. (USN/ Michael R. Morris)

A 1972 line-up in front of CAG, CDR Lowell "Gus" Eggert's assigned F-4J. Curt Dosé is fourth from right (back row). Third from right at the front is Lonny McClung, the 1998 President of the Tailhook Association. (Curt Dosé)

Curt Dosé's self-portrait during a combat mission includes a reflection of his Phantom's instrument panel and windshield. (Curt Dosé)

Red shirted armorers load a VF-96 Phantom with Mk 82s. (Curt Dosé)

in a VF-111 F-4B, saw a MiG-17 following the last A-6A and called down the TARCAP Phantoms. "Devil" Houston credited his RIO with a major role in the subsequent engagement: "Kevin was about 95 per cent responsible for our getting that MiG. He'd reminded me about some armament switches I'd forgotten soon after coast-in, and then he saw the MiG and coaxed me into a good kill position long before I'd spotted it. RIOs don't have to convince me that they earned their keep. I know they did." They turned in behind the MiG, shed the centerline tank, and lit the burners. Closing on the target, Houston saw 37 mm fire from the VPAF fighter as it pursued the third A-6A, the other two having broken away in response to Devil's warning. The lead A-6A, flown by the Air Wing's new CAG, Roger "Blinky" Sheets, inserted himself back between the MiG and the A-6A it was attacking in order to drag it away from its prey. He then proceeded to maneuver his heavy Intruder sufficiently to avoid the MiG's "flaming golf ball" cannon shells. As an experienced fighter pilot, CAG Sheets knew that the MiG driver's con-

trols were ineffective at 500 kts and around 100 ft altitude. Blazing in at 100 kts faster, Jerry Houston and Kevin Moore got a good Sidewinder tone, but couldn't be sure that their missile would not hit CAG rather than the MiG. Houston called the A-6 to break, but at this point his F-4B's "Marine reject" past returned to haunt it.

"The Marines had modified their radios to transmit the Sidewinder growls over the UHF radio when the transmit button was pressed. The transmitted tone overrode all other UHF transmissions and effectively put us without communications during the most important couple of minutes in my airborne career. CAG Sheets decided not to break until he saw my missile fired—the ultimate decoy. It worked, but only because at the last minute (approaching minimum firing range and in total frustration) I decided to fire the missile, despite a chance that it could have been locked on CAG's A-6. I didn't know he couldn't hear my frantic calls to 'Break and get the hell out of there.' All he heard was the transmitted Sidewinder tone."

Hung with Mk 82s and AIM-9s, *Silver Kite 210* awaits a crew. (Curt Dosé)

Curt Dosé (right), Jim McDevitt (left), and their plane captain with F-4J *Silver Kite 211* (BuNo 157269), in which they shot down their MiG on 10 May 1972. (Curt Dosé)

The crew of *Screaming Eagle* 100 (BuNo 150456) then experienced that tiny but seemingly endless delay, during which time it seemed that their AIM-9 had failed to depart the pylon. In fact, after the usual 0.75 seconds the missile left on its corkscrew flight to the control-locked MiG's tailpipe, blowing the rear end of the aircraft off. At around 100 ft altitude the remainder of the wreck impacted the ground almost immediately.

While Jerry was focused on the MiG, his alert RIO had noticed wingman Chuck Schroeder pulling up onto their left flank—with another MiG on his tail and shooting. He veered away to the left, dragging the MiG away. Jerry Houston:

"Because my Sidewinder's loud squealing owned the airwaves, I can't be sure what Chuck's real plan was, other than to get some relief from the bright red 37 mm golf balls whistling by his canopy. Tongue in cheek, I imagine he wouldn't have objected had I rolled in behind the asshole who was peppering his plane, but I knew nothing of his problems. I couldn't even hear Kevin on our ICS [cockpit intercom]. Did the radio problem cost us a second MiG? Who knows? Maybe."

In fact, Schroeder had outmaneuvered the MiG, and it headed back north for home.

At such low altitude the AIM-9 was the only viable missile for the F-4, but it was VF-51's preferred weapon in any case.

"We believed in the Sidewinder. The radars weren't reliable, and we feared that their strobes might have given away our presence and cost us a shot in a MiG engagement. Gunner Thomas and his ordnance gang deserved a world of credit for having achieved the degree of dependability we enjoyed with our missile systems on those old aircraft. Only our radar systems' poor supply support kept us from having equal success with the radar/Sparrow system—that is, having Sparrows that would work. The missile of choice was always the Sidewinder. Always! We would fly into a kill position for the Sidewinder, and once you did that, a kill was practically assured. The RIOs, indignant at first, came to see our point of view, especially given the MiG kills."

After his encounter and VF-51's first kill Jerry took a final sweep of the area and headed back to the *Coral Sea*. "As happy and self-content as I was upon returning to the ship, it was an embarrassing moment because I knew how desperately everyone else wanted their own MiG. Their congratulations had a flavor of a father's congratulations to a new son-in-law. And I understood their feelings." Their F-4B, resplendent then in its CAG decor with multicolored tail-feathers, was itself shot down by a missile, but years later at Point Mugu after it was converted into a QF-4N drone. Ironically, its last assignment prior to drone conversion was back with the Marines at Andrews AFB with VMFA-321 *Hell's Angels*.

VF-51 added three more to their score that year, but there were two more MiGs downed on 6 May near Bai Thuong, marking a return to action for VF-114. CVW-11 on USS *Kitty Hawk* was asked to mount an afternoon Alpha on what was clearly a very active MiG base. LCDR Pete "Viper" Pettigrew, another former Top Gun instructor and later a RADM, switched from BARCAP to MiGCAP after XO CDR John Pitzen, the *Aardvark* MiGCAP leader's radar went down. Vectored by Larry Nowells onto four MiG-21s approaching from the north, Pettigrew and his wingman, Bob Hughes, sought the bandits on radar, and Bob's RIO, LTJG Joe Cruz, got a Doppler (velocity/closing rate) lock-on at about 25 miles. The MiGs were flying a close "v" formation, a familiar tactic to make them appear as only two blips on the Red Crown radar. Pettigrew and his RIO, Mike McCabe, were unable to get their Sparrows to tune, and there were problems with the switchology in tuning their Sidewinders. Bob Hughes took the lead as shooter when they got a visual on the MiGs. As they closed astern the rear pair of MiG-21s saw the F-4Js approach and turned back towards them. Bob Hughes followed the Number 4 MiG, firing an AIM-9 at an angle which was close to the missile's limits. Even so, it made the required turn and exploded close to the MiG's tail, causing severe damage. Pete Pettigrew began to follow it down to confirm the kill as Hughes turned hard after the lead MiG of the pair, firing two AIM-9s unsuccessfully.

LCDR Pettigrew hauled *Linfield* 201 (BuNo 157245) back into the fight, got a better tracking solution on the target, and fired a Sidewinder seconds after Hughes launched his third shot. While Hughes' missile caused damage to the enemy fighter's elevator, Pettigrew's flew up its tailpipe and fragmented the MiG. Meanwhile, both RIOs had been watching the other pair of MiGs returning to the scene and updating their pilots on the threat situation. Finally, with little fuel and only one good missile left, the two Phantoms turned into the oncoming MiGs and headed for the coast. In a mere ninety seconds the *Aardvarks* had scored their fourth and fifth kills, atoning to some extent for the squadron's nine combat losses to ground defenses during the *Kitty Hawk's* three grueling *Rolling Thunder* cruises. Those defenses were even more lethal in 1972.

Helmets off and smiles for the camera from pilot "Soupy" Campbell and RIO Jerry Hill. (Curt Dosé)

Pilot Austin Hawkins and RIO Jerry Hill in *Silver Kite 212*. (Curt Dosé)

XO John Pitzen and his RIO took a SAM three months later, and both men were listed MIA. On the same cruise MiG-killer Roy Cash, by then with VF-213, was shot down by AAA, but recovered with his RIO, Lt R .J. Laib.

May, 1972

These missile fights marked the beginning of a month in which USN Phantoms destroyed more MiGs than they had scored during the whole Vietnam War up to that point. General Giap's "Spring Offensive" invasion of South Vietnam necessitated rapid, massive reinforcement of the diminished USAF and USN air forces in the area, and Operation *Freedom Train*, beginning in April, marked a return to the bombing of the North. On 8 May President Nixon announced that North Vietnam's ports would be sealed off by mines dropped from USN A-6 and A-7 aircraft in an attempt to curtail the flow of war materiel to the North. Bombing of the North's railway network was also re-commenced. A further expansion of the air war was planned to force North Vietnam to negotiate a settlement. *Freedom Train* merged into *Linebacker*, and Operation *Pocket Money*, the mining of Haiphong and other harbors, marked the beginning of a new U.S. determination to end the conflict. A parallel 8 May Alpha strike on a massive truck staging area near Dong Suong included LT Randy Cunningham and LTJG Willy Driscoll in the MiGCAP. As the strike group approached its target, Cunningham and his usual wingman, Brian Grant, "swept" ahead of it for MiGs. Red Crown gave them a vector of 340 [on a compass using Hanoi as its center] at 60 miles for some unidentified bogies. The contact was intermittent. In Brian Grant's estimation, "The second MiG engagement was a classic from the point of view of an overwater CAP being vectored at an overland MiG target. Unfortunately, the Red Crown control vectors seemed 'spotty' to me and, sensing a set-up, I called the flight to make an in-place course reversal, which placed Randy Cunningham behind me in position to down his second MiG, conveniently trapped at my six o'clock. The trap that I had sensed was in fact realized when two other MiG-17s that had been trailing us flew directly over Randy's aircraft due to the course

reversal." The MiG-17 behind Brian's aircraft was firing at him when "Duke" Cunningham saw it and advised Grant to get rid of his centerline tank, unload, and run. The MiG pilot followed up his inaccurate 23 mm "BBs" with an Atoll shot, which Brian Grant outmaneuvered with a 6g turn to port. As the aggressive MiG pilot closed again for another gun attack Cunningham released an AIM-9 at him, knowing it stood little chance of making the required 60 degree turn. It did force the MiG to break off his attack. He reversed in front of Cunningham's *Showtime* 112 (BuNo 157267) just as the two MiGs which Grant had sighted completed their reciprocal turn and entered the fight, firing at Cunningham. His alert RIO, "Irish" Driscoll, spotted the attack as cannon fire started to pursue their Phantom. Cunningham fired an AIM-9, which took the MiG apart. Brian Grant: "Randy engaged the MiG while I went to high cover and jettisoned my centerline fuel tank. Rolling in from a high perch caused them to disengage Randy, go for the low cloud cover and escape. We followed in vain, then egressed to refuel and land. This was as classic a mutual support textbook fight as we practiced in Warning Area W-291, and in hindsight I could not imagine a more successful outcome." Brian had narrowly missed his chance of a MiG but, more importantly, he had protected his flight leader from serious trouble, as the two MiGs had positioned themselves to catch *Showtime* 112 whichever way it turned. The clouds saved the MiGs from Brian's pursuit, but also enabled Cunningham to escape into them and use his burners with relative safety (the clouds masked their heat from Atolls to some extent) and escape, too. On his return trip to the "boat" he used up a remaining AIM-9 to destroy a truck in a convoy they overflew, training its IR seeker onto the vehicle's engine.

Although the pace of events had increased rapidly during the first week in May 1972, it became fevered on May 10th. The carriers *Coral Sea*, *Constellation*, and *Kitty Hawk* all launched Alpha strikes, aimed at targets around Haiphong on a day when 414 USN and USAF sorties were flown against targets in North Vietnam. Air Force Phantoms attacked the monolithic Paul Doumer Bridge in Hanoi with EOGBs and LGBs, rendering it unusable. Four other

Christmas bomb graffiti on an F-4J, 1971. (Curt Dosé)

Christmas bomb graffiti #2. (Curt Dosé)

Bart Flaherty and Russ Ogle pre-flight inspecting their ordnance. (Curt Dosé)

Curt Dosé and Jim McDevitt. (Curt Dosé)

bridges were included on the Navy target list. Railroads, POL areas, and warehouses were also bombed. USS *Constellation's* CVW-9 thirty-three aircraft package launched first. The northern TARCAP was manned by two of Cdr Phil Scott's VF-92 F-4Js, one of which was flown by LT Curt Dosé with his RIO, LCDR Jim "Routeslip" McDevitt. Curt's "Dozo" call-sign had its origins in more pleasant surroundings than *Constellation's* roaring, shaking flight deck environment during the launch:

"These names were picked for you. Although most pilots tried to claim a non-damaging nickname, pre-empting the process, this rarely worked. I had taken the same call-sign that my father used in WWII—"Scarlet." That lasted only until the first time we visited the hotsy bath in Yokosuka, Japan, after I joined VF-92 in 1969. When the young lady handed you the wash-rag toward the end of the bath she would say, 'Hai dozo,' which we figured translated to 'Wash it yourself.' 'Hai dozo' became my nickname, usually shortened to 'Dozo.'"

"An Alpha strike launched everything at once, then made a 'ready deck' for aircraft as they returned. When they were all back

you re-spotted, re-armed, and launched the second strike about two hours later. The pilots were not the same for adjacent Alpha strikes, since the second strike was briefing while the first strike was flown. So, you flew the first and third strikes, as I did on 10 May 1972, or the second strike. As fighter pilots we had the additional responsibility of the Alert fighters. There were always two Alert 5 aircraft on or near the catapults with crews in them ready to launch in five minutes. There were two more in Alert 15 with crews suited up in the ready room, and they would head for their fighters when the Alert 5 launched. A further pair were designated Alert 30, and they suited up and moved into the Alert 15 slot. This was 24 hours a day anytime the carrier was not in port. I slept in the F-4 cockpit as much as I slept in my bunk.

The carrier would generally strike for about six days, then have a stand-down day (Alerts only). This would go on for about four to six weeks, then we would have a week off into Subic Bay or Japan. The normal cruise back then was nine months, but the *Connie's* was extended twice and ended up being almost a year. I would guess about a third of our flights were BARCAP, with the rest divided

Silver Kite 212 with a full eight-missile load. (Curt Dosé)

By 1972 the KA-6D had absorbed some of the Fleet tanking requirement. Each Intruder could usually give away about 12,000-15,000 lbs of fuel. (USN via Curt Dosé)

Gray Ghosts

VF-92's BuNo 155569 takes the barricade, a last resort in the case of severe hydraulic or other problems which prevented a normal arrestment. Item One on the pilot's list in this eventuality was to jettison missiles, avoiding hazards to other aircraft and personnel. (USN via Angelo Romano)

between photo escort (5 per cent), strike (25 per cent), MiGCAP (20 per cent), and SAM suppression (20 per cent). This changed to match the anticipated threat. On Dixie station we were always just strike or BARCAP, although we always carried four Sparrows and four Sidewinders, so we were one button away from being a fighter."

Curt had already made a Westpac deployment with VF-92 *Silver Kings* aboard USS *America* in 1969, a "quiet" period in the war. Before returning to Vietnam in 1971 he attended Top Gun, gaining experience which served him well on 10 May.

"Our CAG-9 fighters were normally very busy with MiGCAP and flak suppression missions on Alpha strikes. The flak suppressers would go in immediately ahead of the strike, accelerating ahead to engage the initial SAMs with Rockeye, then remaining in the area for follow-up and MiGCAP. The assigned MiGCAP would have sectors of responsibility toward all expected air-to-air threat directions. We had wide latitude in these sectors, but had to remain in position to respond to the strike force as needed. Our first May 10th strike MiGCAP was unique in that the strike was close to the coast, so the strike aircraft were quickly on and off the target, then 'feet wet.' We still had lots of gas when the 'blue bandit' guard calls came from threats to the north. Our initial turn to intercept this threat was proper and expected. It was partly coincidence that we ended up heading right for Kep. Only after noting our proximity did we continue for a look-see."

Leading the CAP was LT Austin Hawkins with RIO LT Charles Tinker, and they watched the explosions from the Alpha strike ten miles east of them. The "blue bandit" (MiG-21) call from Red Crown placed the enemy aircraft some twenty-five miles north-west of the CAP, near Kep airfield. Whether "Hawk" Hawkins and Curt should have headed off with such alacrity to this major VPAF base is open to dispute. Certainly, both crews were eager to get a MiG, particularly LT Hawkins, who was nearing the end of his tour.

On 7 May 1972 Curt had written home, saying:

"I take pen in hand again (brought to you live from the Alert 5 cockpit). We have a big one tomorrow—the biggest yet. A three-

carrier operation against Son Toi, a trans-shipment point 20 miles west of Hanoi. MiG action has been very heavy lately. We (Navy pilots off Kitty Hawk and Coral Sea) shot down two '21s and a '17 yesterday on strikes to Bai Thuong. And our target is centered between the major MiG airfields of Kep, Dong Suong, and Yen Bai, and is about 5 nm away from *the* major MiG base in NVN, Phuc Yen. The fact that they have given the Navy this target instead of the Air Force, who could more easily come in the back, leads me to believe that MiGs are really the primary objective and not the 500 trucks.

Hawkins and I should actually have the best opportunity for MiG encounters—to the degree that its almost a certainty!! We will be over the beach for about a 200 nm round trip, and the MiGs are sure to go for the A-7s and A-6s. I really feel ready for this one. We've had a lot of strikes up North now—some milk-runs and some super-hot, but this should be the real one. I feel well-trained, well-equipped, and absolutely ready. I would like to get at least one MiG myself, but I especially want VF-92 to get a couple."

In fact, the next day's victory went to the other squadron; the second for Cunningham and Driscoll, but on 10 May Curt was writing home to Jacksonville again:

"Wahoo!! Scratch one! Just a second between flights—I'm going back to Haiphong this afternoon—but I bagged one MiG and damaged another with Sidewinders this morning over Kep airfield while on TARCAP for a strike on Haiphong. It is VF-92's first MiG kill, and it was beautiful!

After the strike started exiting we took the TARCAP up to the north-west towards Kep after some MiG calls on 'guard.' Over the airfield at 5,000 ft we saw the MiG-17s in revetments on the end of the field, then two MiG-21s holding short of the northern end, and *then* two MiG-21s taking off in section in the opposite direction. We went burner and pulled around, coming down the runway at about 1,000 ft and 600 KIAS. We caught the MiGs about 3 miles off the runway at 500 ft, pulling up and left at about 500 kts. I fired a 'winder at the right one, which went off under him, and I fired

LT Randall H. Cunningham and LTJG William P. Driscoll. (National Archives)

158

Showtime 100 releases its six Mk 82s in unison with other *Fighting Falcons*. BuNo 155792 has its chaff door open, indicating that the strike group has come under SAM attack. BuNo 155800 already had 1815 hours on its airframe when it was lost after its triple MiG victory on 10 May 1972. (U.S. Navy)

VF-161 were noted for their svelte adaptation of the "black tail" pattern in 1972-73. This F-4B (BuNo 150996) also has the smaller 30 inch, rather than 45 inch national insignia on its wings. Assigned RIO Ken Crandall was wingman to LCDR Mugs McKeown when he shot down two MiG-17s on 23 May 1972. (R. Besecker collection via Norman Taylor)

again and the 'winder flew right up his tailpipe and he exploded. I then pulled to the left MiG and fired again. The missile *just* missed (about 10 degrees behind). My last 'winder did not come off the rail.

Hawkins fired three 'winders at the second MiG also while I was shooting, but they were also close misses. Two more MiG-21s jumped us, and we left the area (out of missiles) with Atolls going off around us. I will write again soon—got to go get another!"

It was thought at the time that their MiG had been flown by the legendary Colonel Ton (or Tomb), otherwise known as Nguyen Hong Nhi, with eight aerial victories, but other VPAF records indicate that the pilot was Nguyen Van Ngai of the 421st FR with Dang Ngoc Ngu (credited with seven U.S. aircraft) as leader. Some reports suggested the MiG was flown by a Russian PVO Colonel ("Top Gunsky") who was training the VPAF pilots on a new MiG variant. "Whatever, the pilot we shot down did not survive. The fireball slammed into a hillside." *Silverkite* 211 (BuNo 157269)

was supersonic as they dived towards Kep runway, and had to shed some speed in the turn after their quarry, dropping to below 100 ft as they tailed the MiGs, but still running at Mach 1.1. When Jim McDevitt, RIO in *Silverkite* 211, first noticed the MiGs on the runway he was helped by their gleaming silver finish. Curt noticed their, "pure red stars—Russian, not NVN markings. My best guess now [in 1998] is that they had just been delivered." The possibility that Russian instructors and advisors were still working at North Vietnam's airfields was one reason why Kep and other MiG bases had been off limits for so long. The MiGs were MiG-21MF (MiG-21 PFMA Fishbed J) models, with which the 921st FR had been re-equipped at the beginning of 1972. With a more powerful engine, doubled missile armament, and an effective radar, the "J" mode was capable of 702 kts at sea level, nearly 100 kts faster than earlier variants. Curt Dosé:

"We were the first ones to face the new MF with low-altitude supersonic capability (quite a surprise to us, almost a fatal one for

The F-4B assigned to "Devil" Houston in 1971, though not the aircraft flown on his MiG kill mission. (C. Moggeridge)

Allocated to LT Winston "Mad Dog" Copeland (spelled "Copelands" on the canopy) and LTJG Dale Arends, this sea-worn F-4B was flown by Lt Ken Cannon and LT Roy Morris on their 10 May 1972 MiG kill mission. The small red flag and the name "Ragin Cajun" on the splitter plate record Cannon's victory. (via C. Moggeridge)

Copeland and Arends were previously assigned to F-4B BuNo 150417, which has their names spelled correctly. (via C. Moggeridge)

Austin Hawkins and Jay Tinker). We certainly adjusted to the new observed speed of the MF. It became more important than ever to get rid of the F-4's centerline drop tank in a low-altitude fight. This was difficult, because it had to be dropped below a certain speed [560 kts, wings level and in 1g flight with the tank either full or totally empty to avoid aircraft damage and severe control problems, according to the NATOPs manual] that went by very fast when getting aggressive in a developing situation. You couldn't drop those 600 gallon tanks every flight—we would quickly have run out. If dropped faster than the limit it could take your horizontal tail off—not desirable over enemy territory. The MiG-21MF also re-emphasized the need for our coordinated 'loose deuce' tactics. Your wingman kept bandits from slipping up behind you—as I did for Hawk."

As he pursued Ngai's MiG-21, Curt's first missile, fired at extreme low level and around half a mile from the fleeing, jinking MiG's afterburner, struck the ground. His second shot went up the tailpipe of the sharply-turning fighter. Dang Ngoc Nhu managed to evade all three of Hawkins's AIM-9s by some hard maneuvering close to the ground; so close that Curt's last missile also hit a hill-top as it snaked towards its target. Unable to use Sparrows at that low altitude and lacking a gun, the two Navy pilots were forced to let the prey escape as two more MiG-21MFs closed on them. Although the Phantoms were scorching along at Mach 1.15 their crews were alarmed to notice that they were being overtaken. Curt called a hard turn as the MiG leader launched an Atoll at Hawkins' aircraft from less than a mile astern. Fortunately, the Atoll couldn't match the turn and the two *Silver Kings* were left to make a safe return to the boat. Their excursion to Kep had caused some anxiety, not least when they dropped below Red Crown's radar coverage for several minutes. However, the *Constellation's* official release to the media on 12 May firmly associated their success with the others from 10 May as having, "clearly buoyed the spirits of crew members all over the ship." Curt missed the second strike of the

When he assumed command of VF-31, Geno Lund's F-4J carried his MiG kill symbol from October 1967. The 1972 *Tomcatters* pose in front of his Phantom. (via E. P. Lund)

F-4J BuNo 155529 of VF-31 turns over a choppy sea. (USN via Angelo Romano)

F-4J BuNo 155841 on the point of launching. (Angelo Romano)

day, but his next letter home was composed in a rather different mood:

"Things took a bad turn to the great mood prevailing at the time of my [previous] letter. On the second strike, to a target west of Haiphong [Hai Duong Railyard], our XO [CDR Harry 'Habu' Blackburn] and his RIO [LT Steve 'Sar' Rudloff] took a direct hit by 85 mm over the target area, 15 nm inland. Their F-4J burst into flames, and they were forced to eject. His wingman (Rod Dilworth/ Jerry Hill) was also hit by the well-aimed barrage and lost their starboard engine immediately. They circled the area single-engined until the two good chutes reached the ground, and then made it out for a perfect single-engine landing back at the ship. There was no rescue effort possible for Habu and Sar due to their location."

Reflecting on his MiG kill the previous day in a calmer mood, Curt wrote:

"Routeslip and I really made our MiG kill. It may have been foolish, but at that point the MiGs were not coming to us and the only way to get them was to go after them. Well, we drove 100 miles through hostile territory, dodging SAMs and flak to get to Kep, and sure enough a wish came true (still don't know how we missed the second one). The way out was a bit worse, as the NVN were not very pleased with our little escapade. In other words, we really pissed them off!! If not a noble departure, it was at least a successful one at near Mach below 50 ft with MiG-21s behind us firing Atoll missiles."

Steve Rudloff was another Top Gun graduate who had found that he was, "able to adapt pretty well to the type of combat in S. E. Asia," partly because of that specific training.

"Dropping bombs, we had gone to Fallon and El Centro with the RAG and VF-154, so I was pretty well prepared, more as a semi-bombardier/navigator than as an actual RIO. I spent most of my time with my eyes out of the cockpit. When I was in VF-92 I was selected to go to Top Gun, and it was a fantastic experience—I really locked into it. The RIO's role was really enhanced by Top Gun doctrines. We had our role in terms of air-to-air combat much

more clearly defined, and a great deal more responsibility was cast on the RIO than we had when I first went through the RAG."

Steve almost had the chance to score a MiG in an earlier engagement while he was with VF-154. Having made contact with a MiG near Kep on 8 January 1968 and tracked it on approach, he had sought permission from the controlling USN destroyer to attack it.

"They never bothered to give it to us. Will Haff (the best pilot I ever flew with) went up to the destroyer that evening via helo to talk to the Admiral, and when he came back he said they never saw the aircraft so they couldn't give us clearance to fire. Even though it was a bad weather day and there were no U.S. aircraft doing much except weather patrols like us, the odds were that whoever I had locked up was definitely an enemy aircraft."

On 10 May 1972 Blackburn and Rudloff in *Silver Kite* 212 (BuNo 155797) were orbiting northwest of the target, and Steve was getting MiG contacts on his radar once again:

VF-102 *Diamondbacks* made their 1968 Westpac cruise with VF-33. Their only combat loss was their CO, CDR Wilber (POW), and his RIO, LTJG Rupinski (KIA), to a MiG-21, allegedly flown by Dinh Ton of the 921st Regiment, VPAF. Here, BuNo 155534 leaves the deck with six Mk82SE bombs and six missiles. (Norman Taylor collection)

"We were in a port turn with Rod Dilworth and his RIO Jerry Hill on our left wing. The strike force had already departed the target area. It was my recommendation to CDR Blackburn that we stay in the target area so that we could intercept any MiGs exiting or entering the area. Our mission that day had been TARCAP, so we thought we could leave the strike force because the flak suppression Phantoms could give them fighter cover. We were on our second orbit, and we had extended ourselves a little bit further out than the target area. We were aware of getting extensive flak; jinking to starboard I could see flak on that side. I was about to say 'Break port' to the XO when he reversed [the turn], and I saw a wall of flak on the port side, too. I knew we were going to get hit, they had us zeroed in. It looked to me like barrage fire. I didn't have anything on my EW equipment indicating we had been locked on by AAA. We had dumped quite a bit of chaff. I don't know if we were still dumping chaff at the time we were hit, or whether we had emptied out."

The flak barrage turned the F-4J into an inferno, and an electrical explosion in the cockpit temporarily blinded Stephen Rudloff as he called for an ejection. An extract from a tape transcript of the radio traffic during this incident (provided for this book by Matt Connelly) gives an idea of how rapidly events multiplied.

SILVER KITE 207 (Dilworth/Hill) I've shut down my starboard engine...I've taken a hit...I've got a fire warning light.

Ah, Roger. POUNCER 1 (A-7E Shrike carrier) has started back—enroute to the beach—say your position.

SILVER KITE 207 I'm on the 340 [degrees relative to USS *Chicago*] 57 miles off of 20.

My wingman was hit.

POUNCER 1 Understand your wingman was hit?

SILVER KITE 207 Rog - he was hit by flak and he went in...two good chutes, two good chutes.

HONEYBEE (CAG Gus Eggert) That's 212. Silver Kite 212 went in...two good chutes. OK, mark the position. The downed aircraft was directly over the target area.

JASON-X (LCDR Mike Gravely, VA-147 A-7E) OK, I just passed two Phantoms headed back inbound [actually MiG-21s]. Is that you with the hit, 207?

SILVER KITE 207 That's affirm.... OK, I'm coming back around.

Black Knight BuNo 158377 at Luke AFB in November 1975. (James E. Rotramel)

An impressive line-up of VF-154 F-4Js at Miramar. (Simon Watson)

SHOWTIME 106 (Matt Connelly/LT Tom Blonski, VF-96) Honeybee - Showtime 106, we'll be feet wet in about two minutes. There's a Phantom [SHOWTIME 100, Cunningham/Driscoll, VF-96] feet wet, burning...

Roger....

Get out! Get out! Get out!

Punch out! Punch out!!

Who's that for?

DEEP SEA (offshore destroyer) SAM, SAM, vicinity of Haiphong.

Punch out! Punch out!

Get out of that airplane!!

Who you talking to?

HONEYBEE (CAG) F-4 is going down in a spin off the beach [SHOWTIME 100]

- How far? What's the radial?

-OK, three- three-two-five, no DME (rescue beacons) about five miles off the beach. There's two good chutes [Cunningham/Driscoll]

In the midst of all this Steve Rudloff, still sightless, drifted down on his parachute towards captivity.

"I knew I was being shot at in the chute as I came down, as I could hear the impact of bullets on the parachute above me and whistling sounds, which I presumed were bullets coming through the air. Once I hit the ground and was about to be captured I could hear an aircraft approaching [Mike Gravely, VA-147 in an Iron Hand A-7E], and a lot of small-arms fire began as he got closer. I was lying on the ground face down. I had released my parachute. I tried to stand up, as I wanted to wave and let them know I was OK. When I tried to put weight on my right ankle, which I had landed on when I hit the ground, I collapsed back down, so I don't know if that pilot ever saw me stand up so he could report that I was alive. I do know that Harry [Blackburn] and I were declared 'presumed captured.' Harry, of course, was in the cell next to mine in Hai Duong being severely beaten that afternoon, and I believe that's when he died."

Curt Dosé's account of the day in his letter home continued:

"The second strike was memorable, not only for the XO's loss, but also because it ushered in a new phase of the war. For the first time the MiGs came in force to repel a strike. As the strike aircraft

VF-92 revised their color scheme post-war, as seen on this F-4J in January, 1975. (James E. Rotramel)

"...and make sure the coffee's ready." Last-minute conversations between LT John Kelly, LT Dewaine Cherry, and their plane captain. (David Daniels)

started exiting the area they were hit by at least fifteen MiG-17s and four MiG-21s. I've always dreaded that possibility, thinking we would lose several A-7s and A-6s, but they were fantastic. They were yelling like plucked chickens and swerving around a bit aimlessly, but kept the MiGs off long enough for the F-4s to help them.

The only fighter cover left was VF-96's TARCAP (the XO and Dilly [Dilworth] had been our TARCAP) and a section VF-96 had in as flak suppression. Randy Cunningham (familiar name?) was there and got three '17s, Matt Connelly got two MiG-21s, and Steve Shoemaker got a 'probable' [later confirmed] on a '17. It sounds like the biggest MiG engagement of the war (our own 'Turkey Shoot'). Randy is now the first Vietnam ace, along with RIO Willy Driscoll. All I need is one or two gaggles like that second strike, and I'll get the four more I need. Stay tuned!"

Brian "Bulldog" Grant didn't attend Top Gun until September 1972:

"Top Gun was a turning point for me as a fighter pilot and, although I had the confidence in my skill before the course, I know the cruise outcome would have been much different had I received the training prior to it. Top Gun provided me and others with what Cunningham knew instinctively—being able to focus the development of a multiple-bogey, multi-dimensional dogfight in slow motion, predict all the planes' flight patterns, and make your own aircraft coincide with a firing solution. Randy Cunningham was a very instinctive fighter pilot and a very brave man who additionally worked harder at his craft and more often than any other pilot that I have ever known."

May 10th provided plenty of opportunities for both pilots to demonstrate those skills. Cunningham and Driscoll were added to the afternoon flak suppression flight at the last moment, flying the CAG F-4J *Showtime* 100 (BuNo 155800). They launched with a heavy load of Rockeyes (in addition to their missiles) to knock out

F-4J 157305 was assigned to LCDR Gene Tucker and carried his MiG kill on the splitter, though F-4J BuNo 157299 was the aircraft used for the *Sluggers* MiG kill. (via C. Moggeridge)

CDR Tucker carried his MiG kill over to VF-74 when he assumed command of the squadron. (Angelo Romano)

VF-51's NF 101 overflies some famous Californian architecture. (via Jerry B. Houston)

anti-aircraft batteries. Not perhaps the most obvious start to an ace-creating MiG battle sortie, but certainly a tribute to the versatility of the F-4 and its flyers. Approaching their Hai Duong target area the flak suppressers could detect no obvious AAA sites, so Cunningham and Grant (in *Showtime* 110) bombed a warehouse complex in the main target area. Brian Grant: "As we rolled into our dives SAM and AAA opposition was intense. Close by our position VF-92 TARCAP lost one airplane [Cdr Blackburn]. As Randy and I pulled off target I called him to 'Break port, MiGs at six o'clock.'" Two MiG-17s overshot *Showtime* 110 after shooting at Cunningham, who broke towards the enemy. While Brian Grant kept the wingman at bay, Cunningham put a Sidewinder into the lead MiG as it ran out in front of them, and they scored their third kill. "Approaching the MiGs threatening Randy, my RIO reported that a MiG was firing guns at us." Meanwhile, Cunningham was "dragging" the second MiG in front of Grant to give him a shot, but Brian was too busy with the MiGs following his Phantom. "We disengaged, did defensive maneuvers, and evaded the MiGs to rejoin the fight in progress."

Below them at 10,000 ft was a defensive "wagon wheel" formation of about eight MiGs. This tactic was designed to lure a Phantom into the circle behind a MiG target, so that the following MiG could turn in towards the F-4 and shoot it. Cunningham observed at least two F-4s already in the circle and resolved to go down and engage. At that moment the *Fighting Falcons* XO, CDR Dwight Timm in *Showtime* 112, came under attack by three more MiGs, including one in his blind spot beneath the belly of his turning F-4J. With yet more MiGs approaching from above and others breaking out of the "wheel" and climbing to engage, Cunningham shouted to his XO, "Alright, Showtime, reverse starboard, God dammit!" Wary of the two MiGs he could see, and unaware of the one hiding beneath him, Timm eventually reversed his turn as Cunningham, concerned that his AIM-9 might home on Timm's' afterburner, transmitted, "I don't know how to shoot at this f****** airplane!" "I don't either," replied Brian Grant, "but you got him!"

Throughout this grueling pursuit, RIO Willy Driscoll watched a quartet of MiG-19s which were attacking them from above, calling to his pilot to "Break" each time a MiG got into a firing position. As CDR Timm finally broke away, Cunningham called "Fox two" (for a Sidewinder launch), and his missile blew the "hidden" MiG-17 apart—their fourth victory. Timm extended away safely from his other pursuers.

Sundowners **F-4N BuNo 152293 with full tail art restored, and enhanced for their bicentennial cruise on USS *F. D. Roosevelt*. (Norman Taylor)**

F-4J BuNo 153862 from VF-151 on approach. (Simon Watson)

In another part of the crowded sky Brian Grant came tantalizingly close to his own MiG kill:

"I found myself approaching a simultaneous firing position with Steve Shoemaker (later credited with a kill) and broke off that engagement, fearing a mid-air. I eventually found myself in an extremely nose-high firing position on a MiG-17, but the missile did not appear to guide, possibly due to too high an angle or 'sun capture' [homing onto the sun as a heat source]. At this time gas was low, people had started to bug out, and I followed suit, hitting the tanker with only 500 lbs fuel remaining. The closest I came to Randy after our target pull-off was simultaneously crossing the beach as he ejected."

It was fairly inevitable that flight-leaders and wingmen would become separated in such a melée, particularly with such an unusually large VPAF presence. Having lost sight of Brian Grant during his third encounter, Cunningham and Driscoll came under attack by several more MiG-17s, which got his F-4 in gun range, and a quartet of MiG-21s. Cunningham maneuvered desperately to evade them, finally turning in towards them and then reversing back out towards the coast, climbing steadily. En route, alone, he saw another solo MiG-17, and they met head-on, with the MiG pilot's guns pumping shells. Cunningham fell back on a basic tenet of Top Gun teaching—fight in the vertical—and zoom-climbed his Phantom. To his surprise the MiG pilot did the same. MiG-17s usually flew horizontal, turning fights, but Showtim*e* 100's adversary was clearly a skilled practitioner, and he fired again as the Phantom's climb peaked out just above the stubby MiG. After an inconclusive "rolling scissors" and two more vertical climbs it was clear to Cunningham that something drastic was needed to break the deadlock.

VF-143's distinctive *Pukin' Dog* insignia on F-4J BuNo 155739, straining at the leash in March 1974. The squadron made seven war cruises between 1964 and 1973 before joining CVW-8 in the Atlantic Fleet at the beginning of 1974. (Angelo Romano)

During the third vertical maneuver he suddenly chopped the throttles to idle and popped his speed brakes, forcing the MiG out in front of him. As the two aircraft rolled out at the top of their climb Cunningham was able to push in behind the VPAF fighter and heavily damage it with an AIM-9 so that it flew into the ground. The identity of the pilot remains in doubt. For many years he was the strongest contender for the mythical Colonel Toon (or Toomb), but no such name appears in Vietnamese records. Other evidence points to Nguyen Van Bay, who had claimed seven kills and a "damaged" B-52, though the Commander of the 923rd Fighter Regiment, whose identity remains obscure, is another possibility. Whoever he was, his skill brought him close to preventing Cunningham and Driscoll attaining "ace" status (the only Navy crew to do so), and it was not the last MiG that *Showtime* 100 met that day.

Tom Blonski and Matt Connelly in *Showtime* 106 (BuNo 155769) came close to matching their squadron mates' score that day, shooting down two MiG-17s. Placed on TARCAP at 18,000 ft over Hai Duong, Connelly sighted an A-7E being pursued by two MiG-17s and dove after it, setting up a Sparrow launch as he did so. However, this could not be achieved: "During my initial roll-in, after sighting the two MiGs, both our radar scopes in the aircraft went black. We had lost all video." Matt switched to "heat" and fired a Sidewinder, at the edge of its performance envelope—or a little beyond. The MiG-17 turned back into its attackers, almost hitting them. "My wingman, Aaron Campbell, had a near-miss with the MiG we initially attacked." Campbell, in *Showtime* 113, then departed the battle.

Seeing a number of MiGs circling above him, apparently holding for intercept instructions from their GCI, Connelly climbed after another MiG-17, which entered a tight defensive turn and then, unusually, reversed its turn, allowing another AIM-9 from *Showtime* 106 to acquire it at 1,000 ft and blow it apart. On his combat tape, just as Cunningham was trying to make CDR Timm turn and expose his MiG pursuer, Matt Connelly's voice announced laconically, "OK. Splash one." As Cunningham finally "smoked" Timm's shadower, Matt got his second of the day, also. "The second MiG kill was a random aircraft simply flying by. I turned in pursuit. He saw me and attempted to break with me, but lost sight after two reversals. He rolled wings level, and I fired a Sidewinder to a near-

VMFA-333's base code DN appears on BuNo 155526 in 1970. This aircraft remained with the squadron and was used by "Bear" Lasseter and "Lil' John" Cummings for their September 1972 MiG kill. (Duane Kasulka via Norman Taylor)

miss. The missile exploded on proximity [fuze] and cut off the whole tail assembly of the MiG-17 with its expanding-rod warhead. As I flew past the unflyable MiG the pilot ejected, allowing me to watch the full ejection sequence. I was surprised to see their parachute canopies were square in shape." The unfortunate Vietnamese pilot was nearly "run down" by *Showtime* 100 as LT Cunningham headed out for the beach.

Meanwhile, *Showtime* 111 (BuNo 155749), with LT Steven Shoemaker and LTJG Keith "Cannonball" (a reference to his robust stature) Crenshaw aboard, had made its third run through the dogfight, having used a ballistic Sidewinder shot on his second pass to scare one of the MiG-17s which were harassing LT Dave Erickson (*Showtime* 102), Timm's' wingman. Although the fight so far had consisted of a series of turning or vertical engagements, often with multiple missile shots, Curt Dosé pointed out that, "USN tactics were always for single missile launches, then fly to position for the next shot or next target. That's the way we trained and fought." On the vertical tactic which Cunningham and Driscoll had used to their advantage, Curt concurred: "We emphasized the vertical as an advantage for the F-4, although you had to be careful not to bury the nose. The F-4 took 8,000 ft to do a 'split S,' not a foot less, as learned by many unfortunate F-4 pilots, including the *Blue Angels* and *Thunderbirds* diamonds."

Steve Shoemaker, a former *Blue Angel* himself, made another attack run towards the fight and sighted another MiG-17, but it was following an F-4 (*Showtime* 100) in its blind spot as it climbed out of the bearpit. LT Shoemaker radioed a warning and headed after the MiG, which saw him and drew him into a descending, turning fight. As the MiG gradually gained a turn advantage on him, Shoemaker released an AIM-9 and broke away to ensure he wasn't being followed. Returning to their previous course at low altitude, LTJG Crenshaw noticed a big explosion on the ground, later confirmed as "their" MiG.

Showtime 100 was still not out of trouble. Yet another MiG headed for it, and this time it was Matt Connelly who spotted the bogey, just as he and Tom Blonski were pulling out of a situation which could have provided their third MiG. Matt explained:

A VF-102 night-launch. (USN via Angelo Romano)

Shamrock 202 on patrol with a full warload of eight missiles during the 1972 USS *America* cruise, the first by a USMC F-4 squadron. (via MGEN Paul Fratarangelo)

VMFA-333's *Trip Trey* shamrock design is repeated on the centerline tank of 214 A.J. (via MGEN Paul Fratarangelo)

"My third engagement was a MiG making an attack from my right, 3 o'clock. I turned into the MiG, forcing him to overshoot and pass directly under my tail. At this point I pulled the nose up and rolled over the top of the MiG, which was now on my left at 9 o'clock. So I was in a horizontal engagement, at about 140 kts with half flaps and full afterburner. Way too slow in an F-4. The aircraft promptly departed flight and went into a post-stall gyration. Both Tom Blonski, my RIO, and I lost sight of the MiG during the gyration, and we never saw him again. After my second kill I had expended all my available Sidewinders, and with no video for the radar had no close engagement capability."

There was still a use for Matt's Sparrow armament, even so. When he saw the MiG following Cunningham's Phantom he shouted a warning to keep *Showtime* 100 out of gun range, but he felt that something more tangible was needed. "The MiG definitely had a firing solution [on Cunningham]. Randy was wings level, in basic engine, climbing out of the engagement airspace. I feel this MiG did enough damage to his F-4 to either cripple its ECM capabilities or actually cause it later to become unflyable. I distracted the MiG from Randy's tail with a Sparrow missile, which I fired in boresight mode." Without radar guidance there was little chance of the AIM-7 hitting either the MiG or *Showtime* 100. As Cunningham went into afterburner to escape the MiG banked away, leaving its F-4 target to face its final challenge. Although Cunningham didn't get any SAM alert on his own ECM, possibly due to damage, as Matt conjectured, he heard SAM warnings on his radio, and both crew members stated that they saw a SAM explode about 500 ft from their aircraft. Matt Connelly: "Randy claims he was shot down by an optically guided SAM. I was not aware of any optically guided SAM sites in North Vietnam. It is very possible that a standard SA-2 was fired, and the damage to Randy's ECM meant he did not pick it up. I turned in pursuit of the third MiG, off Randy, but lost sight and turned again to resume my exit from the North. I was behind Randy as we headed 'feet wet.' I saw no SAMs fired."

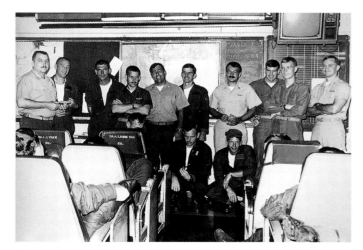

Shamrocks aircrew, all with over 200 landings on USS *America*. Major Lasseter (CO) is on the left. Paul Fratarangelo is fifth from left, and John Cummings is kneeling (right). (via MGEN Paul Fratarangelo)

F-4J BuNo 153904, still wearing DN codes, doing carquals. (via MGEN Paul Fratarangelo)

"Bear" Lasseter and John Cummings make a supersonic pass at low level abeam USS *America*. The shock wave effect on the sea's surface is visible. (via MGEN Paul Fratarangelo)

At any rate, *Showtime* 100 had received mortal damage to its hydraulics, and both PC-1 and PC-2 systems went down to zero. Remembering, like Duke Hernandez in 1968, that the stabilator would swing to the full "nose up" position as the fluid pressure faded, but that control was still possible using rudder and throttles, Cunningham rolled his F-4 using speed brakes and rudder to force the nose down and prevent a stall. He was able to repeat the maneuver several times, buying a few extra miles to take them over the beach and away from captivity. A major fire, fed by hydraulic fluid, had spread through the fuselage and threatened to engulf the rear cockpit, but the crew held out as long as they could before ejecting. As the blazing wreck entered a terminal spin they finally got out and were rescued from the sea, protected by other CVW-9 aircraft as North Vietnamese gunboats began to move out towards them.

The eighth MiG shot down that day fell to *Coral Sea's* VF-51, when LT Ken L. "Ragin Cajun" Cannon and LT Roy "Bud" Morris in F-4B *Screaming Eagle* 111 (BuNo 151398) hit a MiG-17 with their AIM-9 about twenty miles west of the main Hai Duong air battle. The fight was basically a one-on-one. When Ken Cannon described the combat to Jerry Houston the latter believed that Cannon had, "shot down 'Colonel Toomb' rather than Randy Cunningham doing it. Whoever the enemy pilot was he proved a damn tough and experienced opponent. I was proud as hell of Ken and the job he and Bud Morris had done." They were on a MiGCAP with Chuck Schroeder and Dale Arends. Entering over Haiphong at 200 ft they saw a MiG-17 ahead of them in a turn which took it behind Schroeder's Phantom, but into Cannon's field of fire.

It was a day of outstanding success for the Phantom squadrons, a vindication of the Top Gun initiative and a demonstration of the accuracy and reliability of the AIM-9 missile, which was responsible for all of the kills. In May 1972 the Navy claimed sixteen MiGs, and the USAF another eleven. The VPAF claimed seventeen U.S. fighters in return (including six F-4s on 10 May), but U.S. records include only six losses, all from the USAF. It was clear that when the MiGs appeared in substantial numbers the Phantoms now had a distinct advantage. It was several days before they appeared in force again. On that occasion they were met by Phantoms from

another carrier, the USS *Midway*, which had returned to Yankee station on 12 May with VF-151 *Vigilantes* and VF-161 *Chargers*.

On 18 May Top Gun graduate LT Henry "Bart" Bartholomay with LT Oran Brown (*Rock River* 110), and wingman LT Pat Arwood with RIO LT Mike "Taco" Bell (*Rock River* 105) were the MiGCAP for a strike in the Haiphong area. "Bart" Bartholomay: "Our CAP station was at the western edge of the ridge [north of Haiphong] at 15-20,000 ft. It was also about twenty miles east-southeast of Kep airfield." It was Taco's first excursion over North Vietnam (on his previous two cruises he had been confined to Laos and the south), while Pat Arwood had only about 150 hours on F-4s. As the two F-4s arrived on station Taco was checking his radar, hoping to pick up something coming from Kep, as Bart had received a bogey call from Red Crown. Seeing nothing, he stowed the radar console away and resorted to eyeball surveillance. Bart takes up the narrative:

"We had made one race-track pattern over our CAP station in combat spread, three quarters abeam one another with my aircraft on the left. As I looked 'through' Pat's plane to check his below, above and six I saw what appeared to be two aluminum roof tops at about 1,500 ft AGL, seven miles away. As I looked again they had moved, so I made the decision to investigate. As we got closer we could see that they were bandits, but it wasn't until we engaged that

Lil' John (left) and Bear with two MiG kills marked up on the replacement 201 AJ (BuNo 155852). Unofficially, they were allowed 1.5 kills, but officially just one. (via MGEN Paul Fratarangelo)

we identified them as MiG-19s. They were heading into the break at Kep, having come from China, as we later found out. [The VPAF received Chinese Shenyang F-6s—MiG-19 copies—as well as a small number of Russian MiG-19S and MiG-19PM fighters.] Their heading at the time was about 270 degrees. I positioned Pat 3,000 ft above and slightly behind my aircraft. My intention was to engage both MiGs as quickly as possible without bringing both our aircraft down to near ground level. They then turned port to a runway heading of about 190 degrees. By the time they had passed the south end of the airfield I was following behind at 500 ft altitude doing about 600 kts. As I passed over the north end of the runway they broke into me, and the fight was on. My hope was that the lead would break off such that Pat could fall in behind, which he did. The trail MiG and I were engaged."

Taco Bell watched the two MiGs jettison their wing tanks and go into a tight 180 degree turn while Pat Arwood tried to get a Sidewinder tone on the leading fighter. Failing to do so, he launched a ballistic AIM-9 which forced the MiG pilot to break sharply left towards his F-4B in order to avoid the missile, while the second MiG-19 turned away to starboard. Following him, Bart's aircraft lost energy in the turn, so he headed out west of Kep to build up some speed. Arwood and Bell entered a turning fight over Kep airfield with their MiG, which took them on a couple of circuits of the base before the MiG went wings level and left the fight. Rather than following it Pat elected to stay and re-locate Bart, who was on his way in again.

"As I extended west to regain energy I lost sight of my MiG. When I had about 500+ kts I pitched up and reversed back to the east. Oran still had our MiG in sight, and as we got closer I picked him up at about 10 o'clock, slightly low. He was heading for me, and rather than pulling into him to meet head-on, I kept my airspeed up and let him make progress to a point at which I thought I could get him to overshoot. As he closed at 8.30 level I pulled hard into him, pulling up and outside my turn into an outside barrel roll. As he started to overshoot I popped my speed-brakes and pulled power [back]. As soon as we were abeam each other I went to full burner and closed the speed brakes, pulling into him as we both climbed into the vertical. We were no farther apart than a couple of

VMFAT-201 shared USMC F-4 training duties with VMFAT-101 from 1967 at Cherry Point. BuNo 153860 is caught here on 17 May 1971 with empty TERs and spent practice bomb dispenser on the centerline. (R. Besecker via Norman Taylor)

hundred feet, bleeding airspeed like it was going out of style. I could see his black helmet and face as he looked over to check out where I was. But we were getting so slow, almost stalling, that I said to Oran, 'We've got to get out of here!' I knew that we weren't going to get a shot, and would be lucky just to get out with our lives. But MiGs don't fly well low and slow, either, and he had the same idea. As luck would have it, I outlasted him by just a few seconds and was able to fall into a tight 200 ft trail as we both dropped our noses and headed for the deck to gain airspeed. I was too far inside the Sidewinder envelope for any kind of a shot, so I just kept lagging out and pulling back in to gain distance. I would have had the opportunity to use guns four times in our engagement [if the F-4B had been so equipped]. By the time I'd lagged back to 600 ft I felt it was time to give her a try, so I pulled back into the MiG but couldn't hear a Sidewinder tone."

Pat Arwood's MiG-19 had turned back into the scrap, apparently heading for Bart's *Rock River* 110, and Taco surmised that Bart's MiG was trying to drag Bart out so that "theirs" could get behind him. However, the MiG pilot failed to see Arwood and Bell curving in behind him. They got a positive Sidewinder tone, their missile exploded close to the MiG's tail, and it slowly disintegrated as the pilot ejected. Bart recalled:

Shamrock 207 (BuNo 153877) ready for the cat shot. Note the Micky Mouse on the VF-74 aircraft behind. The two F-4 squadrons shared fighter duties on USS *America* in 1972. (via MGEN Paul Fratarangelo)

"I knew Pat had just shot his MiG down, and I knew that we were desperately low on fuel, so I made a decision to try one more time. This time I still couldn't hear an AIM-9 tone, but went ahead and fired. Within a second or two the MiG went into a hard starboard roll, the canopy coming off parallel to the ground and the pilot ejecting momentarily thereafter. The plane went in a second or two later. To this day I don't know whether he was hit by my missile or I just flew him into the ground."

Surprised that they had not encountered SAMs, AAA, or other MiGs, the two *Chargers* headed for the tanker and home, unscathed, though they were told that, "Two MiGs were launched out of Hanoi and actually did get within ten miles of us as we egressed." *Rock River* 105 (BuNo 153915, the last F-4B ordered) later went on display at the Naval Aviation Museum. Once again, careful training and Top Gun tactics had prevailed in a difficult situation. As Bart observed, "Our [VF-161] ACM training was a direct result of Top Gun's influence on re-defining fighter tactics and our squadron's insistence on spending more hours training its aircrews in these tactics than other squadrons did." He reinforced the primacy of the AIM-9 in fights like the one on 23 May: "In almost no situation would I have used a Sparrow in a close-in fight such as this one. We used it for intercepts only. If we didn't have Sidewinders left, we might have used a Sparrow, given a good lock." This view of the AIM-7 in the Vietnam scenario dated back to the earlier Phantom deployments there. Ken Baldry, referring to his in 1966-67, commented, "Nobody that I know of realistically ever expected to use the Sparrow against anything but a PT boat. This led to the concept of lag pursuit, i.e. trying to get the nose behind the opponent and gain space for the Sidewinder to arm."

On using the AIM-9, Bart commented that, "The MiG-17, MiG-19, and -21 all had approximately the same wingspan (within a few feet). Our gunsight had two circles which, although not specifically designed for it, helped the pilot judge the distance to a MiG. If the wingspan extended across the outer circle you were approximately 600 ft behind (the minimum range of the AIM-9D at the time), and if it extended across the inner circle you were approximately two miles behind (approaching the maximum range of the missile)."

Mugs' MiGs

The emphasis on ACM skills in CDR Wayne Connell's VF-161 was naturally encouraged by one of its members in particular. LCDR Ronald "Mugs" McKeown, VX-4's Project Officer for air-to-air and Sparrow tactics, was a founder member of the *Have Drill* project, a former VF-154 F-8 pilot with two combat tours, and a key figure in Top Gun. He had returned for another combat tour to add to a total of combat missions which would eventually reach four hundred. During the 1972 205-day deployment his squadron flew 2,322 sorties and downed five MiGs. Two of them fell to McKeown's *Rock River* 100 on 23 May. Flying an Alpha MiGCAP with LT Mike Rabb and LTJG Ken Crandall on their wing, he and RIO LT Jack Ensch followed the same ingress route that Bart and Pat Arwood had taken to a point near Kep. They received a "bandit" call (meaning a confirmed MiG, not an unidentified "bogie," suggesting that the MiGs' transponders had been interrogated by U.S. ELINT units). As Crandall's APQ-72 had failed, Jack Ensch conducted a radar search, but the two pilots soon found themselves over Kep at about 3,500 ft with two 925th FR MiG-19S fighters. The engagement seemed to be a repeat of Bart and Pat Arwood's fight, as the two F-4Bs turned towards their targets, but they saw about six MiG-17s above them in ambush. The Phantoms entered a frantic "furball" dogfight, narrowly avoiding several mid-airs. Mugs evaded a MiG-17 on his tail by putting his F-4 into a "controlled departure" during a sudden reversal of a tight turn and came out of it behind another MiG-17. He released two AIM-9s at it in a turn which was too tight for the missiles. As another MiG-17 closed on his tail, shooting, McKeown forced it to overshoot by popping his speed brakes and putting negative g on the F-4. Firing his third Sidewinder at it, LCDR McKeown watched the MiG pilot apparently turn to defeat the missile, but then reverse back into its path. His fighter, sans tail, crashed as he ejected. LT Rabb's F-4B also had a MiG-17 behind it and shooting. McKeown called for him to extend away, using his Phantom's superior acceleration, and the MiG turned away, in front of *Rock River* 100. Firing his fourth AIM-9 (which he had previously assumed to be "dead"), Mugs chalked up his second MiG-17, having given the pilots at Kep a master class in how to defeat more nimble opponents who outnumbered the Americans eight-to-one.

Assigned to "Jade" and "Hombre," this F-4J (BuNo 153859) is representative of the many used by VMFA-232 *Red Devils* on its deployments to Chu Lai in 1969 and Nam Phong in 1972-73. (Norman Taylor)

VMFA-451's No 1 F-4J had the nickname *"Fang"* on its splitter plate, and its RIO has been doing some radar training under the "blackout" hood. The *Warlords* flew F-4s from 1978 to 1988 at Beaufort. (J. T. Thompson)

A *Falcons* F-4J (BuNo 155735) rolls out for take-off. (via C. Moggeridge)

Navy MiG-kills continued at a reduced pace up to the end of 1972. "Tooter" Teague got his on 11 June. His aircraft were always named *The Bossier Bearcat*, since Teague had attended Bossier High School in 1952 before moving, like several other VF-51 veterans, including "Devil" Houston, to Texas A&M. VF-51 was one of the five commands (including command of the USS *Kitty Hawk*) in his long career. His 11 June mission, on the last day of a line period for *Coral Sea*, was a MiGCAP with RIO Ralph M. Howell. LT Winston "Mad Dog" Copeland and LT Don Bouchoux were wingmen. They encountered MiG-17s, and each Phantom scored one using AIM-9Gs. Both F-4s had unserviceable radars, and Copeland had no radio. Vectored by USS *Long Beach* to a point near Phu Lai they dropped to 50 ft AGL to avoid SAMs and identified the four MiG-17s, which turned to engage them. Copeland's victim was the Number 2 MiG, which was trying to get onto Teague's six o'clock. Tooter continued to follow another MiG, demolishing it with his second Sidewinder. He got a good tone on the leader, which broke hard right just as he released another AIM-9. The other two MiGs hightailed it for home, leaving the Navy aviators to do the same, though Copeland's F-4B, *Screaming Eagle* 113, took a flak hit in the left engine as he egressed at extreme low altitude. He shut the wounded J79 down when it caught fire and managed a single-engined landing on *Coral Sea*. The ship's MiG score had risen to six. Ten days later an Atlantic squadron, VF-31, notched up their sole air-to-air victory.

When USS *Saratoga* left Mayport on 11 April 1972, MiG killer Geno Lund had taken command over of VF-31 from Gene Geronime, and he described this period as the highlight of his career.

"The squadron's departure from NAS Oceana was most abrupt. We were all set to go off on a routine Mediterranean cruise when I got a phone call at home on a Saturday night from the Wing Commander asking me how many planes I could have ready to fly the coming Monday. I thought this was one of those periodic drills we had all the time, but I was quickly informed that this was no drill and that my answer had better be, 'All twelve of them.' Of course, it was, and we did just that. One plane was in 'check,' which means it was just about completely in parts in the hangar, having everything checked out as is routinely done. We even got that one up and ready by Monday, and we flew them all down to Mayport, where the *Saratoga* awaited us, by the middle of the week. We were

pumped up with new flight crews and maintenance people, and off we went to war. The only preparation was carried out in the thirty day transit from Mayport to S. E. Asia. My XO, Jim Flatley III, and I were about the only ones with combat experience, so we had a very intense ground training program to bring everyone up to speed on the situation in Vietnam. Jim's prospective XO, Sam Flynn, was a combat veteran, also. He later shot down a MiG-21 flying with my RIO, Bill John. We had a Top Gun trained crew already on board. Randy Leddy and his back-seater had taken one of our F-4Js out to Miramar earlier in the cycle.

Jim Flatley's orders were to assume command in May 1972, so I only got to spend the first line period down south in 1 Corps [Dixie station] before I had to leave the squadron to him [26 May 1972]. We had the MiG-17 [from his October 1967 kill] painted on F-4J AC 101 all the time when I was in the squadron, and on AC 102 when I was XO."

As soon as J. H. Flatley had taken over as "Top Cat," VF-31 prepared to conduct Alpha strikes. On 21 June CDR Sam Flynn and LT Bill John were MiGCAP with LTJG Nick Strelchek and LT Dave Arnold. They dodged four SAMs as they approached the target and immediately had to contend with a pair of MiG-21s diving towards them. CDR Flynn (who had for some time advocated a greater training emphasis on ACM) managed to get into an AIM-7 firing position on one of the aircraft, and LT John locked it up, only to have the Sparrow fail to leave its well. He then noticed that his wingman's F-4J was being fired on with an Atoll by another MiG-21 and went to the rescue, firing three AIM-9Ds before getting a conclusive hit with the third. In his Command History for the cruise CDR Flatley noted that it was a "first for Saratoga and for Air Wing Three." It was also a point at which the Navy could rack up a 21-1 success rate in aerial combat against the MiGs. Some confusion arose later over the identity of CDR Flynn's aircraft, but he confirmed that he and Bill John flew BuNo 157307 AC106 on the mission, an aircraft normally assigned to USAF exchange pilot Captain Ron Williams. Their MiG kill marking appeared on their own assigned F-4J, *Bandwagon* 102 (BuNo 157280).

The *Tomcatters'* sister squadron for the cruise was VF-103 *Sluggers*. On the evening of 10 August 1972 LCDR Gene Tucker and LTJG Bruce Edens found themselves on Alert 5 and scrambled after an intermittent contact on a MiG in the area of Vinh. Sharing the search with a *Tomcatter* Alert F-4J who headed south, LCDR Tucker got another vector onto the MiG which was heading back north. He dropped his centerline tank and some empty TERs to gain speed, and got to within two miles astern the MiG-21 at 3,500 ft before launching a good AIM-7, followed by a second. The MiG exploded in a brilliant fireball, the only one to fall to a Navy F-4 at night. Tucker and Eden's usual F-4J, *Clubleaf* 203 (BuNo 157305), bore their kill marking, won in 157299 *Clubleaf* 206.

As *Linebacker II* operations reached a climax with the huge B-52 onslaught of Christmas 1972, MiG activity gradually diminished as their airfields were pounded by U.S. aircraft. VF-142 returned to the line for the final stages, and LTJG Scott Davis with RIO LTJG Jeff Ulrich destroyed a MiG-21 with an AIM-9 on 28 December during a MiGCAP sortie. LCDR Tucker's kill was the only successful Navy AIM-7 attack since 9 May 1968, when USAF exchange pilot Major John P. Hefferman and LTJG Frank Schumacher of VF-96 scored a MiG-21 in F-4B BuNo 153036. Paradoxically, two naval aviators on an exchange with the USAF's 58th TFS did succeed with a Sparrow from an F-4E Phantom. Captain Larry Richard (USMC) and RIO LCDR Mike Ettel (USN) were flying a 432nd TRW weather recce on 12 August 1972 when they were intercepted by a MiG-21 and destroyed it. Another Marine, Captain Doyle Baker, shared a MiG-17 with 1LT John Ryan, USAF, when their 13th TFS F-4D attacked the aircraft with the less-than-reliable AIM-4D Falcon missile. The final USN kill of the war, on 12 January 1973, went to the *Rock Rivers* once again (their sixth) with a Sidewinder. LT Vic Kovaleski and LTJG James Wise took advantage of a relaxation in the rules of engagement, which enabled them to chase MiGs beyond the 20th parallel during a BARCAP. An AIM-9 from their VF-161 F-4B (BuNo 153045, 102 NF) sent a MiG-17 tumbling down. However, the engagement was not without problems. Squadron-mate Bart Bartholomay recalled that it took, "three or four passes for the controller to get him hooked up on the MiG." Kovaleski was shot down two days later with his RIO, ENS D. H. Plautz, on a *Blue Tree*, but both were recovered. Theirs was the last USN combat loss over the North. There was a brief return to action in May 1975 when CVW-15 was called to the Gulf of Siam following the seizure by the Khmer Rouge of the container ship SS *Mayaguez*. VF-51 and VF-111 used their F-4N Phantoms to cover two strikes by A-7Es and A-6As on the Cambodian mainland, hitting Ream airfield and an oil depot. After its last F-4 deployment

CVW-15 returned to San Diego in July 1975, and its fighter squadrons began transition to the F-14A Tomcat.

While digging out some squadron memorabilia for this book, Curt Dosé came across his VF-92 songbook. "We were very impressive in the bar when not on the line, and were renowned for our enthusiastic singing. A psychologist would have a field-day with these songs, I suspect, but it was our way of coping with stress and death." These extracts come from, "one of VF-92's best."

> While flying 'round Quan Lang, your husband's plane went bang,
> And started falling, little pieces through the sky.
> It might have been a SAM. Who really gives a damn?
> He almost got his MiG before he had to die.
> Fly Navy - and be a hero!
>
> Dad entered to the break; he didn't hesitate.
> The airplane snapped up, hauled 'round, pulling the g.
> But just one problem there; your Pop ran out of air,
> And flew his F-4J into the cold gray sea.
> Fly Navy - impress your friends!
>
> It was a black-ass night. There was no moon in sight
> And came your brother's turn to land aboard the ship.
> He got a little low. There was a bright red glow.
> Now all that's left of him is a dent upon the ramp.
> Fly Navy - but not at night!
>
> While rolling down the cat, his flaps blew up to half.
> But your son said he could hack it anyhow.
> The RIO gave a shout and then he punched right out
> And sonny stalled the jet and dived into the sea.
> Fly Navy - and save a plane!

Marines at Sea

Apart from their two "exchange" MiG kills, the nature of USMC F-4 operations seldom took them into the hottest MiG territory. An exception was the Westpac cruise by VMFA-333 *Shamrocks* on USS *America*, with VF-74 as sister squadron. Their cruise book for 1972-73 was sub-titled "Med Cruise 72?." MGEN Paul Fratarangelo, who flew with the squadron in 1971-73, explained:

Falcons **fixers work on an F-4J's fuel system. (via James Sherman)**

A division of *Falcons* **hot-refueling. (via James Sherman)**

"VMFA-333 was the first Marine F-4 squadron to embark aboard a Navy carrier. Up to that time the Marines were committed completely to supporting USMC ground combat ops from Da Nang and Chu Lai. the Navy was skeptical of our ability to field a carrier-capable F-4 squadron. However, we enjoyed the support of LGEN George Axtell, CGFMFLant, CG 2nd MAW and CO, MAG-31. *Trip Trey* [VMFA-333] fielded some of the most experienced pilots, RIOs, and F-4 mechanics and radar technicians in the aircraft Group, and we enjoyed enormously successful Med and Westpac deployments. We made our first cruise to the Mediterranean and would have made our second [1972-73] there also, but President Nixon was determined to bring about closure of the S. E. Asia conflict, and the USS *America* was "chopped" from the Sixth Fleet to the Seventh. We then saw that there were, in fact, two different U.S. Navies with regard to CVW tactics and operating procedures.

One sea story I will never forget was the night I took a no-flaps F-4 cat shot. We were flying back aboard after an in-port period, and Air Wing pilots were getting their carrier qualifications 're-freshed.' I had completed my day of refresher landings and was getting two to four night traps to regain my night CQ currency. Since we were flying back aboard for an at-sea period my cockpit was stuffed with my dop kit, clothes, etc., and in the back seat was the squadron Flight Surgeon. The *America* usually shot all F-4s off the waist cats, but because of refresher ops I was taxied to the bow catapults. I went into the 'hold-back' a little hard, and the cat officer signaled for a push-back. I raised the flaps for the push-back, and everyone (me and the deck-crew) forgot to lower them prior to the cat shot. Because we were at 'carqual' weight, I took the shot at military power [no afterburner]. As soon as the cat released me, the nose of my aircraft over-rotated to approximately 60 degrees nose-up. My regular RIO would have ejected us, but good ol' Doc was just along for the ride. I jammed the throttles to full afterburner, and the wonderful J79 turbojets responded immediately. I ruddered my F-4 out of wing-rock, raised the gear, and when I found the flap handle 'up' I finally realized what had happened. Most Navy guys don't believe this story, saying no-one could prevent an F-4 from going into the water following a no-flap cat shot."

Among the *Shamrocks* were Captain John D. Cummings and Major Lee T. "Bear" Lasseter. Lil' John Cummings had already made quite an impression on Marine aviation since his time at Da Nang in 1965. In 1971 he received the USMC Naval Flight Officer of the Year Award for developing, "Aircrew coordination techniques which allow the [F-4] RIO to virtually fly the aircraft through directions to the pilot, when the pilot is unable to visually acquire the target," and "Tactics which allow fighter aircraft to approach enemy aircraft undetected with all weapons simultaneously brought to bear." In addition, his published articles on fighter tactics were considered, "The 2nd MAW's basis for the tactical deployment of the F-4 on long-range escort, night escort, BARCAP, and TARCAP missions." John Cummings modestly acknowledged that, "Bear and I were instructors at MAWTU Lant in the 1969-70 timeframe and helped to give a minimum of preparation to aircrews headed for Vietnam. There wasn't a lot of time to give them much more, so we did what we could. After the war the powers-that-be allowed the creation of MAWTS-1, which has helped somewhat." MGEN Fratarangelo made clear that Bear and Lil' John were vital figures in the establishment of this organization, also. Their impact on *Trip Trey* squadron operations was considerable, too: "The F-4J ushered in the AWG-10 radar, which was very difficult to maintain. VMFA-333, due to Lil' John Cummings' efforts and dedicated factory rep support, had the most reliable radar (pulse and pulse-Doppler), ECM, and D/L that I experienced in Marine F-4s." John Cummings remembered that:

"We had some problems at first because the AWG-10 pushed the limits of electronic engineering in those days. Some company tech reps were with us on USS *America*, and between them and our own radar techs we were able to make the system work. I did lose a shot at a MiG on 10 September because my damn radar quit, but these things happen. We got a call that MiGs were airborne north of us. Bear snapped the airplane in a northerly direction, and I buried myself in the radar trying to find them. A few seconds later our wingman reported that he was receiving SAM indications, and both Bear and I looked behind our wingman for SAMs, but none were on the way. We received another call from the PIRAZ ship that the MiGs were at our 8 o'clock at four miles. Bear called for an in-place turn to the left (our wingman was on the right), and then called to drop tanks while we were both in the turn. We were over Hanoi when we dropped tanks. As we came out of the turn I had radar contact with the MiGs about 20 miles southwest of us and at about 20,000 ft. We were at about 3,000 ft. I locked one up, and we took a heading to intercept. Shortly thereafter my radar died. Our wingman, who did not have radar contact, was low on gas, so we had to head back to the ship."

F-4J BuNo 155563 returning to the fuel pits after a BARCAP. (via James Sherman)

Commenting on the RIO's task, he said: "With some aviators you were a full-time systems operator handling the radar, ECM gear, etc. A few pilots needed a little more help (because of lack of experience or other factors), and your duties expanded to whatever needed to be done. On daylight missions in high-threat areas most of our attention was outside the aircraft looking for flak, SAMs, MiGs, and the target." A Vietnam deployment also pre-supposed capability in air combat, which the standard intercept-plus-CAS syllabus would not normally have emphasized. Paul Fratarangelo:

"I was at Beaufort in 1969 when we began dissimilar ACM (DACM) training against our sister A-4 squadron. What a surprise to see the Skyhawks stay on the inside of all our turns and lead-turn us. Up until about 1969 USAF, USN, and USMC [F-4] pilots were woefully deficient in ACM. Top Gun and MAWTU [Marine Air Weapon Training Unit] were responsible for the change. John Boyd's 'Energy Maneuverability' and USN tactical manuals were coming out with tactics to fight low wing-loaded aircraft. Bear started out as VMFA-333 'S-3,' and along with Lil' John led our ACM training effort. We did a lot of 1 v 1, 2 v 1, and 2 v 2 DACM. Once we joined CVW-8 the Navy kept emphasizing ACM during ORIs. We were greatly concerned with the lack of ACM training opportunities due to shortage of gas during cyclic operations. Bear convinced the ship to support what we called a Tiger Cycle. We launched ACM missions on the first cycle and recovered about thirty minutes later."

On 11 September 1972, Bear and Lil' John were on a MiGCAP, flying F-4J 201 AJ (BuNo 155526). For John, it was the second mission of the day; earlier he had taken LT Eric Denkewalter on his first hop into RP 6. His account of the second mission records their MiG kill (possibly two), and illustrates the complexity and hazards of strike operations from the MiGCAP viewpoint:

"That afternoon I was scheduled to fly with Bear on a MiGCAP for a strike force hitting a fairly 'easy' target near the coast, north of Haiphong. Because it was an easy target Bear elected to take Captain 'Scotty' Dudley and his RIO Captain 'Diamond' Jim Brady as our wingman. It was their first mission into RP 6, also. The mission was briefed by the Strike leader for a radio-silent rendezvous, and Bear and I briefed the A-6 tanker crew on the refueling point for the MiGCAP prior to preceding the strike force over the target. (We were also the weather recce).

At about 1700 we launched, joined with our wingman, and proceeded to meet the A-6 tanker to top off our fuel before going

B-52 escort, one of the *Falcons* duties. *Falcon 4* rides MiGCAP on B-52D-5-BW 55-0054 on an *Arc Light* mission. (via James Sherman)

Falcons at rest on beautiful Hawaii en route to Da Nang. (via James Sherman)

over the beach. It was at this time that we encountered the first of several glitches. The A-6 wasn't there! We looked for it for what seemed a long time before I broke radio silence to ask his position. The A-6 was about 20 miles from us. (After the mission I asked the pilot why he was not at the briefed rendezvous point, and he said he was told by the PIRAZ to orbit a different point. Since the controller was on another ship and not at our briefing, the fallacy of blindly complying with radio instructions without letting us know of the change should have been obvious.) Bear and I refueled, and Scotty was refueling when I picked up the strike force on my radar. They had rendezvoused and were heading for the target to make their target time. We had to stop Scotty's refueling in order to be in the MiGCAP position before the bombers got to the target. On the way to the coast-in point Bear and I discussed whether Scotty had time to get enough fuel for the mission and decided that he had. After all, the target (a suspected SAM assembly area) was close to the coast, and our assigned CAP station was just a little further inland. After crossing the coast northeast of Haiphong we turned to a westerly heading and passed abeam the primary and alternate targets. We then broadcast the code word that target weather was good and headed for our CAP station.

Before we got to CAP station the MiGCAP controller (a Navy Chief named Dutch Schultz) gave us a vector for Bandits 61 miles west of us. The MiGs were circling Phuc Yen airfield (about ten miles north north-east of Hanoi), and the controller continued giving us information which confirmed this. On the way to the engagement Bear and I both discussed Scotty's fuel again, and I was also apprehensive about being sucked off CAP station and possibly allowing other MiGs to come at the strike force from one of the airfields to the north, like Kep. But there was another section of fighters also on MiGCAP using a different radio frequency, and I thought that they would be kept close to the strike force since we were being vectored to intercept. Besides, Bear and I had been flying fighters a long time, and we really wanted to have a go at the MiGs.

The controller kept giving us information, and I started picking up the MiGs at 19 miles. Our ingress attitude was about 2-3,000 ft. The controller and I later figured that the MiGs were about 20,000 ft when he gave us our first vector, and that they spiraled down to about 1,000 ft, which was about where they were when I picked them up. Because of their low altitude I had a hard time maintaining radar contact, and the controller was no longer holding them on his scope, either. Visual contact was not made any easier by the fact that we were looking into the sun. The MiG-21s were in modified trail formation, with the trailer about three miles behind and to the right. I still had an intermittent contact until about six or seven miles, which was when Scotty picked them up visually. I locked onto the aircraft that was in the lead. The MiG that I had locked up was shiny silver in color and began tightening his left turn. The MiG on the right (Bear told me it was light blue. I never saw it except on radar) reversed his turn, went to the right and disappeared to the north.

Lil John's tape recorder caught the conversation:

OSWALD: (GCI destroyer) Looks like he's headed back to the Northwest. OK, there's definitely more than one of them now. I've got another contact now of 265 at 19.

CUMMINGS: OK, I've got the first one you've got the second. There's one in trail on him at about three miles [To Red Two: Dudley and Brady]

OSWALD: Do you have video [radar] on him?

CUMMINGS: That's affirm, I've got a slight radar contact on him. It's breaking off now.

OSWALD: Roger I hold him 24 - 256 at 12 now.

CUMMINGS: I lost him, Bear.

OSWALD: OK, let's bring it hard around now. Port to 220.

CUMMINGS: Rog. Just a sec. OK, Stick [pilot], dead ahead at about 7.

LASSETER: OK John, how far is he?

CUMMINGS: About seven miles. There's two of them. He's at 11:30 now about six miles. I just lost him.

RED TWO: Talley ho! 12 o'clock - keep going straight! [visual sighting].

LASSETER: OK, John. Go boresight.

CUMMINGS: OK, baby.

LASSETER: Boresight now.

CUMMINGS: OK, I got it. There's your settling time [system is ready to fire missiles].

Bear made sure that he was cleared to fire and let go with two Sparrows. When I felt the missiles come off I looked forward over Bear's shoulder, expecting to see debris. I was shocked to see that the MiG was still flying and that he had not only dodged our missiles, but had also gained angles on us using that magic turn that MiGs are famous for. The fight lasted about four and a half minutes, speeds were subsonic, altitude was below 1,000 ft and remained geographically over the eastern end of Phuc Yen runway. Our speed was around 450 kts. Although the MiG had lost some energy, he gained enough angles to go from our 12 to 10 o'clock position. While the MiG was at our 10 o'clock I took a couple of pictures [Lil' John always carried a Canon SLR camera with a 100 mm lens, as well as his harness-mounted tape recorder, on missions. The

former was lost during his subsequent ejection]. Bear did a high yo-yo to gain back some angles, and Scotty pressed the MiG until Bear could get his nose back on the bandit.

Throughout the engagement ground fire and SAM warnings were continuous. I have never seen flak that heavy before or since. Unknown to us, the other section of F-4s assigned to MiGCAP (which I thought was protecting the strike force) was vectored to the area after us. While Bear's section was still engaged, the wingman of the second section of F-4s was hit by flak and they both headed for the water. During the melée, Bear took more shots at the MiG—two more Sparrows and two Sidewinders, one of which blew up right in front of us, but the MiG stayed low and played his turn so that he was on the edge of our missile envelopes and was able to tighten his turn whenever we took a shot. The entire fight remained a constant Lufbery to the left.

About four minutes into the fight Scotty called Bingo and turned toward the coast. The MiG for some reason *reversed* his turn, and Bear got his second clean shot of the day; with a Sidewinder this time. (I consider the first Sparrows fired to be a clean shot, also.) The growl of the Sidewinder tone was loud enough to drive you from the cockpit, and it really did a job on that MiG. Everything aft of the cockpit was gone, and what was left was in an almost 90 degree dive for the ground at about 500 ft.

LASSETER: Haa! We got him, John!

CUMMINGS: OK!

LASSETER: OK, splash one MiG-21.

Bear snapped our plane into a sharp right bank so I could get a picture, but by the time I got the camera up he had rolled it back level again. I'm pretty sure the pilot didn't get out.

Our next priority was to get joined with Scotty and get out of there alive. I was trying to get an ADF cut when Bear picked him up visually. Bear also picked up a black MiG-21 which had apparently passed us and was making a run on our wingman from his deep six o'clock. Scotty at first thought the MiG was us, and was giving him directions ('OK, I'm on your nose now...') when Bear recognized the situation and told Scotty to BREAK PORT! When

Letting down to land at Wake Island during the long transpac deployment. (via James Sherman)

Scotty broke, the MiG overshot. (I know MiGs aren't supposed to overshoot Phantoms, but this one did.) Bear called for me to lock on, and he fired the remaining Sidewinder. We were, of course, out of Sparrows, and there was no need for a lock, but it was a habit from training. The MiG popped a flare and broke right, but the Sidewinder still guided on him. We last saw him at our 4 o'clock, three to four miles away, heading west-south-west, trailing a thin wisp of smoke.

Right after this last MiG made his run on us, a very authoritative voice came up on the radio and told us to 'Knock it off, check states!' or get out of there now. At this time we didn't need much coaxing, so Authoritative Voice was redundant. A few minutes later he became an outright nuisance. After the MiG had turned away we joined our wingman, who was dangerously low on fuel. We had ingressed over the rugged territory NE of Haiphong, but because of fuel we had to cut that corner and fly nearly over Haiphong. We climbed to about 14,000 ft and slowed to about 400 kts, also to save gas (our state was about 4,200 lbs, but Scotty had 1,700 lbs) and still retain enough energy to dodge SAMs. 'Authoritative Voice' then began vectoring aircraft. While this was going on my ECM gear was noisily giving SAM and AAA indications as it had during the engagement. We continued to clear each others' six visually, and Diamond Jim gave us a call of 'SAM! SAM!' just before we got hit. Unfortunately, the SAM call was given at the same time 'Authoritative Voice' was talking, and Bear and I missed it.

To complicate the problem, 'Authoritative Voice' was talking to us and the A-6 tanker, but looking at the other section of MiGCAP fighters on his radar. They of course beat us to the coast by a few minutes and crossed the beach several miles south of us and on a different frequency. Thus, when we (they) did not respond to his vectors, 'Authoritative Voice' directed the tanker towards what he thought was the fighter low on fuel but *away* from the F-4 that really needed gas! Bear and I were too busy to catch this latest glitch. Dutch Schultz came back on the air to pass intelligence information that there were MiGs five kilometers from us. The telepanel and all our fire lights were lit up, and both wings were full of holes. Scotty notified us that we were 'burning real bad.' The airplane was uncontrollable at first, but Bear stabilized it using full back stick. We stayed with the bird as long as we could in order to get as far out to sea as possible. We stayed too long.

CUMMINGS: We just took a hit!

RED TWO: Roger, keep going!

CUMMINGS: I've got a wing full of holes, but I think we can fly it.

RED TWO: OK, keep your present heading. There's no more coming now.

CUMMINGS: Thank God!

CUMMINGS: We've got a low-level fuel light.

LASSETER: Well, we're fixing to flame out.

CUMMINGS: OK, this is Red One, 201. I'm afraid we're going to have to jump here. We're about out of gas. Roger, what's your state now?

LASSETER: We don't know, its spinning and the gauge is spinning, indicating zero. OK, Red Two, what's your fuel state right now?

RED TWO: Indicating zero. You're burning real bad.

LASSETER: Let's go, John.

The plane pitched forward into a nose-down inverted spiral with increasing negative g forces. We were both pressed against our canopies and unable to muster the force necessary to pull either ejection handle against g. [Paul Fratarangelo and others later listened to John's 'hair-raising' tape of himself and Bear trying to eject under 5 or 6 negative g's'. At the time of their return he was, 'standing watch in Pri-Fly when all this happened; very exciting.']

Bear reached behind his helmet and began threading his face curtain out an inch at a time. I got my face curtain handle past my helmet and pulled upward and forward along the canopy with everything I had (I cracked the cartilage in my rib cage doing it). Neither of us was ever sure who actually triggered the ejection. An armed 'Big Mutha' helicopter was there to pick us up and give us a swig of Cold Duck as we rode to spend the night aboard the USS *England*, the ship that had controlled us. On the *England* we met Dutch Schultz, dried out the tape in my tape recorder, and pieced together a lot of what happened. We also found out our wingman didn't make it to the tanker. [Dudley's F-4J BuNo 154784, AJ 206, flamed out from fuel starvation and its crew were recovered]

The next day we joined Scotty and Diamond Jim aboard *America* and debriefed each other and the intelligence officers. Bear and I looked and felt like the dogfight had been a physical encounter with a motorcycle gang. We had numerous cuts and scratches, our backs were stiff from the ejection, and the whites of our eyes were cherry red from broken blood vessels caused by the negative g.

The following day we were summoned to the *Saratoga*, flagship of Admiral Christianson, for a personal visit with him and his staff. After all, you don't lose two planes, even in combat, without getting to talk to men in high places. The Admiral started the interview on a friendly note, but his staff, particularly the CIC Officer, soon took the role of prosecuting attorneys. Somewhere during the course of his harangue the CIC Officer brought out a map with a series of lines and tight circles showing our flight plan and location of the engagement. He was wrong by about 15 miles, and I pointed this out without the politeness I would normally have used when

A cascade of Mk 82s falls from *Falcons* F-4Js in an above-cloud attack. (via James Sherman)

Graffiti on a *Falcon* F-4J. "Bula Bula" indicates a kill on a BQM target drone during missile practice. (via James Sherman)

talking to an officer of his rank. When I produced and played our tape of the engagement the accusations being obliquely directed at us almost disappeared. The one charge that neither we nor the tape could easily answer was that we had taken a 60 mile vector toward MiGs that did not threaten the strike force we were assigned to protect. Bear and I agreed that was in fact true. I then asked if they were suggesting that we go back to our squadron and tell the aircrews that they could choose which vectors they were going to take. I followed up this logic by pointing out that when aircraft in our squadron were given a vector that we assumed we were part of a bigger picture and that someone higher than a Navy Chief was directing the air battle. Case dismissed! Kill confirmed! Where do you want your medals sent?

I think the North Vietnamese used our aggressiveness and desire to bag a MiG to lure us into a trap consisting of AAA, SAMs, and MiGs. If their coordination and marksmanship had been better they would have got us, saved one of their pilots, and one, possibly two MiG-21s. I don't have a good explanation about what caused the first MiG to reverse his turn and give Bear a clean shot. I do know that the AWG-10 radar and Sparrow missile combination were not reliable in those days. This caused most Navy squadrons, including our sister squadron aboard USS *America* [VF-74], to carry a missile load of four Sidewinders and two Sparrows. VMFA-333 carried four [AIM-7] and four [AIM-9] on MiGCAP. After our sixth missile had been fired the MiG pilot may have figured we were 'out of Schlitz' and went after Scotty. His Intelligence may also have picked up our wigman's bingo call.

If the MiG pilot's mission was to lure us into a flak trap and keep us there it would account for the fight remaining in a confined area. If he was receiving coaching from the ground, it would also account for his being able to fly that low and successfully stay out of our missile envelopes. My ECM gear was busy the entire time we were over North Vietnam, especially during the engagement and just prior to being hit. It was quiet 5-7 seconds before the missile hit. This leads me to believe that the SAM that bagged us probably had IR terminal guidance. The chaff we carried helped, but we were out before the end of the engagement."

When the *Shamrocks'* CO, LTCOL John Cochran, was shot down on a 23 December recce escort he sustained back injuries in the ejection (which left him with a limp), and he opted to retire shortly thereafter. Major Lasseter took over as CO, and the cruise continued until March 1973, giving him nearly two years as skipper. He died eight years after the war during routine surgery, but he is remembered by his former RIO as a, "gentle, kind, and generous man. People have a hard time visualizing a gentle warrior, but he was one." During his command of the *Shamrocks*, as Lil' John noted, the squadron had, "no diverts to land bases, and our only accident was a hard landing by the Navy CAG flying one of our planes. We were the only combat squadron aboard *America* to return from the Westpac cruise without losing any aircrews."

One of the more unusual tactics employed by the squadron during its strike operations was the rigging of its F-4Js with SUU-44/A flare pods, loaded with Mk 24 Mod 4 paraflares, on the lower and outboard TER stations. These were carried, Paul Fratarangelo stated, "on the lead F-4, and we dropped our own flares on two-plane tac recce missions. I recall several I flew both as lead and wingman with excellent results."

Although the *Linebacker* period was a time of comparatively intense MiG activity for the Marine and Navy F-4 squadrons, the majority of their sorties were still strike or BARCAP. A glance at VF-31's mission totals from 18 May 1972 to the end of the year shows that most crews averaged 130 missions each, of which (typically) ten were Alpha strikes, five armed recce, eight photo escort, roughly fourteen MiGCAP/FORCECAP (this varied greatly, with the CO and XO taking up to 29 each), and around 100 BARCAP and strike. In many cases pilots flew around a third of their missions as BARCAP and over a third on strike. An analysis of Curt Dosés logbook with VF-92 for November 1971 to June 1972 showed that he flew 56 per cent of his missions on strike, 28 per cent on BARCAP, 8 per cent escort, and 8 per cent "miscellaneous." During this time he made 180 flights in virtually every aircraft in the squadron, including twenty-four in his assigned "MiG killer," BuNo 157269, and four in an OV-10A FAC. On strike or BARCAP missions MiGs did not normally challenge the Phantoms, but, as Jerry Houston said of the war in general, "MiGs were funny: you ran into them or you didn't. Many great fighter pilots (for example, 'Blackjack' Finley) never saw one. But never ask a fighter pilot what he'd give to get a kill. You might get embarrassed. RIOs feel the same."

To the Rose Garden
The renewal of the air war in 1972 brought Marine F-4 units back into action and prompted the development of a new forward base at Nam Phong, Thailand. While it was being completed by "Seabees," F-4s from VMFA-115 *Silver Eagles* under LT COL Kent McFerren and VMFA-232 *Red Devils* commanded by LT COL Joe L. Gregorcyk flew in on 6 April from Iwakuni and operated from Da Nang in an attempt to frustrate the North Vietnamese Spring invasion. They were joined four days later by a dozen VMFA-212 *Lancers* F-4Js from Kaneohe Bay (including a couple on loan from VMFA-235) and commanded by LT COL Dick Revie. Although

the crews were inexperienced on the whole, they were immediately thrown into support of a vicious land war in which the NVA rapidly drove back the South Vietnamese defenders and took over their armour. Learning to use their RHAW and ECCM gear was a new experience for most on the deployment as they set about delivering their loads (typically twelve Mk 82s) with OV-10A Bronco or F-4E Phantom "Fast FAC" guidance. Unusually, they flew initially without missiles, finally receiving a token AIM-7 to satisfy their fighter-pilot image. Two *Lancer* F-4s were shot down, and only one of the crew was rescued. Back at Da Nang they also came under attack from Viet Cong rockets. In all they flew 863 missions in about three months, dropping 2,500 tons of ordnance before being ordered back to Kaneohe on 20 June.

The other two squadrons, the *Red Devils* and *Silver Eagles*, were sent on to Nam Phong, sardonically referred to as "The Rose Garden" by its occupants. There they had the benefit of a 10,000 ft runway and "wonder arch" aircraft shelters in an otherwise very basic facility with temperatures up to 110 degrees F. The location enabled the Phantoms to operate in the covert war against the Khmer Rouge in Cambodia into the summer of 1973, flying DAS and other strike missions against an elusive enemy. In September the two Nam Phong units were among the last to leave the combat zone when they withdrew to Iwakuni.

VMFA-334 patch.

8

Spooks at Sea

Although much of the F-4's time as the prime USN and USMC fighter was spent in South East Asia, other Navy Phantom squadrons patrolled the seas throughout the world and continued to do so for another twelve years after the Vietnam War. The Marines hung on to their Phantoms even longer, retiring the last in the early 1990s.

Flying from aircraft carriers has always been an extremely demanding task, particularly at night, in war or peace. Simply finding the carrier on return from a sortie can be a problem. Steve Rudloff found that one of the main uses for his radar could be for locating the home deck. "Ninety-nine per cent of the time the TACAN was working, but just for practice I would try to zero in on the carrier itself, lock it up, and let them break lock with their EW equipment. It was just to be able to know I could find the ship out there in the event of something going wrong." On his second cruise an F-4J pilot and his wingman inadvertently made a landing approach to a destroyer instead of the carrier, and two Phantoms ended up in the sea as a result. The squadron Ops Officer and his RIO had failed to switch TACAN stations properly and were running out of fuel as they attempted an approach to the destroyer. "His wingman was NORDO [had no radio], so he was exonerated. The other crew were immediately transferred from the cruise when they had been plucked from the sea.

Carrier flying had its lighter side, too, as Steve demonstrated. "I was flying with P. J. Scott close to midnight one Christmas Eve. Some guy came over the radio on guard frequency and said 'This is the Lord. It's a boy.' I just cracked up." Lynn Batterman tells the tale of Hardy McAllister, one of VF-96's more colorful personalities, who decided to liven up a BARCAP by announcing on his radio, 'Red Crown, we've got a bogie on radar bearing 270 at 20 miles and he's high.' Red Crown replied, 'Negative contact. Vector for bogie.' 'Roger,' replied Hardy, 'bearing 270 at 15 miles now, and he's high!' Red Crown advised, 'Continue.' 'Roger, bearing 270 at 10 miles,' asserted the wily Phantom driver. 'He's real high...5 miles...4 miles, 3...2...1 mile. Oh my God, it's so bright! The brilliance! It's unbelievable!' At this point Hardy switched in his cassette recorder and broadcast a tape of the Tabernacle Choir singing 'The Hallelujah Chorus.'

"YF-4S" 153088 was painted as VX-4's "bicentennial bird" in 1976. The aircraft tested a variety of leading-edge slat configurations (fixed slats in this case) at Patuxent River. Built as an F-4S, it acquired an F-4B radome, which helped with the "eagle" image. (James E. Rotramel)

VF-21 F-4J BuNo 158368 queues for the cat on USS *Ranger* in 1974. (Jan Jacobs)

A *Swordsmen* F-4B (BuNo 151400) ready to launch from USS *John F. Kennedy* with CVW-1 in the Mediterranean, December 1969. (Rasmussen/USN via CDR James Carlton)

Bicentennial markings on VF-74 F-4Js aboard USS *Nimitz*. (U.S. Navy)

The Nature of the Beast

Among the hazards of operational flying in the F-4 were some which derived from the aircraft itself, and throughout the 1960s crews continued to discover, and learned to manage, the mighty Phantom's vicissitudes. William Greer found a couple while he was instructing at VF-121 in 1967.

"While at the bottom of a low yo-yo my aircraft suddenly whipped into three rapid rolls to the left before I could regain control. Investigation revealed that the rudder was hard over to the left, where it would remain for the rest of the flight. The F-4's yaw/roll coupling at high angles of attack had caused the rolls. Use of the rudder pedals had no effect on rudder position. Being uncertain of how the nose gear steering would react when on the ground, I chose to land into the arresting gear at NAS Miramar. Approach required 15 to 18 degrees right wing down to maintain heading, but was effected easily. I heard later that the Air Force lost an aircraft from a similar failure.

Correct use of nosewheel steering was always a point to watch. The rudder pedals had to be centered before nosegear steering could be selected, otherwise the nosegear would be skewed off center when the aircraft touched down, causing lost tires or a slew off course. Other pitfalls in the F-4J/S included remembering that, in the rear seat a pedal on the right operated the radio and the left pedal was for the internal comms system. It was important to remember who your message was intended for! Similarly, the left canopy eject lever would blow the whole canopy off, while the lever on the right sill was a manual release to merely unlock the canopy."

CDR Greer went on to command VF-11 *Red Rippers*, a squadron which made one Westpac cruise, terminated after only five days when a fire devastated USS *Forrestal* on 29 July 1967. The carrier had a strike force ready to launch when a short circuit occurred in the pylon of F-4B BuNo 153061, possibly due to stray voltage as the left engine was started or to the fact that a Zuni pack had been

An overall gray paint-job on this VF-11 F-4J didn't mean toned down markings; these came later. However, little of the original plethora of maintenance stenciling remained. (Norman Taylor)

Black Aces F-4Ns overfly their "base," the USS *F. D. Roosevelt*. The nearest aircraft, BuNo 150442, was later preserved at NAS Memphis. (USN via Norman Taylor)

"Bicentennial" F-4Js of VF-11, lined up on USS *Forrestal* in June 1976. (via Peter B. Mersky)

The red-nosed F-4J (BuNo 153870) assigned to *Pacemakers* skipper, CDR Hank Halleland. (Author's collection)

plugged in too soon before the "safe" time to do so, i.e. when the aircraft was on the cat and pointing out to sea. A Zuni roared from the rocket pod of F-4B BuNo 153061 and ruptured the fuel tank of an A-4E, waiting for launch 100 ft away. The subsequent fire torched 40,000 gallons of aviation fuel, cooked off ordnance, and engulfed the flight deck. The death toll was 134, and 21 aircraft were destroyed, including seven F-4Bs. A horribly similar accident befell USS *Enterprise* on 14 January 1969 near Honolulu. As a cyclical launch was being prepared an MD-3A "huffer" engine starter was inadvertently parked with its hot exhaust blasting over the warhead of a Zuni only two feet away under the starboard wing of F-4J BuNo155804. Once again, the explosion penetrated fuel tanks, causing blazing fuel and detonating ordnance, killing 27 men and injuring 350 others. On that occasion eight F-4Js were among the fifteen aircraft whose charred remains were subsequently bulldozed into the sea.

All VF-11's subsequent deployments were with the Sixth Fleet. One difference CDR Greer noted between Vietnam and 6th Fleet operations was, "the infrequency of tanker support. F-4 crews had

to spend an inordinate amount of time at minimum fuel consumption, since the ship was invariably exercised if you returned with less than maximum allowed landing fuel. This was particularly disheartening in trying to impart some knowledge of ACM to flight crews trained largely on radar intercepts. It seemed that Atlantic Fleet ears were deaf to pleas for more realistic training for two or three years more. In fact, I later heard COMNAVAIRLANT himself tell his Chief of Staff that we should quit ACM because we lost too many aircraft."

Soon after joining the squadron on *Forrestal* in 1968 CDR Greer had to land much earlier than was scheduled and had to burn off fuel to get down to landing weight:

"Using afterburner and a 5 or 6g turn to hold the speed down was doing so nicely when there was a sudden thump and a great deal of wallowing about. Slowing down, I found the aircraft felt a bit unstable, though I was unable to find anything wrong. Inspection by another aircraft also found nothing, so I recovered aboard normally. Visual inspection showed that the outer panel of the left stabilator was missing. Correspondence with the Bureau of Aero-

David Daniels' jet, minus its left outer horizontal stabilizer, at Miramar in 1976. (David Daniels)

Another angle on the abbreviated tail feathers of David Daniels' F-4J. (David Daniels)

VF-101 *Grim Reapers* F-4Js from Oceana. BuNo 155539 was also *Vandy 1* with VX-4 later in its career. (McDonnell Douglas via CDR James Carlton)

VF-171 was tasked with F-4 training when the *Grim Reapers* switched to F-14 Tomcat RAG duties in 1977. Flying from Oceana, it continued in this role until June 1984. (via Peter B. Mersky)

nautics revealed that, had I not caught the arresting wire there would have been insufficient pitch control available to raise the nose back to a flying attitude. Good fortune indeed, although a replacement stabilator was months in arriving, and we badly missed use of the aircraft."

In 1976 LT Dave "Skinny" Daniels had much the same thing happen to "his" VF-21 F-4J when it was flown by LT "Dart" McCarty. On this occasion the landing was a safe one at Miramar, with the whole of the outer white part of the left stabilator cleanly broken off. Possibly the stab failures were fatigue-related. "Smoke" Wilson had one break off during a cruise with VF-142:

"During the 1971 Indian Ocean operations [during which one of his CAP crews intercepted the Senior Admiral of the Indian Navy in his DH Dove transport!] we kept up our proficiency by doing ACM within the Air Wing on routine night operations. On one mission my wingman and I were engaging two A-7s. As we passed head-on I went into a sharp turn, and one of the pilots said, 'Hey, *Dakota*! Something came off your airplane.' My wingman joined

up and checked me over. To this day I remember his words: 'Looks good to me, Skipper.' By now it was getting dark, so we headed for night recovery. As I was slowing down to approach speed I noticed I had to put in more left stick and nose-up trim. By the time I arrested I had almost full left stick in to hold the wings level. When I shut down, the plane captain came up to me and said, 'Jesus, Skipper, you lost half your stabilator!' Sure enough, the outboard honeycomb section of the starboard stabilator was gone. I hate to think what might have happened if I had boltered or waved off..."

Safe carrier flying, whether in combat or not, depended on maintaining proficiency. Paradoxically, operational losses (to human error or mechanical difficulties) tended to be less on a combat cruise. Frank Curcio, flying with Air Wing 5 in 1965, noted that *Coral Sea* had no losses to pilot error in around 15,000 sorties off Vietnam, "whereas a Mediterranean cruise with just hundreds of sorties would get two or three losses from lack of proficiency." Catapult launch and recovery were often the times when the rather brutal mechanics of both processes took a toll. "Smoke" Wilson's

Another VF-101 F-4J showing the squadron decor. (Simon Watson)

A VF-11 F-4B being guided onto the catapult track aboard USS *Forrestal* during a "Med" cruise. This aircraft eventually passed to VF-171. (via Peter B. Mersky)

A pair of VF-11 *Red Rippers* on a CAP mission. (via Peter B. Mersky)

Armed with LAU-10 Zuni pods (made from processed paper with an aluminum skin), each weighing around 650 lbs and holding four five-inch rockets, this F-4B overflies the USS *Forrestal* in June 1968. VF-11 Phantoms patrolled the Atlantic and Mediterranean for over twelve years. (USN via Peter B. Mersky)

wingman, Bob Cross, and his RIO Dean Hutchinson were on the starboard cat of the *F. D. Roosevelt* on a Subic Bay visit. As Chuck D'Ambrosia observed, "When they launched, the catapult shuttle broke and flipped up, rupturing the centerline tank and causing an enormous explosion. While we were all looking for a plane in the water, Bob was driving around wondering what the bump was that he felt on launch. Unfortunately, the bow runner [deck crewman] was blown over the side and severely burned."

Deck landing presented its own hazardous and curious situations. One of the oddest was a case Dave Daniels observed where a disoriented F-4 pilot attempted a landing approach on a carrier from the bow end. A member of each squadron was designated as Landing Signal Officer, and when he wasn't flying he would stand on the LSO platform on the port side of the ship, abeam the landing area during all recoveries of the squadron's Phantoms. Dave Daniels described the routine:

"The LSO carefully watched each approach and shouted comments to an assistant, who recorded everything in a small green notebook for use later. The LSO would immediately transmit instructions if he thought the approaching pilot needed it. It was usually hot, humid work, and extremely windy and noisy. Not only was every approach carefully scrutinized by the LSO, two TV cameras were also recording all the action. When the crews returned to their ready room the tape was run on the closed circuit TV, not just for that one squadron, but the entire recovery: every approach of every aircraft was shown. If you screwed up it was there for everyone to see at least once as a re-run. If a pilot boltered or touched the deck but the tailhook didn't catch a wire, that stayed on the tape, and every one got to see them try again. After everyone had seen the tape the LSO de-briefed each pilot on his approach. He graded the start, the middle, in close, the touch-down, which wire was engaged, and gave an overall grade. He even gave a grade for a pass or approach that resulted in a bolter. The very best grade was 'OK,' the next best was called a 'Fair' and written as 'OK'; worse was a 'No Grade,' and the most shameful of all was the 'Cut Pass,' which was sure to bring hoots of derision and plenty of comments from pilots in the other fighter squadrons."

A VF-32 Swordsmen *Phantom* takes the wire during FCLPs. (USN via Peter B. Mersky)

Chained securely to USS *Independence's* deck, a VF-102 F-4J visits port. (Simon Watson)

VF-102 was teamed with VF-33 for a series of Mediterranean cruises in the 1970s. (via C. Moggeridge)

VF-33 and VF-102 F-4Js aboard USS *Independence* in July 1977. (Angelo Romano)

There were eight squadrons on the ship, and each squadron had a 'greenie board' attached to a bulkhead in a prominent location in their ready room. Every night, after the ship secured from flight ops, the LSO would sit down in the ready room and update the greenie board. It started out as a white sheet of paper with the names of all seventeen pilots in the squadron listed by seniority down the left side. Everyone was on the list, from the skipper right down to the most junior pilot. Horizontally, next to each name there were perhaps thirty empty squares. Each square represented one approach, not necessarily a successful landing, by that pilot. Each night the LSO would get out his grade book, grab a handful of colored markers, and spend half an hour putting each pilot's grades on the greenie board. An OK earned a solid green square. The goal was a row of solid green. If the approach actually resulted in an arrested landing, the wire engaged was recorded as a small number

in the lower right hand corner. There were four wires in the landing area [the number one wire being nearest the stern], and the number three wire was the target. Everyone was aiming for a solid green square with a '3' in the corner. If the approach was at night, a large black 'N' was placed in the center of the square. Each pilot was required to have at least 30 per cent of his flight time at night. A 'Fair' pass earned an amber square, and a 'No Grade' or 'Cut' was a glaring red acre of embarrassment. There is nothing like having one's work videotaped, critiqued, graded, and having those grades posted each day on the 'greenie board.' You could talk all day about being the world's best (simulated) MiG-killer, but if you couldn't consistently plant the plane back aboard the ship, day or night, right in front of the number three wire, all the talk didn't go far."

CVW-1's Mediterranean Cruise in 1967 had its interesting moments in the recovery phases. "Smoke" Wilson:

VF-11 provides some definitive formation flying. (USN via CDR James Carlton)

A RA-5C's cameras captured this VF-33 F-4J escorting a Badger away from the USS *Independence* in September 1971. (USN via Fred Staudenmayer)

One of VF-41's Alert F-4Bs takes a good look at a Badger in 1964. (USN via Fred Staudenmayer)

Another Tu-95 Bear D intercept by VF-41, on 22 September 1973. (USN via CDR Jim Carlton)

"Johnny Barto made the record book for landing the furthest aft on the *FDR* and flying away. During my tour XO Jack Koach was the first record holder when he hit his tailhook about two feet down the ramp on a night recovery. Then VF-32 *Swordsmen's* XO put his main gear tire tracks about the same distance down the ramp, also in a night recovery. CDR Burt's record held until Johnny Barto made two new marks in the ramp with the stabilator of his F-4B, also at night (yes, the night carrier landing is rather challenging). After he hit the ramp the drooped stabilator was only partially drooped. The outboard portions were then bent upward in an inverted gull-wing configuration like the wings of a F4U Corsair fighter. When he got airborne after the ramp strike he was diverted to the Naval Rework facility at Naples. They could replace the stabilator, but there was a problem: the F-4's radar was classified, so it had to be removed before the Italians could work on the plane. Steve Paul, a fine maintenance officer, had the radar removed and put in a secure Naval facility. Everything OK? Right? Wrong! The automatic sprinkler system went off and flooded the radar. It was flown back to the ship, dried out, bench-checked, and some black

Down but not out. A preserved VF-14 awaits further calls to duty in the AMARC sun. Sadly, BuNo 149460 was not recalled for F-4N or QF-4 conversion. (via Paul Smith)

Bear interception often involved getting "up close and personal" to photograph all those minor structural modifications. (USN via Fred Staudenmayer)

Launch time for this VF-171 F-4J during a carqual session. (USN via Peter B. Mersky)

VF-154's BuNo 158362 eases in towards the basket of a VA-145 KA-6D near the Philippines in 1974. (Jan Jacobs)

VF-21 and VF-154 F-4Js share the USS *Ranger's* deck in July 1975. The yellow shirted plane director gives a "hard right" signal as he guides an aircraft forward for launch. The breaks in the black walkway over its intakes are the result of adding the ALQ-126 (code-named *Pride*) antennas. At the same time, the aircraft received the ALQ-45 (*Charger Blue*) equipment during a comprehensive ECM update in late 1974 to mid-1975. (Jan Jacobs)

boxes were replaced. Because of the weight and balance considerations it had to be re-installed in the airplane before it flew—so, back to Naples it went. Believe it or not, everything, including the radar, checked '4.0' on the flight back to the ship."

"Smoke" also described the "hairiest night operation" he had ever participated in:

"We were operating south of Sardinia and were scheduled to take part in a U.S./Italian anti-PT boat exercise. The weather was *terrible*, and the deck was pitching like I had never seen it before or since. The ramp was going up *and* down fifteen feet. On a normal approach to the ship, if the aircraft is exactly on glide-path there will be ten feet of clearance between the ramp and the tailhook. I recommended to CO Rudy Krause we should not fly. He recom-

mended to CAG, CAG to the Captain, and he went to the Admiral, who said, 'No, we can't back out at this late hour. We'll lose face with the Italians.' They compromised: cancel the A-4 Skyhawks, since they have no all-weather capability, anyway. We'll send the F-4s with the best pilots, and they can use that tactic developed by Smoke Wilson. Now it's my fault!

The anti-PT boat tactic was that the RIO would detect and lock onto a surface target. At a specific radar antenna depression angle approximating the desired dive angle the pilot would push over and center the missile steering cue and, at the prescribed altitude and slant range, salvo six 19-shot rocket pods. That's one hundred and fourteen 2.75 inch rockets aimed at the one spot. The theory was likened to both barrels of a shotgun; at least one BB must hit the target.

F-4J BuNo 158373 had an extra "0" added to its Modex to mark the Air Wing Commander, CDR Massey's 1,000th arrested landing. Wing tanks were fitted for the Air Wing's after-cruise fly-in to Miramar in order to get the tanks off the ship before it entered the yards in October 1974. They contained no fuel, and were virtually never used by CVW-2 F-4s. (Jan Jacobs)

David Daniels recalled that his whole body shook as he took this shot of a *Freelancers* F-4J in afterburner about to launch from USS *Ranger* in 1976. Note the extended nose gear, full flaps, and stabilator leading-edge full down. (David Daniels)

A VF-33 F-4J (BuNo 153892) showing the changes brought by revised ideas on naval aircraft color schemes. In March 1980 the yellow logo at least remained intact. (James E. Rotramel)

F-4B BuNo 150475 first flew on 31 January 1963, was recycled as an F-4N ten years later, and then entered a third "life" as a QF-4N over a decade on. By 1981 it had over 5,000 hours on its airframe. In this view its glossy gray paint includes the much-reduced livery of VF-201. (Simon Watson)

CDR Krause took the first section with two birds from VF-32. A half-hour later I launched with my wingman, LCDR Steve Paul, and two planes from VF-32 flown by LCDR Don Primeau and his wingman LT Don McCrory. We flew down to the Gulf of Taranto, and on arrival learned that the ceiling was below 1,000 ft and the Italian air controllers would not let us down unless we were 'visual.' Remember, its night and raining. In sum, the whole evolution was a waste. Back to the ship we came, to learn that the normal landing aid, the Fresnel lens, was out of commission. It was stuck in the limit stops of the stabilization system because the ship was pitching so violently. We were using the manual back-up system. The VF-32 guys were first down the chute. Primeau got aboard, but McCrory made a very long touch-down and was diverted 'bingo' to the airfield at Decimomannu on Sardinia. I made it aboard OK,

and as I was arrested I heard Steve Paul 'call the ball,' which meant he was close behind. Not wanting to create a fouled deck situation [aircraft or obstruction in the landing area] for him I added a whole bunch of power and rapidly spun out of the landing area, skidding to a stop only inches from a parked airplane. Then I heard Steve over the radio say;

'Hey, Paddles. Where'd the ball go?'

LSO [landing signals officer] Keep it coming. You're looking good.

AIR BOSS Wave him off.

STEVE Roger that, Paddles.

AIR BOSS Wave him off!

LSO You're looking good, Camelot [VF-14 call-sign]. There are no wave-off lights.

VF-74, the first operational F-4 unit, continued to operate Phantoms into the "lo viz" 1980s. BuNo 153864 off USS *Forrestal* rolls out at Oceana in October 1980. (James E. Rotramel)

VF-102 kept some of their squadron markings, but lost their color even on the CAG airplane. BuNo 153850 graces the hardstanding at Oceana in May 1980. This Vietnam veteran F-4J was one of the batch which the UK bought in 1984. From AMARC (where it was labeled 8F102) it became ZE 356 with 74 Tiger Squadron, RAF. (James E. Rotramel)

Dumping fuel to get down to the required 5,100 lbs of fuel on landing, a flight of F-4Js enters the USS *Ranger's* landing pattern in the southern California Operating Area (Socal Oparea) in July 1975. (Jan Jacobs)

VF-154 made two deployments on USS *Coral Sea* in 1980-82, but had to "de-convert" from the F-4S to the lighter F-4N to meet the carrier's weight limits. This F-4N (BuNo 153053) has VTAS units fitted to the pilot's canopy rails. (Simon Watson)

VF-103 *Sluggers* kept a trace of color in their tail markings when the tactical paint scheme was introduced. This F-4S was used for their final Phantom cruise, with CVW-17 on USS *Forrestal*. (James E. Rotramel)

Turning above San Diego's suburbs, a *Black Knights* F-4J returns from an exercise. (Jan Jacobs)

VF-191 *Satan's Kittens* transitioned to the F-4 comparatively late, exchanging their F-8 Crusaders for F-4Js in 1976. They deployed once on USS *Coral Sea* and were decommissioned in 1978. (via Peter B. Mersky)

BuNo 158372 makes firm contact with the deck and arresting cable, late in 1974. The folding ACLS antenna is deployed. (Jan Jacobs)

The neatly-designed folding ACLS antenna, deployed on a VF-74 F-4J, above the "red-jacket's" head (Angelo Romano)

AIR BOSS I'm securing the deck lights.

STEVE Falcon 01! First the ball's gone, now the ship's gone!

He waved off and was sent to Deci with a number of others. What had happened was that in my haste to clear the landing area, I cleared it all right! I managed to blow the temporary landing aid over the side, hence no ball for Steve and no wave-off lights. I later learned that both Rudy Krause and Don McCrory had landed so long that their main landing gear hit the deck, but the nose-gear missed the ship! The long landing was because the bow of the ship was going down as they landed. When they left the ship their airplanes were pitching down instead of up. When last seen disappearing into that dark and rainy night their afterburners were blowing rooster tails in the Tyrrenian Sea. They came within inches of hitting the water. When Steve arrived at Deci an F-4B was off the side of the runway with a blown tire. An A-3B that didn't even try to get aboard was off the end of the runway in the over-run with another F-4 close by. All in all, not a very productive evening, except for scaring the crap out of otherwise intrepid airmen. I have often wondered whether the Admiral considered the debacle at Deci as 'losing face' with the Italians."

Bears and Badgers

Keeping up appearances in the Atlantic and Mediterranean also involved making sure that all airborne intruders were met well within the Carrier Group's outer air defense perimeter. LCDR Fred Staudenmayer commanded VF-33 *Tarsiers* with its F-4Js on one of USS *Independence's* Mediterranean cruises (21 June 1973 to 19 January, 1974).

"Carrier Group Commanders dreaded that a picture would appear in the national press of an unescorted Soviet bomber over the Fleet, and particularly over the carrier (a picture no doubt conceived by the jealous U.S. Air Force!). This was much more important than any tactical or intelligence consideration, or any concern for conflict breaking out! I once launched against a Soviet Tu-95 Bear that was almost upon the carrier when initially detected by our pathetic ship's radar. As soon as we were in the air we were vectored 180 degrees from launch heading. I had the radar operating [as the first RIO to command an East Coast operational USN squadron] and detected a huge radar blip at about twelve miles, followed right away by a visual, and we were able to join up on his wing before he passed over the carrier at about 500 ft. This was always the goal

VF-11 F-4Js, in use with the squadron from 1973 onwards. (via Peter B. Mersky)

On USS *Forrestal's* crowded deck an F-4J crew check out their aircraft in May 1981. (Angelo Romano)

Leaving little doubt as to their identity, the *Be-devilers* wore this bold bicentennial scheme in 1976. (USN via Angelo Romano)

F-4N Phantoms of VF-41 and VF-84, closely packed on deck in June 1975. (Angelo Romano)

and the politically correct thing; be on the Bear or Badger's wing, showing the world that you were escorting these uninvited visitors.

When the Bear was in the air armed fighters were put on Alert, either five minutes with crews in cockpits, airplane behind the catapults and plugged into electric and air start carts, or in one minute with engines turning and broken down from chains and ready to go on the cats. Upon launch, crews would use full afterburner climbs to try and catch the target aircraft, usually Bears, at the maximum distance from the carrier. In the 1960s this was nominally at 200 miles, about the maximum range of the air-to-surface missiles carried by the Badgers, which would be sent in to the carriers in descending flight, simulating the missile.

Once joined up on the 'baddies' the RIOs would take pictures of various radomes, weapons, chaff tubes, structural surprises, and anything else the Intelligence types might want, with hand-held 35 mm cameras. These were always carried when manning the Alert aircraft. During a cruise in the Bay of Biscay in USS *Independence* we had a large number of Soviet over-flights, thirty of forty as I recall, and we intercepted all of them (with assistance from sensors

external to the Fleet!). As a general rule, our attack profile started from a low or mid CAP station (5,000 or 15,000 ft), and depending on ranges, etc. we would be in a climbing attack, usually trying to attack from below. Not too much thought was given to vertical separation, sun position, hiding in the clouds, etc. These were all-weather attack profiles.

As the Soviet air-to-surface missiles got faster and more formidable our CAP stations got pushed further and further out. The goal was to be in a position to destroy the targeting or launch aircraft prior to missile release. Nevertheless, we usually trained against descending supersonic missile simulation, not having much faith that we would ever get clearance to fire on targeting or launch aircraft. We always thought we had a pretty good capability against such missiles, and an outstanding capability against Bears and Badgers. Before the Vietnam War, emphasis on coordinated Alpha strikes was limited, with each squadron doing their own thing. The attack aircraft had long-range missions, and the fighters' Fleet air defense was evolving into anti-Bear or Badger tactics as the Soviets simulated air-to-surface missile launches, particularly during the carriers' transits in and out of the Mediterranean.

VF-161 F-4N BuNo 151006 from USS *Midway* in March 1975. (Masumi Wada via Angelo Romano)

F-4J BuNo 155835, seen in September 1977 in VF-161's sleek decor. (Masumi Wada via Angelo Romano)

VF-103's mid-1970s markings are displayed on BuNo 155864 in October 1977. (Angelo Romano)

VF-154's BuNo 158347 takes a close-in wave-off for another landing approach. (Jan Jacobs)

With regard to tactics simulation [in training] we would usually use a wingman to fly a descending profile closely resembling the Badger's descent from formation with a Bear, starting about 200 nm from the carrier from a general threat direction. Threat aircraft would typically be at 20-30,000 ft and Mach 0.7 to 0.8. Our attacks would always start head-on, then, after contact, maneuver to a forward-quarter Sparrow firing position (120 to 140 degrees crossing angle). This was in order to be in place should we be granted a 'clear to fire' without (or with) visual acquisition and still be in a position to convert to a pure Sidewinder attack when the inevitable lack of clearance [to fire] occurred. When intercepting a profile aircraft, we would start low, accelerate to 450 to 500 kts and then start a climb, picking up 0.9 Mach to maintain a reasonably high energy level, subject always to fuel considerations. If an actual Bear or Badger was inbound we would set up an ID pass with a pure pursuit approach to the rear, converted from a head-on starting position. If we had the luxury of having a wingman he would maneuver to a one or two mile trail position to be able to launch a missile at the bogey being identified by the lead. Our computer didn't tell

us that, of course. It could only compute a type of lead pursuit for the Sparrow (radar mode), or tell us which way to go to point the radar at the bogey ('heat' mode or 'pure pursuit'). Most of our training intercepts prior to ACM days were of the forward-quartering approach, designed to get a Sparrow (all-weather) shot off first, then get into position for a Sidewinder. Intercepts were usually conducted at Mach 0.8 to 0.9, with fuel conservation being the overriding consideration when embarked."

The *Red Rippers* made many interceptions on their cruises, as William Greer explained:

"The Badger was the most prevalent. Most of them bore Egyptian markings and Arabic numerals. The main requirement laid on from the Staff was to ensure that no picture of the ship could be taken without an F-4 appearing in the foreground. Many intercepts were run at night, and the Badger would frequently shine a rather bright and distracting light at the escorting Phantom pilot. VF-11 rigged up a very strong spotlight, powered from the Phantom's electrical system, and the first time we hit the Badger with that their performance became somewhat more restrained. I once intercepted

For their final cruise with F-4s VF-21 sailed on USS *Coral* Sea with F-4Ns in place of their newly-acquired F-4S Phantoms, including BuNo 150464. (U.S. Navy)

VF-151 *Vigilantes* operated Phantoms right up to March 1986, making the last carrier trap by a Fleet Phantom. (via CDR James Carlton)

A VF-154 crew mans up for a patrol mission over the Tonkin Gulf in March 1973, just after the war's end. The pilot, LTJG Jim "Ox" Van Hoften, is wearing a first-generation VTAS helmet sight. "Ox" later went on to become a Space Shuttle astronaut and participated in a satellite rescue in the mid-1980s. (Jan Jacobs)

The later-style VTAS equipment is visible, mounted inside the pilot's canopy side-rails on this February 1975 Miramar flight-line. LT Ron Grubb is starting F-4J 204 NE, which also carries an ACMR pod on its left Sidewinder station, indicating a mission over the instrumented range at MCAS Yuma. (Jan Jacobs)

a Bear while returning from my cruise in USS *Enterprise*, and with the aid of my two years of Russian at the Naval Academy, some white cards and a grease pencil, exchanged brief notes with the crewman occupying the rear gun sighting station."

Steve Rudloff's experience of Cold War Tu-95 intercepts involved a rather more intimate form of contact after a Bear rear gunner had teased an F-4 crew by waving a bottle of vodka at them.

"On Alert 5 aircraft for a brief time the back seat was equipped with a copy of Playboy magazine. I took off and intercepted a Bear, and in retaliation for the vodka I flashed the magazine centerfold, getting a hearty smile and a thumbs-up in response. We were always taking pictures of them, and vice versa. We were more than willing to take our oxygen masks off and let them get pictures. There was a point on one of my cruises where we actually spoke to some of the Bear crew members. We indicated which frequency we were on and talked to a crew member who spoke some English. He told us he lived outside Moscow. Suddenly there was some talk

in the background in Russian, and the conversation ceased, even though we tried to raise him again."

Captain William Knutson's personal calendar for his 1967 CVW-6 cruise aboard USS *America* gives an idea of the range of activities involved in a "Med" cruise at the time, though part of it coincided with the Six Day War in the Mid-East.

JAN 10, 1967. Depart Norfolk, VA, on USS *America* direct for Mediterranean. Refresher landings 11th through 13th.

JAN 17. Fly and set Bear overflight CAP.

Bear C and D overflew carrier 19 JAN, 400 nm South of Azores.

JAN 21 Pass Gibraltar.

'Quickdraw' Exercise with Italian Navy.

FEB 4-11. Athens, in port. FEB 12-14. Fly independent ops.

FEB 15-17. Anchorage for Administrative Inspection. [With Vietnam draining operating funds we often anchored so that we conserved steaming and flying fuel.]

FEB 17-20. Fly independent ops. FEB 21-26. Naples, in port.

A *Vigilantes* F-4S on approach to Atsugi. (via Simon Watson)

VF-21's CAG bird overflies Lake Tahoe in September 1975, crewed by LT Ron Grubb and LTJG Bill Martin. (Jan Jacobs)

Another view of the CAG F-4 BuNo 158378, near NAS Fallon. The wrap-around black nose scheme was the idea of CDR "Devil" Houston when he took over the squadron from CDR Doug Clower in 1975. (Jan Jacobs)

VF-171 F-4J BuNo 153773 "hot refueling" in May 1981. In its long career this Phantom flew with VF-31, VF-33, VMFA 251, VMFA-122, and the Royal Air Force as ZE 351. (James E. Rotramel)

FEB 27-MARCH 5. Fly independent ops. MARCH 6-14. Valencia, in port.

MARCH 15-23. Fly independent ops. Lost George Jones on a catapult shot on March 15. The right catapult hook in the F-4B was torn out of its mount due to cracks in the metal. The F-4Bs were grounded for two weeks.

MARCH 24-30. Taranto and Naples, in port.

MARCH 31-APRIL 4. Fly independent ops after F-4B cat hooks were inspected and dye-penetrated.

APRIL 5-9. Malta, in port. APRIL 10-17. Fly independent ops.

APRIL 18-20. Aranci Bay amphibious operation and support with strikes and close air support.

APRIL 20-21. Fleet exchange. Flash orders to get underway and proceed to Greece area. Helos were loaded.

APRIL 21-30. At sea flying on Alert status, but nothing happened.

MAY 1-7. Taranto in port. MAY 8-14. Fly independent ops.
MAY 15-21. Livorno in port.

MAY 22-25. Poopdeck Exercise with the Spanish. Flash message received May 24 to depart Exercise and proceed to Eastern Med at best speed. Flying in Alert status, arriving Aegean Sea, May 26.

MAY 25-27. Israeli/Jordan/UAR crisis. Gulf of Aqaba mined.

MAY 28-JUNE 7. Stationed north of Crete. Planning for all contingencies [due to] UAR/Jordan alliance and tension in Middle east. VF-33 flying CAP and standing 15 minute Alert. All forces in the Sixth Fleet congregate in the Task Force.

JUNE 8. USS *Liberty* is attacked by Israeli torpedo boats and aircraft. I launched on Alert when the attack message was received at 1405, but USS *America* was too far away for us to render any assistance. Task Force headed south.

JUNE 9. Flew CAP over USS *Liberty* and escorted her back to Souda Bay, Crete. [34 U.S. crewmen were killed and 164 injured in

F-4J BuNo 157293 of VF-102 suns itself on Oceana's crowded flightline in 1976. Lacking many of the squadron's original markings, it compensates with a few "zaps," possibly from the Royal Navy's 892 (Phantom) Squadron. (via B. Pickering)

VF-21 F-4Js over San Diego's Coronado Bridge during "Devil" Houston's command of the squadron. (via Jerry B. Houston)

Going bombing from Miramar with small Mk 76 "blue bombs" on the rack. The pilot is believed to be LCDR Brian "Bulldog" Grant (Randy Cunningham's wingman in 1972), with LTJG Greg "Shifter" Hurst as RIO. (David Daniels)

LT John Kelly and LT Dewaine Cherry, ready to fly at Miramar. (David Daniels)

the attack, for which Israel later apologized, though the motive for the assault remains mysterious}

JUNE 10-20. Remained at sea in Crete area. VF-33 had major missile-firing exercise JUNE 17-18. JUNE 21-25 Istanbul, in port. I assumed command of VF-33, JUNE 24.

JUNE 26-30 Fly independent ops. JUNE 30-JULY 6 Thessalonica, in port.

JULY 7-16. Fly independent ops. JULY 17-23 Athens, in port.

JULY 24-28. Task Force ops and open sea Missilex. JULY 29-AUG 2 Genoa, in port.

AUG 3-15 Fly support for Sub Phiblex 2-68. AUG 16-21 Genoa, in port.

AUG 21-SEPT 1 Task Force Ops, training, anchorage, independent ops.

SEPT 2-9 Valencia, in port. SEPT 10-19 Transit to CONUS. Air Wing fly-off.

Crews averaged about 25 flying hours per month. Everyone made at least 100 landings by the end of the cruise. It was our practice to fly whatever mission we had, but to save gas so we could engage in ACM at the end of the cycle before landing."

CDR Greer found this kind of schedule very different from what he had experienced on four Westpac deployments:

"During my nine months on USS *Enterprise* (Westpac) we had about 24 days in port. In the Med it was common to spend about half the time in port. Frequently we were able to put a detachment ashore for those periods so we did at least some flying."

Cooperative exercises with other NATO forces were part of most Mediterranean or Atlantic exercises. Fred Staudenmayer:

"Working with NATO ships never got very sophisticated. It was difficult just initiating communications, though we often worked with HMS *Ark Royal* and other British air controllers with ease, marveling at the professionalism of the controllers. I cross-decked

A Tu-95 Bear intercept on 25 August 1976. Keeping Russian aircraft under escort near the Fleet formed a major part of the fighter squadron's task in peacetime. (David Daniels)

USS *Ranger* leaves Pearl Harbor on 1 September 1976, en route for the USA. (David Daniels)

Freelancer NE201 off Luzon, Philippines, in 1976. (David Daniels)

LT David "Skinny" Daniels with his F-4J (BuNo 158365) after the 1976 cruise. (David Daniels)

F-4J BuNo 158359 in March 1974. (Jan Jacobs)

A ramp strike (every pilot's nightmare) caused this damage to USS *Ranger* on 20 March 1976 near Cubi Point. Both the pilot, LT Gary Caswell, and RIO, LT Herb Jones, were recovered. On USS *Ranger* an F-4's tailhook-to-ramp clearance was a mere eleven feet if the plane was exactly on glideslope and the deck wasn't pitching. (David Daniels)

CDR Doug Clower and LTJG Pete Covey fly high above USS *Ranger* in the South China Sea. (Jan Jacobs)

BuNo 158358 in tension for a cat shot in 1974. (Jan Jacobs)

A sunset landing for an F-4J. (David Daniels)

on the *Ark*, which gave a great sense of improvisation. We envied the night-flying conducted by the Royal Navy—now that was the proper way to conduct flight ops!"

Airshows and firepower demonstrations for foreign forces were sometimes part of the NATO cooperative effort. VF-14 put on one for some senior Greek officers which included a supersonic two-ship pass led by "Smoke" Wilson and his RIO Ron Bird, with Bruce Cobi and Fred Rauscher in the other F-4B. At the start of their 100 ft altitude pass "Smoke" received a shock:

"As we lined up about four miles astern the ship, we had just gone supersonic, and I looked over to check that Bruce was in position. About the time I passed the ship Bruce came up on the radio yelling 'My RIO has ejected himself and I can't see!' Bruce's problem was the sudden decompression. The inrush of supersonic air tore loose the sound-proofing and cockpit insulation material (mostly fiberglass), and his eyes were full of the stuff. He managed to recover aboard an hour later. The CAP and I spent the next hour looking for any sign of Fred. Later, one of our destroyers recovered Fred's helmet and a piece of his life-jacket."

Chuck D'Ambrosia, who investigated this tragic event, concluded that it was a bizarre accident:

"The recovered helmet had two dents that corresponded to the rubber bumpers on the metal crosspiece for the face curtain [on the ejection seat]. That meant that the face curtain was in place [i.e. not pulled down to initiate ejection] when the ejection sequence started. In order to pull the ejection handle between your legs you had to put the guard down and pull the handle up several inches to make it fire. Even with the guard down and a hand resting on the handle, if the aircraft bumped the most you would expect was to dislodge the handle [not initiate ejection]. I don't know of anyone this ever happened to, even on rough cat shots where every forward-thinking RIO had one hand on the ejection handle with the guard down."

Suspecting an accidental rear canopy loss was involved, Chuck's team got a list from the Naval Aviation Safety Center showing a number of inadvertent rear canopy losses, mostly due to a failed sear pin where the canopy attached to its actuator, which in turn operated the canopy locking mechanism. They established that the sear pin probably failed when Fred closed the canopy, and the extra g during the low-level supersonic bouncing about could have caused the actuator, now free of the sear pin, to move downwards and unlock the canopy.

The VF-154 CAG aircraft makes a play for the deck. Its nose has dropped, and it contacts the deck right wing down. The hook is between the "2" and "3" wires. (Jan Jacobs)

Deck launch time for NE207, replacement aircraft for the one lost in the 20 March ramp strike. (David Daniels)

Pictured during their 1976 Westpac cruise are (left to right); LT Dave Daniels, LCDR "Woody" Woodbury (his RIO), LCDR Larry "Buzzard" Urbik, and LTJG Greg "Shifter" Hurst. (David Daniels)

A F-4J smokes in to rejoin the Air Group's A-7, RA-5C, and E-2A complement. (David Daniels)

"We had a Martin Baker [ejection seat manufacturer] rep come out to the ship. We demonstrated to him that the locking mechanism for the face curtain could easily be defeated by rapping on the rubber bumpers on the face curtain crosspiece (representing Fred's helmet hitting the bumpers), causing the face curtain to become dislodged. If this happened the flexible handles [black/orange striped 'pull-rings' at the top of the seat] on the face curtain could be pulled backwards over the seat by wind-blast, firing the seat."

Chuck encountered the forces of nature in a very different way when the ship ran into a giant waterspout while aircraft were being launched at sea between Malta and Sicily:

"There was a large black cloud in front of the ship. I was with Bud Lineburger, and we were the second aircraft launched. The ship was steaming directly into this monstrous waterspout that looked about a quarter-mile in diameter. I called the ship and suggested that they not drive into it. 'Roger'—but they continued on course and launched ten or fifteen airplanes. From what I understood from the people on the ship they drove the ship right into the waterspout. Fish and seaweed were falling on the flight-deck. A

Inferno 107, a VF-301 F-4S in an early Heater-Ferris camouflage scheme (including a fake canopy and helmets under its nose), returns from ACM over the Pacific with an A-4E and TA-4J from VF-126 early in 1982. (Jan Jacobs)

The classic lines of an F-4N (BuNo 152278) from VF-301, returning to Miramar in April 1979. Like many of the "Beeline" aircraft, it accumulated well over 5,000 airframe hours by 1981. (Jan Jacobs)

Inferno 110 shows the upper surface Heater-Ferris pattern. This F-4 was used by LT Steve Shoemaker and LTJG Keith Crenshaw for their 10 May 1972 MiG-17 kill with VF-96, recorded on the rear fuselage. (Bruce Thorkelson)

This view of BuNo 155749 shows the soft outlines of the Heater-Ferris scheme, applied to all VF-301's Phantoms until their conversion to the F-14A in October 1984. The revised scheme has the darker grays towards the nose. Note the MiG kill markings. CDR Bruce Thorkelson and Bob Shaw made the squadron's last operational F-4 flight in September 1984.(Bruce Thorkelson)

lightning bolt hit one of VF-32's airplanes on the cat, at which point the launch was canceled. Bud and I climbed to about 65,000 ft in burner to get over the top of the monster."

Chuck almost had to fly another hazardous VF-14 mission which would have been a one-way trip.

"We were at anchor in Naples when the balloon went up. There was a ship off the coast of Algeria that was in danger of attack. Since we were out of range for the F-4 the plan was to steam for a few hours, launch the A-3 tanker, and send it ahead. Sometime later we were we were to launch the F-4s, refuel from the tanker, and proceed to the Algerian coast to find and defend our ship against Soviet-style patrol boats (we were going to use Sparrows against them). Considering the range we were at we figured that we would have about five minutes of time on station before flame-out. We were supposed to attack the boats and then punch out and wait for the ship to arrive. Bruce and I decided that this plan had a low

possibility of survival, so we found a couple of small islands nearby on a map and decided we were going to punch out over the islands instead and wait for the cavalry to arrive. We were strapped into the airplanes on the flight deck waiting when the ship went to GQ, and they proceeded to launch the heavy tanker. My thoughts were not very positive as we were positioned on the catapult. At that point the launch was canceled."

After the squadron's return to Oceana there was another near miss. On the way back from ordnance practice near Cherry Point with pilot Gus Watters, Chuck called his wingman to move under his F-4B and take a look at his extended landing gear, as he suspected a blown tire. "The next thing we felt was a large bump. Fred had come up beneath us and then tried to get out of the position by moving sideways. Our outer wing panels collided. Fred was a little ahead of us, and his outer wing panel then ripped up our leading edge flaps. Then the aircraft did a kind of canopy-to-canopy look at

F-4S BuNo 153874 low and fast over the Eastern Californian desert in April 1984. (Jan Jacobs)

F-4S BuNo 153856, one of the first aircraft to test the Heater-Ferris scheme. The first to carry it was BuNo 153814, also ND 101 at the time. (Jan Jacobs)

Another *Devil's Disciples* F-4S (BuNo 155893) on a low-level sortie, showing the early Heater-Ferris pattern. (Jan Jacobs)

F-4S BuNo 155842 returns to NAS Fallon after a practice bombing sortie. (Jan Jacobs)

each other. I could read RIO Bob Graham's kneeboard. Both aircraft separated. I asked Gus if he could still fly the airplane, and he said 'Yes.' I knew we could not survive in the water, as nobody had poopy suits on. Fred could also control his F-4, so we flew back to Oceana and took the arresting gear. Both planes were a mess, but repairable."

Some of the Sixth Fleet exercises in the Mediterranean were large-scale strikes involving whole Air Wings, and "Smoke" Wilson took part in several:

"We had excellent exercises with the French (at least from our point of view). On our first attack on their air base near Marseilles we employed the 'mirror' of the classic Alpha strike, and they took the bait. We rendezvoused all the fighters at high altitude in what appeared to be a strike group, while the real strike A-4s went in

low. We had a grand melée; Phantoms versus Mirages and Mystéres, while the A-4s sneaked in and 'destroyed' their air base. This was a new tactic, and it worked so well because the French were watching our Alpha strike tactics á la Vietnam and expected the same. Not so. Too bad."

William Greer also enjoyed exercises with the French Navy:

"I was once jumped by an F-8 from the carrier *Arromanche*s, who apparently thought I did not see him. Since I didn't have any fuel to play with I held my course until he was a thousand yards or so behind, then did a high-g roll underneath and was delighted to see him dash out in front of me. One of my more frustrating moments came during one such exercise when the Admiral aboard *Forrestal* had promised a bottle of Napoleon brandy to the crew that could get a picture of their F-4 with at least five French air-

The final users of the Navy F-4 variants had the advantage of a new-generation tanker, the KC-10A, for some of their missions. On this occasion a flight of four VF-301 F-4S Phantoms were engaging USAF F-16s from Luke AFB, and the Navy pilots got a March AFB tanker over Yuma between engagements. (Jan Jacobs)

VF-301 F-4S Phantoms break overhead Miramar for a landing. (Jan Jacobs)

Live Mk 82 SE bombs en route for the Fallon ranges beneath F-4S BuNo 153856. (Jan Jacobs)

VF-302 Stallions flew the F-4N in 1978, including BuNo 151475, seen here on a VFR entry to the Miramar landing pattern. (Jan Jacobs)

craft. Out on CAP alone I spotted a flight of eight Super Étendards inboard to the Fleet. I escorted them some eighty miles to USS *Intrepid* and was never able to get the ship to send another F-4 so we could get the required picture."

Despite the ever-increasing demands on F-4 flyers, there were still sources of entertainment. David Daniels revealed one which relied on the basic visual similarity between the EA-6B Prowler four-seat ECM aircraft and its two-seat KA-6D tanker relative. Feigning radio failure, F-4 pilots would fly abeam a Prowler and extend their refueling probe for a "plug." Puzzled responses from the screen-gazers aboard the EA-6B would be met with more urgent gestures and further waggling of the F-4's fuel probe. While the electronics experts got increasingly insistent, making side-to-side "radar-scanning" gestures with their hands to suggest their true function, the F-4 pilot would draw a little closer and flip open his upper fuselage chaff and flare compartment door. Painted on its

inner surface was a conspicuous "rigid digit" silhouette, whose effect would be all too visible on the grinning faces of the fighter crew as they hit burners and pulled away.

Phixers

At all times the maintainers below deck had to respond to the demands of the Air Wing as they managed their elaborate jigsaw of aircraft in the hangars, fitting them in for attention and then extricating them for lifting back up to the flight deck. Rod Preston:

"On board the ship there were not many jobs that could be considered routine, because many of the tasks that were easy on land sometimes became a nightmare at sea. One of these was simply jacking the F-4 for maintenance, such as a landing gear problem. On land you basically just put your jacks under the plane, but not at sea! You had to use a series of jacks that were chained to the deck 'padeyes,' and a number of additional chains to keep the air-

VF-202 *Superheats* exchanged their F-8H Crusaders for F-4Ns in 1976 within CVWR-20. Note the open chaff door on BuNo 152298, named *Bump*. (via Norman Taylor)

A mixed formation from three of the Reserve units. VF-201 *Renegades* flew the F-4N from 1974 for ten years from NAS Dallas. (via Norman Taylor)

VF-301 also flew F-4Ns before receiving the F-4S. This colorful CAG plane was photographed just off the San Diego coastline in April 1978. (Jan Jacobs)

A *Devil's Disciple* F-4N sits at NAS Miramar's hold-short line while two others fly touch and go patterns. (Jan Jacobs)

VF-301 F-4Ns were deployed to McChord AFB, Washington, in 1978 to play "bad guys" for the USS *Constellation's* Battle Group. (Jan Jacobs)

A section of VF-301 Phantoms tank from a VAQ-308 EKA-3B over the W-291 "SoCal Oparea." (Jan Jacobs)

VF-302 F-4Bs lined up at Miramar in September 1974 with BuNo 152970 in the foreground. (via A. Collishaw)

An unusual combination of VF-301 F-4S Phantoms, led by an F4U Corsair and accompanied by a VC-13 TA-4J photo plane. (Jan Jacobs)

The original Ferris "dazzle camouflage" on a VF-194 F-4J in February, 1977. (via C. Moggeridge)

Immaculate formation by the "Blues" in February 1969. (Boeing/ McDonnell Douglas)

craft from falling off the jacks. On top of all this, you had to wait for the carrier to be sailing in a direction that was most stable in case the ship took a bad roll and caused the plane to fall off the jacks during the transition from being on deck to the raised and secured position. Something so simple on land took a lot of collaboration between the squadron and the ship's company at sea."

Corrosion control improved greatly throughout the 1960s, but it required constant surveillance. Towards the end of the Phantom's career one aircraft made a hard deck-landing and some badly corroded centerline splice-plates on the wing spars failed, causing a wing to separate. Only one of its crew ejected from the blazing wreck.

Learners

As each new batch of pilots and RIOs came fresh to the F-4 they quickly learned to appreciate its strengths and manage its minor eccentricities. In Curt Dosé's opinion, "The F-4 was a marvelous fighter, especially for 1970. We learned to fly it to the limits safely,

at any altitude. We used the heavy-g buffet as a 'Q' to energy and angle of attack. You stayed off the ailerons at high angles of attack unless you wanted to depart the aircraft for a snap turn. We lived in it and loved it. In Vietnam with VF-92 we never lost an airplane we were escorting to MiGs." Transitioning to the F-4 for the first time five years later was Dave Daniels. He noted that his first flight was in F-4J BuNo153805 on Tuesday, 11 February 1975, with LCDR Terry Heath in the rear seat. He jotted in his diary, 'It went pretty well. It really has got a lot of power. I did a break and some landings. Very tiring." Later he recollected:

"During engine start and at idle I was surprised at how much those J79s rumbled. They didn't have the silky-smooth whine that I was used to [in the TF-9J Cougar]. I also noticed that even a tiny nudge on the throttles caused a massive increase in noise and a whooshing sound of air going in the intakes just outboard of my elbows. The plane had an empty 600 gallon tank, and that made it 4,000 lbs lighter than normal, so I made the take-off using 'mil' power only. I thought it accelerated down the runway all right, but

The *Blue Angels* taxi past in formation. In all, fourteen F-4Js were converted for use by the team from 1969 to 1973. Dummy Sparrow missiles contained blue and red dyes which were used to emphasize some of the maneuvers. (Simon Watson)

Blue Angel 4 (BuNo 153085) with LT Ernie Christensen at the controls returns from a spectacular display. (via Norman Taylor)

LT (later RADM) Christensen's famous "no-gear" landing at Cedar Rapids, Iowa, in August 1970 with one engine jammed in afterburner. He ejected safely, as did leader Don Bentley during a similar incident at Kingston, North Carolina. (via Norman Taylor)

wasn't anything spectacular. During the flight we did a few acrobatics, and at slightly higher angles of attack the plane shook like it was about to enter a high-speed stall when it still had plenty of lift. It impressed me as something of a truck in the sky.

My second flight the next day rated a few more notes: 'The afterburner take-off really opened my eyes. I went to Mach 1.4, did some acrobatics, half and full-flap landings after a TACAN-GCA approach, and one with a single engine. I distinctly remember the instructor telling me to put the plane into a 300 kts circle pulling 3g, then go to full afterburner. We remained in the 3g turn and quickly picked up 100 kts. I also discovered that the plane could pick up 100 kts every second by going to near zero g, watching the oil pressure drop to almost nothing, and going to full burner. That sort of performance really impressed me. One thing I really liked about the F-4 was the reliability of the engines. I have about 1,760 hours in the Phantom and only had two compressor stalls. Both occurred at low airspeeds and high angles of attack. I only had one engine fire, on engine start. In the mid-1970s the F-14 Tomcat was having a lot of engine stall and fire problems. I was glad I was in the F-4."

William Greer experienced the F-14A Tomcat in the mid-1970s: "I had flown aircraft with high thrust and aircraft with good stability, but the F-4 was the first to have both. While the F-14 was undoubtedly the Rolls-Royce of fighter aircraft, the F-4 was certainly the Land Rover. Powerful, utilitarian, and aggressively ominous; I loved it without reservation." These views were also echoed among the first generation of Navy Phantom pilots, such as Hap Chandler, who considered the F-4, "the finest fighter, along with the Spitfire, that I had the privilege to fly. The F-4's engine response and excess power could get you out of delicate situations." Group Captain Mike Shaw also found it, "a remarkable aircraft. It flew very well when it was supersonic and when it was in the landing configuration, but it was rather barge-like and vague in between. The aircraft's strength was its adaptability. It was truly a multi-role combat aircraft, although to me it seemed a shame to have to use such a capable air-to-air weapons system for mud-moving. All its air-to-ground ordnance had to be suspended on external racks. The drag was significant, and the aircraft's supersonic abilities were then irrelevant.

Nevertheless, the F-4B proved the view that, 'fighters can always be made into useful bombers, but rarely the other way 'round.'" MGEN "Lancer" Sullivan always considered the Phantom his favorite, "mainly because it was such a manly aircraft. It took a strong pilot to be a really good Phantom driver, like the AV-8 Harrier." AMS Rod Preston, fixing Phantoms on USS *Midway*, admitted that, "it was ugly and a big beast, but even today it was the kind of beast I'd want on my side." For Manfred Rietsch, the F-4 was, "The greatest fighter-attack aircraft of its era. Pilots who were privileged to have flown the F-4 fell in love with that brute machine. I suspect that aviation history will eventually judge the F/A-18 Hornet in the same light."

Colonel Rietsch was among an increasing number of F-4 pilots who, from the mid-1970s were in a position to compare the Phantom with its inevitable successors. The first of these was the F-14, which appeared off South Vietnam with USS *Enterprise* in the closing stages of the Vietnam conflict to cover the evacuation of Saigon in 1975. Steve Rudloff was enthusiastic about the new Grumman aircraft's "pure fighter" conception: "The F-4 was certainly a long-range interceptor. It wasn't a very good fighter because of its turning ratio. It wasn't a very decent bombing platform because it had never been designed to be a bombing platform. The roles we undertook during the Vietnam War were, for the most part, not designed for the aircraft we were in. I later flew the F-14A, an absolutely fantastic experience, because it really was a fighter plane, and we were never asked to take over the bombing role." Curt Dosé also transitioned to the Tomcat, flying two tours with VF-2. On one occasion he drag-raced a F-14 on a catapult against a famous ultra-high-performance car for a publicity stunt. The Tomcat won.

Phantom Facelift

At the end of the S. E. Asia conflict the Navy and Marines had twenty-one F-4J squadrons and a total of 448 F-4s of all varieties. The last F-4J was delivered on 7 January 1972, and the F-14 began its first cruise on 7 September 1974, equipping two new squadrons, VF-1 and VF-2. Phantom squadrons, such as VF-142 and VF-143, began to transition to the Tomcat in the same year. VF-14 and VF-

32 had begun their transitions even earlier, at the beginning of 1974. However, it was another ten years before the last F-4 squadron to move to the Tomcat relinquished its Phantoms. During that time many improvements were initiated to keep the F-4 viable.

One proposal to improve its accuracy in the attack role was to use laser-guided munitions, as the Air Force had done with its F-4s. The Paveway LGB was essentially an Air Force initiative, and by late 1972 it had enabled F-4Ds to test KMU-342 precision-guided munitions and carry out precision attacks with the early "GBU" series weapons. At that stage the Navy showed an interest, but it never resulted in combat use of LGBs by Navy Phantoms. Brian Grant was one of a batch of aircrew from Westpac-deployed squadrons who became involved in initial training when it seemed that the weapons might be used in combat: "I had volunteered as the LGB designator from our squadron and went through training at China Lake. This occurred during the work-up to the 1973 cruise, but our Air Wing never deployed this weapon technology." The designator device at the time was a hand-held variant, used by pilot or (more likely) RIO to mark the target. It was occasionally used by F-4s to "lase" LGBs carried by A-6 and A-7 squadrons.

John Nash filled in a little of the politics behind the Navy's rather tardy start in employing these new weapons:

"I was on USS *Enterprise* with VF-142. I flew the Carrier Group Commander to Ubon RTAB, and we had our introduction to the LGB there as the Air Force had started to use them. Our Admiral wanted to see what they were all about, so we spent a couple of days there. Our problem was that we didn't have the laser markers for LGBs, but everything we heard about them was impressive, and the Navy adopted them thereafter. The biggest problem was getting the Navy and Air Force to cooperate on anything. It could have been done, using USAF Fast FAC aircraft [as designators] and Navy F-4s [as LGB droppers] in Vietnam."

Bee Line

A chance to improve both the F-4B and F-4J and to prolong their service life came in 1970. The enormous cost of the South East Asia engagement had drained defense funds, making it imperative for the Navy to get the most out of its existing equipment. At the same time the massively over-budget F-14A was experiencing a delayed development and a reduction of the 772 aircraft originally ordered to only 301. The gap between the planned retirement of the Phantom and its replacement by the F-14 had to be covered, and there were various proposals for a new-generation F-4 to do this. One Congressional group suggested Rolls-Royce powered F-4s similar to those supplied to the UK. These were to have been bought back from the British in part-exchange for a version of the F-4J, which was actually designed as the McDonnell-Douglas "Tomcat alternative" when the F-14 encountered major cost-overrun difficulties. Known as the F-4J(FV)S, it had totally new, F-111-style, shoulder-mounted, variable-sweep wings and a larger, zero-anhedral stabilator, mated to the existing F-4 fuselage, but with the landing gear housed in the center fuselage. General Electric GE1/10 engines were planned, with Rolls-Royce engines for the UK variant (intended as a Panavia Tornado alternative). The "Swing-Wing Phantom" never left the basic drawing board.

More attractive to the Navy's budget controllers was the prospect of refurbishing the weary, ten year old F-4B fleet. Project *Bee Line* followed the very successful conversion in lieu of procurement (CILOP) program for the F-8 Crusader by re-lifing 228 F-4Bs. Airframe life was effectively turned back to zero, and a further 4-5,000 hours per aircraft was anticipated. The "Bee Line" was established at the Naval Air Rework Facility at North Island (NARF NORIS) and Norfolk, Virginia, and the first development aircraft flew on 14 June 1972. Re-designated F-4N but retaining their BuNo identities, the refurbished aircraft provided the Navy and Marines with updated, first-generation Phantom airframes into the 1980s. The aircraft were completely stripped, fatigue-tested, inspected, and repaired where necessary. Wiring was replaced, IFF was updated, the inboard leading-edge flaps were deactivated, and a new Sanders AN/ALQ-126 DECM system was installed. Two of this system's antennas appeared near the intake lips, with their cable conduits "scabbed" onto the intake flanks, and others were in four positions under the intakes at the wing root and under the rear fuselage. The same system had been retro-fitted to many F-4Js, but with shorter cable fairings, as the internal structure could be accessed more easily for re-wiring.

The basic Aero-1A radar/FCS and J79-GE-8B engines had to be retained, but some significant weapons updates were allowed for in the rebuild. One was the Sidewinder expanded acquisition mode (SEAM) AIM-9G, which was able to acquire its target from an off-boresight flank position as well as from the rear. More revolutionary was the Honeywell AN/AVG-8 visual target acquisition system (VTAS). This used a sight mounted on the pilot's helmet,

Renowned for their incredible close formations, the *Blue Angels* made full use of the F-4's exciting power and acceleration in their shows. (Boeing/McDonnell Douglas)

VMFA-531 adapted the fashionable black vertical stabilizer to include their logo. (Simon Watson)

enabling him to track targets visually and use the radar to guide AIM-7 missiles towards them by "sighting" them through his slaved helmet optics.

Looks Could Kill

F-4B RIO Chuck D'Ambrosia went to work for Honeywell's research Group at Minneapolis in January 1969. The company had already received a NAVAIR contract to develop a helmet-mounted sight based on previous work to produce such a system for the Army's canceled AH-56 Cheyenne attack helicopter.

"While the company had good engineers, they lacked somebody with direct experience of fighter aircraft and especially the F-4. By default I became their resource. I was well aware of the shortcomings of the F-4's weapons systems for ACM, and was highly motivated (overly idealistic) to see if I could influence some hoped-for major improvements. As a fighter at medium altitude the F-4 was woefully lacking. Even if it had had a gun it could never have brought one to bear on a target and tracked it, because of its poor turning performance and the extreme buffeting experienced with high angles of attack. Missiles were the inevitable weapons of choice. One of the big problems in Vietnam was the need to do visual ID passes before an engagement to sort out the bad guys from the friendlies. That meant that even the newest Sparrows were useless, because you were either inside minimum range or you could never achieve a lock with the fire control radar. The radar was stabilized to scan horizontally, and there was no good way to put it effectively where you wanted it and quickly get a lock on the correct target. Even with the newest Sidewinders (then AIM-9D) getting a lock was a problem because of this difficulty of tracking the target within the narrow field of vision of the missile.

When I saw what Honeywell was doing I recognized lots of interesting possibilities. What they had was a standard Navy helmet, which was heavy as hell (after thirty years my neck still hurts!), with relatively heavy electronics and two IR sensors mounted on each side of the helmet. The unit projected a reticule on either the visor itself or a small 'granny glass' lens on a stem in front of the

pilot's eye. What I wanted was a light-weight helmet with very light-weight electronics projecting the image on the helmet visor. On each side of the F-4 cockpit, mounted on the sides of the canopy rails were IR projection devices [small rectangular, flat 'boxes,' angled towards the pilot] that put out rotating beams of IR energy. The sensors on the helmet detected the beams of energy and worked out the position of the helmet [i.e. where the pilot was looking]. This calculation was then used to slave the radar antenna to the proper location, where a target lock could be achieved.

The unit was tested by VX-4. One of my squadron mates, 'Speedy' Marv Sealy (Captain, USN, Retd.), had recently been transferred to VX-4 and was involved in the tests. It operated as designed, but the overall system required much more work to make it really effective. I wanted the company to get behind miniaturizing the electronics to greatly reduce the weight on the helmet, and also a new, lightweight helmet that would not 'float' around on the pilot's head during maneuvers. I wanted a system to control the seeker heads on the AIM-9 so that you could slew the seekers to the target and not have to point the aircraft directly at it. (The Soviets watched us, did all of this, and made it part of the MiG-29 weapons system.) I even wanted our Radiation Center to conduct a multi-spectral IR measurement program in conjunction with the Navy in order to develop an IR sensor for the Sidewinder that could be relatively immune to sunlight reflections off clouds and other distracting influences. I was very naive as to what was possible with a profit-making company and a Navy that was not really interested in spending more money. However, I do find it extremely ironic that after thirty years all these things are now being incorporated in new fighter programs. The technology existed thirty years ago, and at relatively modest cost. The Navy did deploy VTAS in limited numbers of F-4s [VF-154 tested the unit in combat], but by that time the Vietnam War was all but over, and costs were being reduced everywhere."

VTAS added considerably to the Phantom's credibility as a fighter for another ten years at least. However, it was the major structural inspections and reinforcements administered under *Bee Line* which really saved the F-4Bs from the boneyard. Fatigue prob-

VMFA-321's bicentennial scheme on F-4B BuNo 151499 (via B. Pickering)

lems had been endemic in some of the Fleet aircraft from early days, as John Nash described:

"We had tech reps around, but they didn't really know what we were doing, though we were given g-limits to observe. When I went through the RAG in 1964 they were starting to see wing cracks and wing-tips were vibrating off. That was during the time they put the big steel strap through the wing. We weren't even doing ACM in 1964—there wasn't an ACM program—and we were limited to 2-3g or so. When they put that strap in they moved the stress points inboard, and by the time I was CO of a squadron (VF-161 in 1976) we had airplanes that were breaking in half down the middle, along the keel, on landing. The stress was moved to the main fuselage. Generally speaking, if we maintained g-limits (6.5 g with less than 10,000 lbs of fuel) there wasn't any short-term damage. Over ten or twelve years you'd start to see it. Also, there were always guys who would over-stress the airplane and wouldn't report it. One 7g pull may be equal to ten years' fatigue life at 6g. If the maintenance

guys found cracks they would notify NARF, and they would tell everyone to check fatigue indicators in that area."

Squadrons began to test the F-4N's improved systems in routine use. VF-111 *Sundowners* returned to Vietnam once again in 1974-75 to cover the final U.S. evacuation of South Vietnam and Cambodia. CDR John "Hot Dog" Brickner, their CO, found that some parts of the package of updates in his squadron's F-4Ns brought real benefits:

"VTAS was one of those new systems that actually worked as advertised. Put the pipper on the target and hit the nose-wheel steering button and, like magic, a radar lock! Of course, it was limited to the gimbal limits of the radar: plus or minus 60 degrees. The sensors on your helmet and the sensors in the cockpit combined with the little monocle 'pipper' eyepiece that flicked down from the helmet, and slaved the radar to the pilot's line of sight. I thought it worked beautifully, though we had to be very careful of the sensors on our helmets: there were no replacements. For some reason, VTAS was removed from our F-4Ns after six months.

Some of VMFA-321's markings lived on when the overall gloss gull gray scheme was introduced. (Simon Watson)

Another *Black Barons* F-4N on return from a sortie in May 1980. Note the braking 'chute draped over the wing. (James E. Rotramel)

SEAM Sidewinder was a real bonus for ACM. It allowed the AIM-9 to lock on to 20 degrees off aircraft boresight, with its seeker head slaved to the radar, and the pilot could lock on to the IR tone and shoot if he thought there was a reasonable chance of a kill. It was still up to the crew to decide if the combination of range, off-angle, closing velocity, and dynamics gave a high kill probability."

The F-4N also received the AN/ASW-25B Link 4 one-way data-link, as used in the F-4J. John Brickner was less enthusiastic about that:

"It sounded great, but it was a one-way, receive-only link. It was VHF, i.e. line of sight, and the same as the good old voice radio. While the sender (E-2 Hawkeye, the aircraft carrier, or a cruiser) could continually send up-dated link information to many aircraft at the same time, there was no real way to determine if the messages were received or verified except by voice. For air-to-air intercepts it was tactically useless. The sender could send information, such as heading to intercept, bearing, distance to target, and relative altitude (assuming you had your IFF switched on). Most information was displayed on the bearing/distance/heading indicator (BDHI). It was almost like a toy. When we flew intercepts at night or were on a CAP station we tried to make it work, or should I say, we tried to make the controller make it work.

Link 4 also provided an automatic landing system, which worked a little better but had its limitations. I remember making the first Link 4 landing aboard USS *Coral Sea* in 1974. Many of the squadron pilots didn't trust it and wouldn't try it. First, you needed a good automatic power compensating system (APCS), coupling engine power with angle of attack to keep you 'on speed.' Second, you needed a great auto-pilot that you could trust. I don't know many fighter pilots who trusted an auto-pilot. Then the carrier would lock-on your beacon with their radar, and the ship's computer would send glide-slope and line-up information to your aircraft. You always had your finger on the disengage switch, just in case. If the weather was good and the deck wasn't pitching it worked pretty good. As for the DECM system; we never really trusted it,

VMFA-321 adopted the Tactical Paint Scheme for their F-4S Phantoms in the 1980s. (via Jerry B. Houston)

This F-4S shows the Tactical Paint Scheme on MG 12 (BuNo 153900). (via B. Pickering)

and you were never sure of it working, but we never really had to use it. Fighter pilots are very distrustful. That way, they live longer."

S for Super

Beefing up the Phantom for a longer life meant several ECPs, which resulted in external straps and plates on wings, stabilators, and lower fuselages, as well as internal reinforcement, adding up to 1,800 lbs to some variants. A second program of major refurbishment was devised for surviving F-4Js as a follow-on to *Bee Line*, and it affected 248 out of a proposed total of 262 aircraft. The resulting aircraft, the F-4S, was so much improved that one F-4 squadron commander assured the author he would have "performed an unnatural act" to get the chance to fly it. F-4S make-overs kept a number of USN squadrons in business while they awaited their delayed Tomcats, and the variant served with the Reserve squadrons until 1985. For the Marines, the F-4S held the fort until the F/A-18 entered service and remained in use as late as 1992. The F-4S service

life extension program (SLEP) included VTAS (for aircraft not already equipped with it) and much the same detailed inspection, rewiring, and strengthening as *Bee Line*, but added a number of radical and long-desired improvements. Its J79-GE-10B engines were genuinely smokeless, virtually removing the F-4's incriminating sky-signature and operating at higher exhaust gas temperatures (EGT). A new digital version of the AWG-10, the -10B, was bought, plus updated TACAN and, at last, provision for two UHF radios (ARC-159). The AWG-10B offered superior reliability and automatically computed ordnance delivery using radar ranging. Pale green electro-luminiscent formation strip lights, similar to those on USAF F-4s, were fixed to the vertical stabilizer, nose, and fuselage flanks. A more important carry-over from the ultimate USAF Phantom, the F-4E (LES), was a set of leading-edge maneuvering slats. These were originally developed under the *Agile Eagle* project and tested on YF-4E 62-12200 (inboard and outboard slats) and YF-4J BuNo 151497 (outboard wing panels only). Tests of various F-4E

For their retirement ceremony on 13 July 1991 a *Black Barons* team under 1 Lt "Bo" Bogaczyk repainted their "Triple Nuts" F-4S in a scheme resembling their original starry pattern. Aviation writer and photographer Don Spering applied the final artwork. (Simon Watson)

slat profiles continued through 1970, including combat testing by the Israeli AF. A contract for a full F-4E slats program was issued in November 1970, and the configuration was hardened by February 1972. Modification kits, including new outer wing panels, were then issued for 304 F-4Es. The thin, hydraulically operated slats improved the F-4E's turn performance by 50 per cent and, in the F-4J, reduced landing speed by 12 kts. CDR Bruce Thorkelson found the slats, "gave the aircraft much improved nose control, especially at high angle of attack. This improvement was most noticeable during ACM engagements. A few A-4 Aggressor pilots from the Adversary squadron were more than a little surprised by the improved control."

Contract go-ahead for slats for the F-4J was given on 9 July 1976. Patuxent River had tested a variety of configurations using F-4J 153088, eventually deciding on a set with a rather more square outboard cross-section and a longer wing fence than the F-4E type. Although two slat sections were installed on each wing leading edge at first, it was found that a single unit gave much the same advantage. It extended automatically at 11.5 units of angle of attack, retracting at 10.5 units.

When the first F-4S (J001, BuNo 155565) was rolled out at NARF/NORIS on 26 May 1978 for NATC testing it lacked the slats, as did the first 47 aircraft, though they were later retro-fitted. The first slatted aircraft (BuNo 155855), which also had reinforcing straps, was the 48th F-4S conversion (J048). VMFA-451 *Warlords* became the first recipients of the F-4S when they took delivery of the first service conversion (BuNo 155899, J050) in November 1979, and VF-21 became the first USN user the following month. In all,

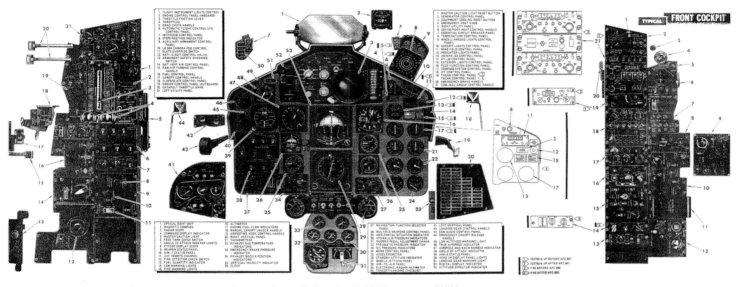

Front and rear cockpit instrumentation and control panels for the F-4S Phantom. (USN)

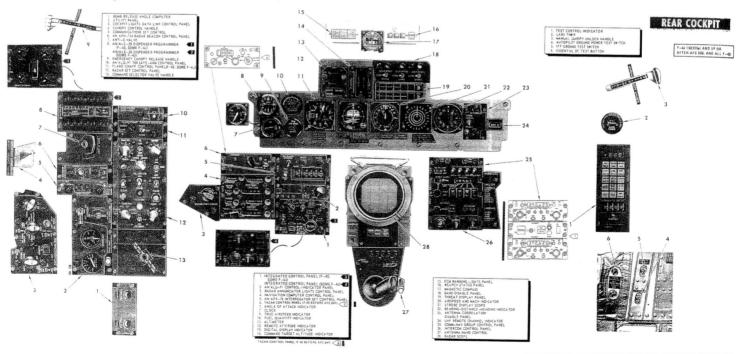

VMFA-312 *Checkerboards* had one of the most colorful Marine paint schemes, but tone-down time shrank the color element severely by August 1981. (James E. Rotramel)

ten USN squadrons (plus test units) and twelve USMC units flew the F-4S, though in some cases for quite short periods. VF-21 and VF-154 went back to the F-4N for their 1980 cruise on USS *Coral Sea* when it was found that the F-4S was a little too hot and heavy on approach to its smaller, *Midway*-class deck. VF-154's compensation was to service-test the AIM-9M all-aspect Sidewinder in 1983, having done the same for the longer-ranging AIM-7F Sparrow in 1977. These missiles added significantly to the Phantom's "fighter" capability by reducing the need for agility on the part of the launch aircraft.

For John Nash the F-4S, which he flew at VX-4, was, "a tremendous airplane. We used the VTAS and improved ACM performance quite a bit, in that you could acquire a target reasonably at about 30 degrees off your nose, off-heading, which would reduce your requirement to turn so much." He had seen Phantoms at St. Louis in 1967 with wooden test slats during the *Agile Eagle* program. Prior to that, "flying the F-4 to the edge of the envelope in ACM was more of an art than a science. In the old hard-wing Phantom, if you were going straight down at 250 kts and tried to pull the nose to the horizon it would just buffet. The wings would rock, and eventually you'd just stall and fall out of the sky. Then, when we got slats you could do the same thing, comfortably pull 3 g, and pull the nose to the horizon and actually maneuver the airplane. There was very little similarity in the flying qualities at high angles of attack for the two airplanes. Performance in a turn was almost identical to the MiG-21. When I left VX-4 I went to VF-161 in 1976, and we had F-4Bs, including one I had flown as an Ensign in 1963. It was hard to resign yourself to flying that airplane after just leaving the F-4S."

This F-4S, photographed later in 1981, displayed nothing to indicate its *Checkerboards* ownership. (via B. Pickering)

The striking color scheme worn by VMFA-451 *Warlords* for their short cruise on USS *Forrestal* in the bicentennial year. (USN via CDR James Carlton)

The *Silver Eagles* stayed with F-4s until the end of 1984, having moved from Iwakuni to Beaufort in 1977. BuNo 153848 appeared at Oceana in May 1980. (James E. Rotramel)

Captain Fred Vogt, USN, flew some of the last block of F-4Js with VTAS. "They were great aircraft, except that McDonnell must have put in all their old stock they had on the shelves to finish the production run! These F-4s were maintenance nightmares. I heard that the F-4S was what the Phantom should have been all along. The weapon system stayed up, it had several good, hard turns before it lost energy, and slow-speed maneuvering was definitely improved."

In Reserve

New Phantom squadrons continued to appear well into the 1970s, partly to compensate for the shortfall in F-14 procurement. Two veteran F-8 Crusader units, VF-191 *Satan's Kittens* and VF-194 *Red Lightnings*, converted to F-4Js as late as June 1976, though they de-commissioned less than two years later after a couple of cruises on USS *Coral Sea*. VF-171 *Aces* formed at Oceana in August 1977 to take over as the East Coast F-4 RAG, VF-101 having moved on to the F-14A. It continued to train F-4S crews until mid-1984, at which time VMFAT-101 *Sharpshooters* took over the role, since the Marines by then had opted to use the F-4 longer than the Navy. A second Marines training unit, VMFAT-201, had formed on F-4Bs in 1967, disestablishing with F-4Js in 1974. The first USMC line squadron to receive the F-4S was VMFA-251 *Thunderbolts*, which had been among the first to re-equip on the F-4 in 1964. The squadron flew "DW" coded Phantoms from MCAS Beaufort for over twenty-one years.

The mid-1970s also saw the F-4 passing into USN Reserve squadrons to replace F-8s. West Coast unit VF-301 *Devil's Disciples* and VF-302 *Stallions* acquired F-4Bs in 1974, graduating to the F-4S (via the F-4N) in 1984. The decision to establish an East Coast Reserve Carrier Air Wing (CVWR-20) in 1970 also resulted in VF-201 *Hunters/Renegades* and VF-202 *Superheats* operating F-4N and F-4S Phantoms in 1974-76 until transitioning to the F-

14A in 1986. The *Superheats* had the distinction of flying the last tactical USN Phantom to storage in May 1987.

USNR use of the Phantom actually began in 1969 when VF-22L1 formed at NAS Los Alamitos, south of Los Angeles. Ken Baldry was its CO, and presided over an accident-free year of operations, albeit the squadron's only year of existence.

"VF-22 was an experiment to see if the Reserves could operate the F-4. Los Alamitos is very close to a retirement community called Leisure World. Those who are familiar with the unearthly howl of the J79 engine as it is accelerated can imagine the consternation of the retired folk as we checked out our airplanes and shot touch-and-goes over their heads. The base telephone operators were shell-shocked, to say the least!

I can say that we were by far the most combat-ready squadron in the entire U. S. Navy, as all our pilots and RIOs had made at least one combat cruise in the F-4, and a number had two or more. Between the pilots and RIOs we had accounted for three MiGs and one An-2 Colt. During the one personnel inspection that was held while the squadron was in existence the Inspecting Admiral walked away, shaking his head at the display of medals.

Unfortunately, we had airplane problems, as they gave us the oldest F-4s in existence at the time, the '148' series machines [e.g. BuNo 148392, coded 7L 108, the first USN Reserve F-4 on strength], and we were at the very end of the supply pipeline. We did a lot of scrounging at Miramar, North Island, and as far away as Alameda and China Lake. On a good day we could plan on flying four aircraft, and on a few red letter days we could get six in the air at one time. There was no shortage of experienced pilots, and RIOs as officers were resigning from the Navy in droves. Flying in VF-22L1 consisted of a lot of intercept flights and service flights, acting as 'targets' for ships based at Long Beach. We also took non-explosive ordnance (i.e. Mk 76 bomblets and non-warhead 2.75 inch rockets) over to various desert targets at Yuma and El Centro.

VMFA-115 deployed on the USS *Forrestal* in 1981 before transitioning to the F-4S. (Simon Watson)

Displaying that distinctive "Sierra Hotel" code in the largest possible lettering, a *Sharpshooters* F-4N (BuNo 152230) rolls out for an ACM sortie from Yuma in February 1976. VTAS units are visible inside the front canopy. (James E. Rotramel)

For obvious reasons they did not want us carrying the heavy stuff out of Los Alamitos! We did make a mini-deployment to Point Mugu, where I had the pleasure of firing the first Reserve Sparrow at a drone for a kill.

One of my more memorable flights was with my XO, CDR Bill Kiper, aiding the Marines in calibrating a new radar. They gave us a GV [KC-130F] tanker loaded with 80,000 lbs of fuel and had us run supersonic head-on passes at each other from about 200 miles apart. Each time we met we 'threw out the boards' [airbrakes] and rendezvoused with the tanker. In two and a half hours flying we sucked the tanker dry. VF-22 became a casualty of a giant re-organization of the Navy when they moved the Reserves to active duty air stations. It was re-named VF-301 and given F-8Js at Miramar [F-4Bs in 1974]. The Phantom was a great Reserve airplane, as far as the pilots were concerned, because it was so comfortable and easy to fly, even with the inevitable lay-offs that Reserve flying is

subject to. Even after 2,500 hours in the Crusader I had a tendency to be a 'passenger' for the first few minutes if any great amount of time had passed since the last time I had flown the beast."

Among the Reservists was David Daniels, formerly one of "Devil" Houston's *Freelancers*, where he often flew with CDR Jan Jacobs [later to become Managing Editor of *The Hook* and *Smoke Trails*] as his RIO.

"I was brand new in VF-21. Once, I was out practicing radar intercepts with Dewaine 'Dewy' Cherry, also a 'nugget,' as RIO. Devil and his RIO were the designated 'bogey.' We were charging through the sky at a tremendous speed, with Dewy pouring 'bogey dope' in my ears. We were doing a stern conversion, so we offset about 40,000 ft laterally so we could do a hard turn and come in behind the 'bogey' to shoot a Sidewinder. We overshot really badly, and saw at the last moment that Devil was at 25,000 ft, but only doing about 200 kts, and he had the gear, flaps, and hook down.

VMFA-251's F-4J (BuNo 155836) leads examples of the Marines' other two principal attack/fighter types, the AV-8A Harrier and A-4M Skyhawk, in September 1971. Harriers replaced Phantoms in two USMC squadrons, VMFA-513 and VMFA-542. (Boeing/McDonnell Douglas)

VMFA-251 flew the F-4B in the 1960s. Captain Jake Albright takes off in his to fly chase for an altitude record attempt by LT COL Gordon Keller on 14 November 1966 at Beaufort. (USMC via Steven Albright)

Gray Ghosts **F-4Ns over the California coast in 1982. (Bud Brown via Steven Albright)**

Showing the Tactical Paint Scheme's poor resistance to wear and aviation fluids, a pair of *Cowboys* **F-4S Phantoms (BuNos 155829 and 153792) await the "go" signal at Cannon AFB in December 1988. (James E. Rotramel)**

That taught Dewy and me to closely watch the 'V sub C,' a notch on the radar scope that showed overtake rate."

One of the MiG-killers who remained in Reserve F-4s "for the duration" was CDR Bill Freckleton, who then moved on to the F-14A, which he flew with VF-302 until August 1993.

"We did have a training syllabus as far back as 1973 in VF-302, but in those days it more or less amounted to ACM. There were required training hops that were tracked, such as air-to-air and air-to-ground gunnery, aerial refueling, and so on. At Fallon, during our annual training the squadron adhered strictly to the training requirements that would lead to us getting the 'E award' in Readiness. All the elements of training were in the syllabus, and we usually tried to check all the boxes. When we got the F-14A in 1986 a very rigid, compartmentalized training syllabus was initiated that went by the book—no more fun days of ACM. Transition to the Tomcat wasn't particularly easy, but most of the aircrews had an

average of more than 1,000 hours of F-4 flight-time. The difference from the back seat of an F-4B, where the RIO didn't even have a fuel gauge (not even a totalizer, just a low fuel light) to an F-14A was night and day. There was so much more to do and be responsible for in the back of a Tomcat."

However, Bill Freckleton approved strongly of the F-4S, which the *Stallions* received towards the end of 1980.

"Compared to the F-4B's Aero-1A radar system, which was pulse-only search and track, the F-4S's AWG-10 was much more capable in being pulse Doppler in both search and track. This allowed for superior look down/shoot down capability. The dual radios also enabled a 'strike common' frequency, as well as a 'tactical/fighter' frequency at the same time, as opposed to being on 'strike common' and having to switch to 'fighter common.' In training situations the F-4s eliminated the necessity to 'go burner' at 18 miles in an attempt to cut down on the smoke trail. The smokeless

Red Devils **F-4Js bask in the Hawaiian sun. After grueling combat service in Vietnam and Thailand the unit remained part of MAG-24 at Kaneohe Bay, flying Phantoms until 1988. (via Simon Watson)**

Jay Kuenzle's VMFA-232 F-4 "Super J" (BuNo 155848) in January 1987 after shipment to the UK. (Author)

Detail of the AN/ALQ-126 DECM antenna on BuNo 155848. (Author)

engines gave us the ability to stay out of burner unless we really needed to get the knots up past 450 or so. With the F-4B you could see the smoke trail for miles and miles. It was actually somewhat stirring to see several smoke trails coming for miles away and know that a formation of Phantoms was inbound.

Miramar Reserve F-4Ss had VTAS, and it worked quite well. I also recall some pilots actually taking telescopic sights from rifles out to the F-4Bs and mounting them along the boresight axis of the aircraft to aid in visual acquisition of targets that had been contacted on radar. We used to place the radar target on boresight and let the pilot use his 'device.' However, I believe the F-4s, or at least some in VF-302, were requisitioned when we needed them for Fighter Derby or TACTS competitions."

CDR Bruce Thorkelson flew two Vietnam combat cruises, with VF-154 on USS *Ranger* and VF-142 on the *Enterprise*, joining VF-301 *Devil's Disciples* in 1977.

"During the late 1970s and early 1980s the entire squadron [VF-301] was made up of combat veterans. Several of these were MiG killers, including Scott Davis and Willie Driscoll, with his wingman Brian Grant. My opinion is that VF-301 was as professional and capable as the best of the Fleet squadrons. For every opening in the squadron there were multiple applications, and only the best were selected. We trained with and against Top Gun, the Air Force, Marines, the Canadians, and other Navy units. In the summer of 1983 VF-301 sent a Detachment to Langley AFB and spent several days fighting 'many vs. many' against the F-15 Eagle. Even though we were outclassed by technology (both airframe and electronic) we held our own against this fine aircraft. Of course, experience, cleverness, and trickery helped.

Our role as a Reserve squadron was to be ready and able to deploy with a minimum of warning and take our place alongside the active duty Fleet. Until the carriers changed over to the F-14 this could have been done relatively easily. Of course, as the Fleet went to the Tomcat the F-4 became obsolete and less viable."

VF-301 had twenty-four accident-free years, with 71,300 flying hours, 50,000 of them on F-4s. After two and a half years in VF-21, some of whose F-4Js had smokeless engines and VTAS,

Dave Daniels moved to the cockpit of an airliner, like many of his colleagues. He maintained his contact with the Phantom at NAS Dallas with VF-201 *Renegades'* F-4Ns.

"When I flew the F-4N the squadron had installed a small, digital backup radio in the RIO's cockpit, since one of the problems with the Phantom was having only the one UHF radio. This was very useful, not only for the ability of the RIO to transmit and receive separately, but also it was great in case of a main radio failure. When we switched from the F-4N to the F-4S we thought we would also have a second radio. No such logic had crept into the system—it had one radio. When we requested that the backup ra-

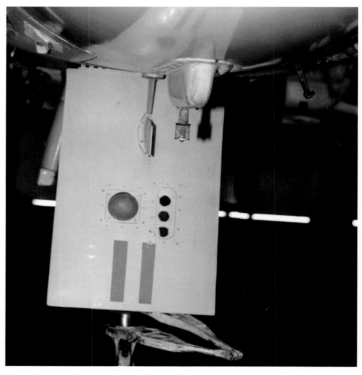

The Navy-style nosewheel door with carrier landing indicator lights and AN/APR-32 antenna beneath the radome distinguishes this F-4J from USAF variants. (Author)

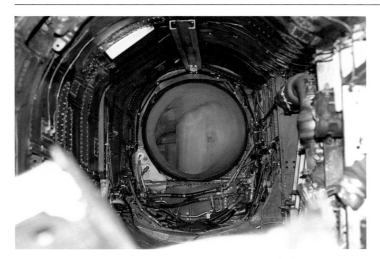

The left engine bay and intake interior of BuNo 155848. (Author)

The *Red Devils* patch was the only area which was repainted after receipt by the Fleet Air Arm Museum in England. (Author)

dios be installed it was denied, even though the radios were at the squadrons.

I quickly came upon a situation where I wished the F-4s had a second radio. At VF-201 we often flew an 'out and in' (flew to another base, landed and refueled, then returned home at night). Late one afternoon, shortly after we got the F-4s in mid-1984, I planned an 'out and in' with my RIO, Dave 'Mick' Jaggers. The plan was to go to NAS Pensacola, land and wolf a cheeseburger while the plane was refueled, then return to NAS Dallas. We got a late start due to a mechanical problem, and it was after dark when we began our approach. We had heard one of the *Blue Angels* land, and he had reported strong cross-winds and some rain. I was flying an enroute descent to a GCA when we lost our one and only radio. We continued to broadcast in the blind, hoping we still had a transmitter. I decided to transition to a TACAN approach and could catch a glimpse of the shoreline. I transmitted my intentions, put 7700 then 7600 in the IFF and confirmed. Right after that it began to rain

very hard, and we went to instruments. Then the TACAN began to spin. No radio, no TACAN, IFR at night: not a good situation. We took a wave-off and a dead-reckoning heading for Eglin AFB. We climbed up, reset the IFF for 7700, and began timing how far we had to go to Eglin. As we neared the base we found a hole in the clouds and let down. The TACAN recovered sufficiently to point to Eglin, and we were feeling better. We picked out a runway (turned out to be the wrong one; heck of a crosswind) and landed. When I popped the drag chute the crosswind just about swung us round, but we managed OK. It just reinforced my notion that we should have had two radios."

The Phantom could certainly absorb some systems failures and still survive in skillful hands. A VF-194 R*ed Lightnings* F-4J on a night intercept progressively lost its starboard generator, radio, cockpit intercom (ICS), and TACAN. On approach to the carrier the other generator also failed, and the pilot experienced vertigo, losing several hundred feet in altitude before he regained control and

Right airbrake and hydraulic strut. BuNo 155848 also flew with VMFA-235 (07DB) and VMFA-212 before its *Red Devils* service, accumulating several hundred missions in South East Asia. It was placed in open storage at NARF and was replaced in the squadron by BuNo 157255 at "WT 11." It received the "straps" update, but not slats, though it was scheduled to do so. (Author)

VMFA-323 was one of two USMC squadrons on USS *Coral Sea*, which were ready to provide air cover for the abortive rescue attempt for the U.S. hostages in Iran. ID bands were painted to avoid confusion with Iranian F-4D Phantoms. (Angelo Romano)

VMFA-212 included F-4J BuNo 155559 in its number during its final Phantom years at "K-Bay." (via Simon Watson)

An early and unadopted color scheme which VMFA-212 tried out on BuNo 153862 when they first received F-4Js in 1968. (LT COL Tom F. Rochford, USMC (Ret) via Steven Albright)

re-set the generator. He boltered twice, and lost the second generator again on his third bolter, leaving the aircraft without electricity and cockpit lighting as it headed into the night once again just above the sea. With the RAT providing limited power the crew then had to make a refueling connection with a KA-6D tanker. The pilot's vertigo deepened again, and his aircraft was diverted ashore, where its generator failed once more on landing.

New Hues
USNR squadrons participated in a series of camouflage scheme trials in the early 1980s. Sea-going Phantoms had retained their basic Light Gull Gray (FS 36440) topsides and Gloss Insignia White (FS 17875) underside colors from 1958 until February 1977. At that point CNO changed the fighter scheme to overall Gloss Light Gull Gray (16440). After twenty-two years it was decided that the white undersides of Naval aircraft made it easier to see their intended direction of turn in a fight. At the same time, CNO instigated tests to establish a new Tactical Paint Scheme for Naval air-

craft. Both moves were accompanied by an increasing emphasis on low-visibility markings and the toning down or removal of colorful unit insignia, which had always distinguished Naval and Marine fighter units like no others in the world. The importance of this heraldry to individual squadrons was emphasized by AMS Rodney D. Preston, who was with VF-161 on USS *Midway's* 1976-78 cruises with the F-4J and F-4N. His squadron's prominent black and red scheme was among the most attractive.

"The job of corrosion control techs was to remove and treat any corrosion found before it destroyed an area of the airframe. They were also the guys who put the fancy paint jobs on the aircraft which, before they went to a low visibility type, were the pride of many squadrons. A unit's logo is very important to it, and squadrons competed with each other to see who could make their planes look best. That changed with the enactment of a 'lo-viz' paint scheme that was basically shades of gray."

One of the experimental schemes applied to Reserve Phantoms was devised by aviation artist Keith Ferris and comprised a

Death Angels BuNo 157246 at the Kaneohe Bay refueling pit. This F-4J was eventually put on display at El Toro in VMFA-134 markings. (via Peter B. Mersky)

Thunderbolt 10 on approach in reduced visibility markings (via Peter B. Mersky)

F-4S BuNo 153824 in a distinctly blue variation on the Tactical Paint Scheme in October 1987. (via B. Pickering)

hard-edged geometric, disruptive pattern. This was modified by LCDR Chuck "Heater" Heatley at NFWS, using a "false canopy" (painted on the underside in a position equating to the real canopy) and less contrasting colors. Two VF-33 F-4Js were used to test this scheme in 1980 using water-based paints. Both VF-301 and VF-302 painted up sample aircraft also, and then adopted the so-called "Heater-Ferris" scheme as standard until 1984-85. For the Marines, VMFA-312 *Checkerboards* ran trials with the scheme in 1977. USMC Reserve squadrons VMFA-112 and VMFA-321 *Hell's Angels* also "went gray." Four shades of gray were used, starting with a light shade (FS 35307) at the tail and on the right wing, and working through FS 36375 and FS 35237 to a dark gray, FS35164, on the nose and left outer wing. Initially, the pattern worked in the opposite order of gradation. CDR Bruce Thorkelson was with VF-301 at the time. "The paint scheme looked sharp, and I believed it was effective in confusing visual reference to opposing aircrew. I personally experienced portions of the opposing aircraft 'disappearing' during ACM engagements. This was the intent in using four shades of gray. Against almost any background, a part of the aircraft closely matched that background." Other paint schemes, including a disruptive pattern of blue shades were tested by VX-4, and another scheme using three shades of gray (FS 35307, 36320, and 36375) was approved on 3 February 1982. VF-21 and VF-154 were among the last squadrons still operating the F-4 when the new scheme was applied to some of their F-4Ns on the USS *Coral Sea's* last Pacific Fleet deployment in 1983, before transitioning to the F-14A. For Navy, but mainly Marine F-4S aircraft, the official Tactical Paint Scheme used a more conventional pattern of Dark Gray (35237) upper surfaces, Medium Gray (36320) sides, and Light Gray (36495) undersides, though the Medium Gray was later extended to the undersides, too.

Blue Angels

The most distinctive paint scheme of all was the blue and gold of the *Blue Angels* Flight Demonstration squadron, which brought the Navy's F-4 to the attention of a huge audience world-wide. In 1968 the team retired their neat Grumman F11F Tigers, considered switching to the F-8 Crusader, and made the brave decision to work with the F-4J. A number of early development airframes had their radars removed and were modified for the team's use (including BuNos 153072/075/078/079/080 through 086), and the first of these to be readied, 153079, was flown to Pensacola on 23 December 1968. The first seven aircraft had J79-GE-8 engines, but subsequent additions had the more powerful Dash 10 engines, and these were used for the solo displays.

There were other minor modifications to the airframes, as former team member Bill "Burner" Beardsley explained:

"We had the 89 per cent afterburner index [operating at 89 per cent engine power, rather than the standard 94 per cent], and a constant thirty pounds of stick-pull force (with full nose-down trim) through a system of springs and cams, and that's what made the flying so physical." The idea was to remove some of the sensitivity from the control process and prevent minor pilot-induced oscillations by making the controls so "solid." With only two or three feet of separation between aircraft in many of the maneuvers it was important to damp out any sudden, small control movements. This arrangement made the aircraft very tiring to fly, and real strength was needed to hold position and yank the plane into sustained aerobatics, especially when the airflow disturbance between such large fighters in close proximity was constantly tending to force them apart. Pilots adjusted their seats and pedals so that they could keep their right arm braced against the leg to give the required stability and leverage on the control column. Also, as 'Burner' recalled, "G-suits really weren't necessary, especially for pilots in the Diamond formation. We never pulled over 5g, and the suit inflating and deflating would affect the 'platform' made by our thigh and forearm pivot point." A further engine modification was made to BuNo 153082, flown by the 1970 team leader, CDR Harley Hall: "He had a physical block on the throttles so that he couldn't select the fourth stage of afterburner, in order to give the wingmen something to play with."

BuNo 149434, one of the original batch of twenty-five QF-4B conversions. Above and below the nose are UHF command/control link antennas. (Simon Watson)

After two months of hard training in their new aircraft at the start of 1969 LCDR Bill Wheat's team evolved a program which adapted traditional *Blue Angel* routines for the hot and heavy F-4J. Several new maneuvers were added to show off the Phantom's power, particularly in the vertical plane. The first show was flown successfully on 15 March 1969, and the *Blue Angels* entered a hectic schedule, requiring almost 3,000 flight hours in their first season. CDR Harley Hall took over as the team's much-respected leader for the 1970-71 period, adding more new variations for their 88-show season in the continental USA, Hawaii, Puerto Rico, Canada, and South America. Their 1971 tour extended to five Far Eastern countries, which "Burner" found "very demanding. We did twenty-six shows in less than a month." CDR Hall returned to operational flying in 1972 and tragically became the last Navy pilot to be lost in Vietnam. LCDR Don Bently took over and led the 1971-72 displays, which mingled triumph with tragedy. During training before the 1972 season the Number 7 Phantom crashed inverted in the desert, killing the team's back-up pilot, LT Larry Watters. The rest of the 1972 season took the team on another extensive and successful U.S. tour, culminating in the award of the Meritorious Unit Citation for their first two years on the F-4.

Sadly, the 1973 season began with another accident in training, when three aircraft in the basic Diamond touched in flight. All three Phantoms went out of control, and their pilots abandoned ship at low altitude. Don Bently suffered back injuries and had to give up the leadership. His successor, LCDR "Skip" Umstead, led the *Blue Angels* to dazzling displays in Europe and Iran, but on their return to the USA another accident occurred at NAS Lakehurst when the Number 4 "slot" position aircraft collided with Umstead's lead Phantom. Of the two pilots and two crew chiefs aboard only one was able to eject. CDR Umstead and Captain Mike Murphy, flying Number 4, both died.

After the loss of six aircraft and four team members in a short time the Navy had to cancel the rest of the season, and the team's future was in jeopardy for a time. Eventually it was decided to re-

place the F-4J with the more agile and economical A-4F Skyhawk.

During its comparatively short time with the F-4 the team undoubtedly offered the most stirring and exacting formation flying that any crowd could have seen. Spectators would long remember the mighty sound of eight J79s afterburning in formation, and the sight of a tight Diamond formation with two of its components flying inverted. It was a powerful publicity device, and the *Blue Angels'* ethos inspired its audiences and evoked devoted support from all concerned with the team's operation. A *Blue Angels* crew chief described to the author how the jets were completely washed and waxed every week, and all aluminum brightwork was buffed with polishing compound. The aircraft were repainted each winter at El Centro by the squadron's own maintainers, and the paint jobs were constantly touched up, for example, after flying through a rainstorm.

Semper Fi

VF-161 and sister squadron VF-151 continued to cruise aboard USS *Midway*, making three Indian Ocean deployments in 1979-80 and transitioning to the F-4S in February 1981. Their last deployment was in 1986 as the final front-line USN Phantom operators, and the last cat shot occurred on 24 March of that year. In the same year CDR "Black George" Kraus, XO of VF-202 *Superheats*, a Dallas-based Reserve unit, made the last carrier trap by a USN Phantom from an operational unit on 21 October aboard USS *America*. With the F-4N, VF-202 had specialized in ACM, receiving the Combat Efficiency "E" Award in 1981, 1982, and 1985. Their "AF" coded Phantoms deployed to the Sixth Fleet in April 1983 (operating from Sigonella), and they moved to the F-4S in 1984. At the time of their final deployment the F-4's old partner in combat, the RF-8A Crusader was also making its final deployment. VF-171 Aces took to the F-4S in April 1981 and served as the East Coast RAG right up to June 1983, remaining busy to the last. In February 1980 they "carqualled" the first all-USAF F-4 crew aboard USS *Eisenhower*, and the first all-RAF crew to fly from a West Coast carrier, on USS *Kitty Hawk*. In Spring 1983 they scored a "first and last" for an F-4

RAG when eight F-4S Phantoms provided fighter cover for USS *Saratoga*, logging 244 traps during this last East Coast F-4 deployment.

Whereas the Navy followed a fighter/interceptor procurement policy, which led from the F-4 through the canceled F-111B to the F-14A, the Marines remained committed to the F-4. When the overall USN F-14 purchase was projected it included aircraft for four USMC squadrons, too. Originally the Marines had opted for a further 138 F-4Js rather than Tomcats at well over twice the price, but the F-14's superior ACM performance won them over in 1974. VMFA-122 was scheduled to transition to the Tomcat in late 1975, but the funding for the whole Marine F-14 program was canceled, and the Corps eventually ordered F/A-18As for the four projected Tomcat squadrons instead. Until the F/A-18 began to reach Marine units in 1983 the F-4 continued to serve in fourteen Fleet Marine Force and Reserve squadrons. The run-down of U.S. forces in Vietnam had resulted in the loss of several units. VMFA-542 and VMFA-513, both pioneer Phantom users, closed down F-4 operations in 1970 and switched to the AV-8A Harrier. VMFA-334 *Falcons* disestablished the following year. The shortest-lived squadron was VMFA-351, which formed as a Reserve F-4N unit at Atlanta in 1977 only to disestablish during the following year.

Other veteran F-4 outfits continued to refine their craft, including the *Gray Ghosts*. Eugene R. Hamamoto was a RIO with VMFA-531, and he described parts of the training routine at El Toro:

"My deployments were mainly to Yuma, and we ran missions in the Chocolate Mountain area. We used the Mk 4 gunpod and adjusted our bomb sights accordingly. We would practice the gun pattern where one of the birds would be the 'tractor' and drag a banner, while the rest of the flight made passes at it. I can remember on a two-week deployment there were zero hits on the banner, but one round hit the engine exhaust 'turkey feathers' of the towing F-4B, and three or four pilots argued to take credit for this solitary hit. The Mk 4 was added as an afterthought, and with the inadequate sighting system it was only effective in strafing where you could 'walk' the rounds towards the target.

My job as RIO in the bombing run was mainly to monitor speed, dive angle, and 'pickle' altitude. As we approached 'pickle' height I would call out, 'Stand by, stand by...Mark' to signal the correct drop altitude. If I didn't feel the g-forces come on very soon after the call I started reaching for my ejection face curtain! Really, the pilots were on the ball. If they needed to press a bit they would let me know immediately after the call. I remember one incident where we were commencing our bombing run and I called the speed 'too slow.' About a quarter of the way down I called that we were still too slow. At this point I felt a jolt as the pilot went into afterburners coming down the slide! It surprised the hell out of me, but this old salt knew exactly where we were at and wanted to have some fun with me.

Dropping bombs sometimes seemed so sterile. You'd pick out the target, determine run-in heading, dive angle, airspeed, pickle altitude, and pull-off heading. As you watched the bomb impact you'd see the silent flash and smoke, do it again, and then go home. You'd deal with the chatter on the airways and talk to the FAC, if there was one, while the oxygen mask wheezed in the background. One day I had a rude awakening when I was sent to FAC school in Coronado. Now I could see what it looked like from the ground.

As I watched the aircraft roll in and I gave the 'clear to fire' I marveled at how pretty and serene the scene was. The hills of the desert were silent, reminding me of Disney's 'Living Desert.' I watched as the speck dove out of the sky. Half way down, an aileron roll—what a hot dog! Soon after that the bomb came off, briefly flying formation with the plane. Poetry in motion. Then, at impact—Booom! I jumped out of my skivvies, went on my knees, and felt the concussion. It was the loudest noise I had ever heard in my life. Then I remembered why I signed up to be in the air instead."

K-Bay

One of the main USMC Phantom bases outside the continental USA was Kaneohe Bay on Hawaii's Mokapu Peninsula, home of MAG-24, which included VMFA squadrons 212 *Lancers*, -232 *Red Devils*, and -235 *Death Angels*. The three squadrons acquired F-4Js in 1968 to replace F-8Es, though the *Red Devils* transitioned a year

QF-4N BuNo 153039 departs Runway 21 at NAWS Point Mugu on a manned sortie. (USN/Vance Vasquez)

A QF-4N from Point Mugu takes on fuel from a USAF KC-135E over San Nichols Island during a Tomahawk missile test operation. (USN/Vance Vasquez)

The first F-4B converted to QF-4B configuration was BuNo 148365, formerly with VMFAT-201. It was delivered as a QF-4B to Point Mugu in March 1972 and shot down on 31 January 1974. (via C. Moggeridge)

F-4J (UK) ZE 356/Q of 74 Squadron RAF returns from ACM practice in August 1986. As BuNo 153850 the aircraft had flown with VF-33, VF-101, VF-74 (including 99.4 hours during *Linebacker II* and many sorties over Cambodia and Laos), VF-31, VF-102, and VMFAT-101. (Author)

earlier at El Toro, spent eight years forward-deployed in Japan, South Vietnam, and Thailand, and became the last USMC F-4 squadron to withdraw from S. E. Asia, in September 1973 before settling back at "K-Bay." All three squadrons, which were essentially training units, took part in rotational deployments (Unit Deployment Program, or UDP) to Iwakuni, as did ten F-4 units in total. Of these, VMFA-312 *Checkerboards* were the first squadron in the 2nd MAW to participate, in 1979. Two others, VMFA-513 and -531 deployed once to Atsugi. A further squadron which spent a brief period at K-Bay was VMFA-122 *Crusaders*, one of the earliest Marine units to receive Phantoms. After two Vietnam tours the squadron re-deployed to Hawaii in 1970 and prepared itself to transition to the F-14A, but was inactivated a few months later when the Corps withdrew from the Tomcat program. The unit re-established on the F-4J at Beaufort in 1975, remaining with the F-4J/S for nine more years.

According to LT COL Charlie "Burner" Mitchell, assigned to VMFA-232 from June 1982 to June 1985 (and later the CO of the *Red Devils* in 1997-98 in F/A-18s), the rotation program with jets to Iwakuni, "is like a revolving door. The three years I spent in VMFA-232 we would deploy to Iwakuni for normally six months and then return to K Bay for eighteen months. Prior to this, in the Vietnam era, squadrons were permanently stationed at Iwakuni. Today, Iwakuni has a mix of both; one squadron is permanently stationed there, and other squadrons move in and out on a rotation basis."

LT COL John Jay Kuenzle arrived at LT COL Gus "Rebel" Fitch's VMFA-232 in Hawaii in July 1982 to fly the F-4J/S:

"This differed from a normal F-4J because it had an upgraded radar (the AWG-10B), upgraded solid-state radios, and an 1,100 lbs belly strap which helped to hold the wings on. We were part of the unit rotation program which had begun in the late eighties when VMFA-232 was no longer permanently stationed at MCAS Iwakuni. The flying in Hawaii was very pleasant, with the weather being virtually identical day in and day out. Typically, there were seventeen pilots in a squadron, which meant you flew three times a week. During my four and a half years there I averaged about fifteen hours

a month. If you were flying a two-ship (called a section) mission, which was about half the time, you would fly either radar intercepts or 1 vs 1 ACM. The radar intercept missions were called 'attack/re-attack' because we would fire an AIM-7 at range and then re-attack with an AIM-9 in the rear quarter. Since we were flying an 'all-weather interceptor' the RIOs were supposedly directing the pilots around during these intercepts to simulate doing them in IFR conditions. The AWG-10B was a good radar for what it was designed to do, but by today's standards, or the F-15's, we often thought it was very primitive. It gave you two modes for doing intercepts; pulse and pulse-Doppler, and most RIOs used a combination of the modes depending on their ability and the stage of the intercept. A third of the time we would go out as a three or four-ship (a division) and practice VIDs [visual identification interceptions]. We did this a lot, because the feeling was that the Rules of Engagement were going to require it. We would get twenty or thirty miles apart and then practice any number of different types of VIDs. They were

F-4J (UK) ZE 354/R at RAF Wattisham in June 1986 with the black fin which became standard. This Phantom had flown with VMFA-232 at Chu Lai, VF-143 on a combat cruise in 1970, VF-121, VF-213, VX-4, and NATC. (Mick Sudds)

actually a lot of fun. We then flew about a tenth of the time against dissimilar bogies, either TA-4Fs from H&MS-24 or the F-4Cs from the Hawaiian ANG, never against the other Marine F-4 unit at Kaneohe.

The remaining flying was limited to air-to-ground. We dropped Mk 76s on one of three ranges, although we had no way of scoring our hits other than doing it ourselves. Then, on a rare occasion we dropped live ordnance, but only on two ranges and never more than one bomb at a time. We normally practiced pop-up attacks and occasionally worked with ground FACs. We did fly in major exercises with the Navy if they were passing through, and we did have an occasional missile shoot, firing AIM-7Fs and AIM-9Ls. I had only been with the squadron two months when we left for the continental USA in mid-September, 1982. We went to MCAS Yuma and bombed our brains out. Then up to Nellis AFB where we painted our F-4s with water-color paints and made them look very sneaky. We flew in Red Flag and participated in a *Constant Peg*. We were scheduled to return to Hawaii in November, but we had a devil of a time getting our jets back, and I didn't make it back until 19 December. On 4 March 1983, we headed for Japan for my first of three Westpac deployments."

Jay Kuenzle's F-4J/S, BuNo 155848, made the Iwakuni trip in 1982, but by September 1984 it was in store at NARF, replaced as "11WT" by BuNo 157255. It made the trans-Atlantic trip to the Royal Naval Air Station at Yeovilton, England, where it was presented to the Fleet Air Arm Museum. At the time of writing it was being transferred to another museum at East Fortune, Scotland.

F-4S models were phased in at MAG-24 between 1979 and 1982, and the squadrons used their "Super Js" (early F-4S prior to the wing-slat retro-fit) and slatted F-4S fighters in some punishing ACM practice. "Burner" Mitchell reckoned that the F-4S, "had a much better instantaneous turn rate than the F-4J, but at the same time it depleted your energy that much quicker. The F-4S was a tad bit slower than the J due to the slats, but the trade-off was well worth it for the better turning rate of the F-4S."

On the verge of their retirement, MAG-24 broke a few regulations and painted a *Red Devils* F-4J (BuNo 157307) in its Vietnam colors and a *Lancers* plane (155805) in a superb dark blue and white scheme. VMFA-235 were the last Kaneohe unit to move to the F/A-18 when its first Hornets began to arrive in 1989. Their CO, LT COL Craig E. "Spot" Sooy (a veteran of over eighteen years on the F-4 with over 100 combat missions), led the last of the "DB" coded F-4S Phantoms out of the base on 31 January 1989, making a final low-level pass along the main runway. In fact, the squadron transferred to Nellis AFB for a final aggressor role exercise with Air Force F-15s and F-16s before disposing of its Phantoms officially on 7 March.

In USMC service the F-4S proved to be a popular steed, and it eventually equipped all twelve F-4 line squadrons and the training unit, VMFAT-101. Two Reserve units, Andrews AFB based VMFA-321 *Hell's Angels* and VMFA-112 *Cowboys*, just made it into the 1990s with the F-4, initiating the aircraft's fourth decade of service. The *Cowboys* gave up their last F-4S at NAS Dallas on 18 January 1992, while the *Angels* moved to the Hornet following Phantom closedown on 13 July 1991. As "Lancer" Sullivan observed, while the F/A-18 began to take over the Marines' fighter role:

"The F-4 was a fine machine, and doing well against the newer fighters, but this was due to tactics and the fact that the F-4 crews were so experienced. As more experience was gained with the F/A-18 it was readily apparent why the F-4 was being replaced. In every category, from acceleration to sustained g, from radar performance and reliability to aircraft maintainability and modern avionics or weapons systems the F/A-18 was superior. The F-4S was the safest Phantom, and I don't believe we ever lost another one to stall/spin in ACM after we got the slats. The AWG-10 finally got to be fairly reliable, but was still a bear to maintain. CCIP was in the aircraft, which made it a better bomber, and we had two UHF radios.

I still preferred the F-4N in ACM, as it weighed several thousand pounds less and the lighter nose didn't 'fall through' as fast as

VX-4's "Black Bunny" F-4S, BuNo 155539, the replacement for F-4J BuNo 153783 as *Vandy 1*, the latter having been sold to the RAF. (Simon Watson)

The "White Bunny" was F-4J BuNo 158350, finished overall in gloss FS 17875 white. National insignia were absent, as were most stencil markings. (Simon Watson)

the F-4S. The 'hard wing' F-4 was definitely not everyone's choice in ACM, as you had to fly it well into heavy buffet and keep your angle of attack under control, even though it was pegged at 30 units. The F-4B/N's APQ-72 radar, though simple, stayed up, and our RIOs were great at working it to its maximum advantage. I believe the strongest point in the Marine ACM program (other than well-trained pilots) was the exceptional qualities of the RIOs in both radar tactics and commentary, but also in the engagement once it became visual and 'heads out of cockpit.' They had the ability to keep a tally-ho, protect your six and let you know when you had to stop working the guy you were trying to shoot and start concentrating on the guy trying to shoot you, as he was beginning to become a serious threat. All in all, the Marines got about everything you could out of the F-4 in both air-to-air and air-to-ground. It wasn't so much the versatility of the F-4, but the dedication of the pilots, RIOs, and mechanics that kept it in the hunt long after there were infinitely better fighter/attack aircraft in operational service. Today's Marine fighter/attack skills and tactics were honed in the F-4 community and MAWTS from the early F-4 days of learning how to do it right through trial, error, and fixes.

As the Phantom's McDonnell-Douglas successor, the Hornet, began to enter squadron service, pilots were able to evaluate it against the F-4. Manfred Rietsch's experience was unusual in that he moved from the Phantom to the Hornet and then back to the F-4 for a while.

"There is such a quantum leap in technology from the 1950s to the 1970s and beyond. I had been flying the F/A-18 for six years, getting very comfortable with all the new technology, when I started flying the F-4 again [with VMFP-3 as Commander of MAG-11]. We had the last F-4s in MAG-11 in 1989—old RF-4Bs with 151 to 153 series Bureau numbers. There were no real F-4 simulators around, so the first hop after six years was refresher in more ways than one. On the take-off roll I wondered...are the burners lit? It seemed like a 'basic engine' take-off in the F/A-18. Why is the cockpit and air conditioner so noisy? Where is my HUD? How can

I fly this thing without a HUD? The fuel gauges must be broken, because it doesn't change...Oh, yes, we are transferring fuel from the centerline, and only the F/A-18 reads internal and external fuel on the gauge. Where is my moving map? How can I find my way? My God, the INS dumped! How will we fly low-level? This refresher was, to me, a very graphic example of the technological changes in fighters.

The hard-wing [non-slatted] F-4 turned well above 420 kts IAS. ACM against better-turning aircraft was a judicious exercise in energy management. And nearly every fighter turned better than the F-4. About the only airplane around that had a higher wing-loading was the AV-8. The only advantage the heavy, high wing-loaded F-4 would have over the F/A-18 was that the F-4 could get down on the deck and ride like a Cadillac. In the F/A-18 we did not look forward to afternoon low-levels over the hot desert, because the aircraft would react to every little turbulent bump and beat the heck out of the pilot: the price you pay for low wing-loading and high-lift devices.

Comparing handling qualities, the Phantom was a very honest airplane. It would tell the experienced pilot exactly what it was going to do. Few pilots flew the hard-wing F-4 *well*. It took experience, balls, and aggressiveness mixed with wily cunning, patience, and energy management. It was not a 'point and pull' aircraft. You had to think two or three turns ahead what you were going to do with the aircraft. If you didn't treat it right it would spin, or at least give you the ride of your life if you ran yourself out of airspeed and ideas. But, if you mastered that beast it could be a magnificent air-to-air machine. The slatted F-4 took away the challenge of flying the aircraft well. It wouldn't spin, but it bled energy at a tremendous rate. They made it safe, but...The F/A-18 is the easiest aircraft I have ever flown. It has great thrust combined with the turning performance of a souped-up A-4. It truly is a point-and-pull aircraft. At 100 kts in level flight you can yank it into the pure vertical.

As for human engineering, the cockpit of the F-4 was like a Cadillac. You had four to six inches between the canopy sill and

Less well-known was VX-4's "Bluebird" F-4J, BuNo 155896, which tested this camouflage in September 1973. (James E. Rotramel)

your shoulders. The stick was long and way up front, with lots of 'throw.' You had to be at least 5 ft 8 inches to fly it, or you had to sit on cushions. A *manly* airplane! On the other hand, you strap on the F/A-18. It has a compact cockpit, and everything is logical and in easy reach. It fits guys from 5 ft 6 inches to 6 ft 4 inches. What a difference the two handles on the canopy bow make to help you turn around in ACM. You fly and fight the aircraft with your hands on the stick and throttle (HOTAS). All your weapons information is at eye level and can be see in the brightest sunlight. You never have to bend over and get vertigo when flying on instruments. It's all in front of you. Switches are designed to feel different to the touch, based on functionality. In the F-4 the gun-charging switch was way on the left console. You had to look to find it. Switching from SW (Sidewinder) to SP (Sparrow) or vice versa, you had to find the second switch from the left on the lower front panel among many other similar switches. We ended up putting a rubber hose on this weapons switch (called a 'donkey dick') to let us find it without looking down. All weapons information and control was inside the cockpit. The TACAN and IFF controls were in the right console, great for vertigo in the 'soup.' There were tiny oil and hydraulic gauges way down between your feet that you couldn't read at night.

It took experience and a certain sense of 'Kentucky windage' to be a very good bomber in the F-4. Later models had computer-aided bombing systems which improved accuracy. In the F/A-18 you have a 'pipper' which is truly a 'death dot.' The bombing system is extremely accurate. Anybody can be a good bomber."

There were still tasks for the Marines' F-4s at sea in the 1980s. When USS *Coral Sea* deployed to the Gulf of Oman in Spring 1980 in response to the Iranian hostage crisis, VMFA-531 *Gray Ghosts* (who had deployed on *Forrestal* in 1972-73) and VMFA-323 *Death Rattlers* (the USMC's premier air-to-air exponents at the time) were the Air Wing's fighter squadrons, with F-4Ns. Had they been called to action in the ill-fated Operation *Eagle Claw*, covering the planned extraction of the U.S. hostages by RH-53D helicopter crews, they

might well have entered Phantom vs. Phantom clashes with the F-4s of the Iranian Air Force. The Marine F-4Ns were given temporary red and black wing stripes to prevent mis-identification in such an eventuality, though their colorful unit markings were probably sufficient for this purpose. Carrier deployment pioneers VMFA-333 made another cruise in 1976 on USS *Nimitz's* maiden voyage, while VMFA-451 *Warlords* sailed the Mediterranean on the *Forrestal* in 1976, followed by VMFA-115 *Silver Eagles* in 1981. In all cases they substituted for squadrons who were in transition to the F-14A.

Recce Run-down

Although the Marines could feel confident in the successor to their F-4 fighters, there was no planned replacement for the dwindling RF-4B force. Following the amalgamation of "Photo-Phantom" assets into a single squadron, VMFP-3 at El Toro, in July 1975, a refurbishment program was initiated for the surviving Phantoms. This was triggered both by the exhausting rate of utilization in Viet-

Another view of the *"Screaming Eagle"* bicentennial bird painted up for VX-4. (Boeing/McDonnell Douglas)

F-4J/YF-4S BuNo 153088, immaculately finished. (Simon Watson)

nam and by perceived deficiencies in the sensor systems. A sensor update and refurbishment effort (SURE) was conducted at NARF North Island in collaboration with McDonnell-Douglas, who provided twenty-eight kits and a pattern aircraft. According to John Harty a number of structural beef-ups were done, the aircraft were re-wired, an F-14A-type INS (AN/ASN-92 carrier alignment inertial navigation system, or CAINS) was installed, plus new passive/active ECM equipment (AN/ALQ-26 and AN/ALR-45/50). Revised sensors included the AN/APD-10 SLAR (side-looking airborne radar/mapping set) and AN/AAD-5 infra-red set. The Phantoms received slotted stabilators, and also the AN/ASW-25B data link.

RF-4B BuNo 153105 was the first to be converted (FR 01) with the kit, and BuNo 157351, along with 157348 acted as verification aircraft for the updates to the other RF-4Bs. With its new APD-10 set the RF-4B gained a high-resolution system offering long-range target stand-off and genuine all-weather surveillance. Its moving target indication (MTI) capability allowed it to detect jeep-sized vehicles on the move, using one of six radar modes.

VMFP-3 *Rhinos* trained its own aircrew and maintained a core of around a dozen RF-4Bs at El Toro, with detachments at Kaneohe Bay, in the Westpac area, and for various USMC exercises within the USA. A further detachment deployed regularly on USS *Midway* from 1975 until F-4 carquals were ended in 1984. Colonel Mike Fagan was the squadron's last Commanding Officer and saw the RF-4B through its final years, when funding to support the benefits of its costly SURE updates was already being reduced.

"To save money on programs such as SURE, maintenance training, spare parts, and manuals are sacrificed when money gets tight. Accordingly, the infra-red and [APD-10] SLAR worked well when they were new, but did not actually see much use after a couple of years due to inadequate spares and maintenance knowledge. I personally had never flown a SLAR mission using the near-real-time data-link [AN/ASW-25B], and I don't remember ever seeing imagery from a SLAR mission. That illustrates how infrequently we used that system. The infra-red was used more, but the resolution was not as good as when the systems were new. No fault of the systems; we just didn't keep them up. There was a switch that would open the door to the IR detector if power was on the plane, and I used to look in at the rotating mirror (which was static when the aircraft was on the ground). I was always disappointed that the mirrors were far from being shiny. They were scratched from improper cleaning, and were the color of tarnished silver. I didn't understand how they could produce any imagery, but they did. I had some conversations with factory reps about improving the quality of the IR imagery, and they described a calibration procedure to some of our technicians. I did see some improvement after that.

When I became CO, improvement in imagery quality was one of my chief objectives. I directed Camera Repair to conduct a resolution test of all our lenses. We had many, as the number of aircraft had decreased, but we retained the cameras and lenses. I knew that we were not getting the results we should, based on informal observation. A camera body (KS-87) was placed on a table and pointed at a resolution target hung on a blast fence. I was interested in the relative performance of the lenses, not the absolute numbers, so the distance to the target was irrelevant. The technicians simply noted the number of the smallest pattern that was resolvable. I had the lenses in the bottom third of performance marked 'never to be used'; the top third marked for use in combat only (which nobody envisaged); and the middle third for training and peacetime requirements."

Originally, RF-4Bs were equipped with the Litton LN-12 (AN/ASN-56), a twin-gyro unit linked to the reconnaissance systems and navigational computer, and controlled by a small navigation control panel with a master mode switch to select standby, align, or navigate. LN-12 was tied into the radar, autopilot, and air data computer. When RF-4Bs deployed on carriers they had to attempt an alignment process for their INS inertial platforms, which equated to the correct process on land. Data had to be relayed from the ship's central navigation system via a "SINS" (ship's INS) cable linked to the RF-4B's INS. Colonel Fagan:

VX-4's "normal" scheme and markings on F-4B BuNo 141439. (MAP)

"On my 1976-77 deployment on the USS *Midway* we used a SINS cable to help align the INS. Ashore, the aircraft had to be perfectly still so that the INS could align itself level with the earth and know which way was true north. The inertial platform also had to know the latitude at which it was being aligned so that it could continuously point to true north during the alignment process. The navigation computer also needed to know about longitude so that it had a place to start navigating." Trying to align the platform on a moving ship made that process impossible without a SINS input. "The system worked well enough. We never expected much from INS anyway, but it did supply the aircraft with primary altitude reference, and it always was handy to have the altitude indicator working, particularly at night. The back-up altitude reference was a separate vertical gyro, and its signal was used whenever the primary wasn't available. The back-up source always drove a smaller altitude indicator we called the 'peanut gyro,' because of its size. It was a good idea to include it in your visual scan for night catapult shots in case the primary went 'spazo.' In practice, though, the SINS didn't always give us good alignment and we were frequently banged off the front of the ship with no primary altitude indicator, planning to get a rougher in-flight alignment. Many folks were less inclined to do this at night, though.

Later, we got CAINS which, while ashore, could be partially aligned and then interrupted for some taxiing before completing alignment. It also had a UHF data link that supplied corrections from the ship's INS for alignment afloat. It worked well, but as the aircraft INS was newer then it was more accurate and reliable. Sometimes, though, confidence in the newer systems was a problem because failures were not as expected as they were with the older system. The CAINS also had the ability to pass twelve waypoints [for planning a navigation route] from the ship to the aircraft during alignment (but not in flight), though I don't believe I ever used this. We were happy with the CAINS data-link alignment feature because of the greater possibility of a good alignment, but the INS system itself continued to bother me because of its sudden failures.

Our pilots were prepared to transition to the 'peanut gyro' quickly, and most had to at some time or another."

RF-4Bs were never again called to war, as VMFP-3 was de-activated on 30 September 1990 after a final ceremony on 10 August. The last five aircraft, including the spectacularly-marked BuNo 157342 (the MAG-11 Commander's aircraft) and the slinky all-black 157351, were flown to Cherry Point on 13 August. Ironically, when Operation *Desert Storm* began six months later the lack of the RF-4B was sorely felt by the Marines as they sought targets and accurate BDA in their attempt to cut down Saddam Hussein's artillery and armor. Col Mike Fagan, who had sent one of the last RF-4Bs to storage on the day Saddam invaded Kuwait, became Targeting Officer for the Marine Expeditionary Force in *Desert Storm*. He reflected on the areas where independent reconnaissance intelligence could well have provided to the Marines' advantage by RF-4Bs.

"The mission type I remember thinking would have been most useful was patrolling the border between 'us' and 'them,' particularly at night. The enemy often used the cover of darkness to reposition and re-supply his forces. My feeling was that this was a period of vulnerability for them, which we could have exploited if only we had better knowledge of their locations and activities. Although the Marines benefited from some J-STARS and U-2 support, I remember that many of us wished we had more of it. I thought our own RF-4Bs could have patrolled the border using their synthetic-aperture radar and data-link. The receiving ground station for the data would have located at 1 MEF HQ, and we could have enjoyed full-time, near real-time (there was a system latency of about a minute) fixed and moving target border surveillance, all night long."

Rhino's Return

With the prospect of many surplus F-4s becoming available, and the need for a realistic, supersonic full scale target drone (FSAT), the Navy began to eye the F-4B as a likely candidate as early as 1970. As Vance Vasquez at Point Mugu Naval Air Warfare Center Weapons Division (NAWCWPNS) explained:

YF-4J BuNo 151473, the "White Ghost," with its center cockpit transparencies filled in for ejection seat trials. (Simon Watson)

The Navy began a new program to begin conversion of a VMFAT-201 F-4B-6-MC to a QF-4B or Navy Agile Target (NAT). It was developed at the then-Naval Air Development Center, Warminster, PA. This QF-4B (BuNo 148315) was flown to the Naval Missile Center during April 1972 and painted in an overall bright red/orange scheme. Carrying the nose-number '40,' this was one of six QF-4Bs to undergo conversion at NARF. Altogether forty-four QF-4Bs were converted, initially taken direct from front-line units, such as VF-14, VF-74, and VMFA-251, and later from AMARC storage. Radar was replaced by ballast, and a UHF command/control link replaced the armament and navigation systems. Remote-control tests began in 1971 with a piloted DF-8L Crusader (later a DF-4J, BuNo 153084) as airborne control vehicle. Later conversions had a variety of countermeasures devices and on-board cameras and electronics to score hits and near-misses by missiles. Roll and pitch maneuvers were limited to 45 degrees and +5g (later 6.5g). QF-4B '40' lasted until 31 January 1974, but by 1987 only one of the QF-4Bs remained intact at NWC, and in that year BuNo 149452 fell to the first YAIM-120 AMRAAM shot. Others went to White Sands to test the Patriot missile system and were blown out of the sky there.

Attention then shifted to later Phantom variants as they began to line up in the Arizona desert, and the F-4N was the obvious next in line. Its great agility was a major consideration as an AAM target. Additional removal of non-essential equipment saved over 2,000 lbs, and the rear cockpit was stripped out to accommodate the AN/ASA-32 analog autopilot. The aircraft could either be flown conventionally by a front seat safety pilot, or NOLO (no live operator aboard). In the former case the pilot was essentially along for the ride unless an emergency occurred, and at the mercy of the ground operator 'pilot,' who flew the aircraft via a small TV screen and a control stick. Safety pilots had to be prepared for some unexpected maneuvers."

In 1999 NWC China Lake's QF-4Ns (over 78 had been converted) are kept in large protective bags, guarded against the depredations of moisture and the sun. Vance Vasquez assessed the operational situation at the time.

"Approximately ninety per cent of the flight time of a QF-4N is manned. At Cherry Point an interface unit is placed in the nose of the Phantom, becoming the nucleus of the conversion to QF-4N and acting as a central point between the ground and the remote target functions of the aircraft. A target control panel is added, with target systems antennas and onboard cameras. When a QF-4B is being prepared for NOLO presentation a pilot sits in a universal control console where he performs a remote check of the aircraft, while the pilot monitors pre-flight instrument readings. While the QF-4N is on the ground, the set-up person leaves the cockpit and the flight termination [self-destruct] charges are set. Control is then turned over to the 'pilot,' who uses the forward-looking TV camera, while looking at a TV monitor, aided by instrument gauges, to 'fly' the QF-4N remotely."

QF-4Ns were also used as test beds for systems such as the airborne turret infra-red measurement system (ATIMS), in an external pod, and various decoy flare systems. They were the only supersonic high-altitude launch vehicle for the AQM-37C target

drone. One QF-4N (BuNo 150465, *Wolverine* 40) became test-bed for the General Electric Warfare Platform, with an AAR-54 modular missile warning system to alert a pilot to IR missiles being aimed at him. ALE-47 countermeasures pods were also carried, and some crucial new EW systems including the AAR-47, ALE-50, ALR-164, and ALR-167 were all tested aboard the aircraft in the mid-1990s. By 1997 NOLO operations accounted for only five per cent of the annual flight time of the Naval Weapons Test squadron, the *Bloodhounds*, the rest being given over to a wide variety of test and Fleet support tasks. However, many QF-4Ns earned extensive "kill marking" score boards of missile stencils on their splitter plates, indicating misses rather than kills by "hostile" fighters. *Bloodhounds* QF-4Ns could be deployed to other missile ranges (at Puerto Rico and Hawaii) for realistic interception training. Two of their number briefly resurrected VF-143 *Pukin' Dogs* as a Phantom squadron when they were hoisted aboard the USS *Independence* to appear in the 1989 film "Flight of the Intruder," though the markings were less than authentic.

Supplies of QF-4Ns were stretched by using other drone vehicles and towed targets, but the intention was to extend the conversion process into F-4S stocks as the Marines switched to the F/A-18 after fifteen years of relying on the F-4S. VX-4 retired their last F-4 "Sierra" in 1990 and VMFA-112 ended USMC use of the type in 1992, releasing quantities of these high-performance aircraft. A single F-4S, BuNo 158358, was pulled out of AMARC and converted by NARF, arriving at Point Mugu on 27 July 1995 as *Bloodhound* 100, piloted by Dave "Fireball" Hayes. The program had been initiated in 1987, but was suspended when the QF-4E/G was selected as the tri-service FSAT and the prototype QF-4S was stored at China Lake. As a drone, it differed from the QF-4N in having an automatic flight-control system which would respond to fifteen "command words," rather than the QF-4N's ten. Roll and rudder trim were provided, as were commands for engine shut-down and pre-planned maneuvers. Dual color video cameras were mounted ahead of the pilot, giving the ground pilot a more sophisticated image of the aircraft's flight. Innate F-4S characteristics were in themselves major advantages, such as smokeless engines and the fifty per cent improvement in turning capability conferred by

EF-4B BuNo 153070 of VAQ-33 *Firebirds* with ALQ-126 DECM. (Simon Watson)

slats. In 1996 ten QF-4S conversions were planned, with conversions beyond that being, "based on customer needs." A total of up to 35 has been proposed. Conversion avionics kits were put together at China Lake and installed at Cherry Point. CDR Al Bradford at Point Mugu explained the rationale behind the QF-4S:

"The F-4S was inducted as a cost reduction because extended scheduled depot level maintenance (SDLM) savings will be realized by converting the F-4S as opposed to F-4N aircraft. The F-4S is relatively young (average storage time is between two and nine years), the alloy used in the F-4S airframe is not susceptible to the extensive metal fatigue associated with the F-4N, and the F-4S is already in the Aircraft Service Period Adjustment Program. In this, if the QF-4S passes acceptance testing after target conversion, twelve months will be added to its service life. The process is repeated until the aircraft fails inspection, at which time it will be returned to the depot for SDLM.

The initial QF-4Bs used radio control very similar to that used by radio control model airplanes, while the QF-4N's use the Integrated Target Control System (ITCS). The QF-4S configuration has been designed to minimally change the hardware and software from the current QF-4N system while incorporating unique QF-4S characteristics. The system modification has six key elements: the aircraft interface unit (AIU) provides command uplink controls and telemetry downlink flight data through the AN/DKW-3A/B transponder. The AIU, which replaces the existing F-4S automatic flight control system, also contains system-control logic and automatic flight-stabilization and control functions. The Target Control Panel (TCP) provides power distribution and circuit breaker protection for the target systems, and switch commands to the AIU, it displays necessary aircraft and target systems parameters for the drone safety pilot, and also provides a safety interlock for target system disengagement. Dual forward-looking video cameras give the remote pilot a visual reference for takeoff and landing, and are used during remote flights for formation flight reference (in multiple NOLO presentations) and as a secondary attitude-reference source." Like the QF-4N, the "S" model is deprived of its radar and parts of the weapons control system. All circuit breakers in the back cockpit are moved to the front, and the QF-4S becomes a single-seat Phan-

One of two EF-4Js, BuNo 153084, flew with VAQ-33 on a wide variety of electronic warfare development tasks. (Simon Watson)

tom. More recent additions, in the QF-4S-Plus, include a Trimble GPS set and a new EW suite.

"The aircraft is flown to the range using radar track or Extended Area Test System (EATS). Prior to NOLO flight the aircraft is checked out with a safety pilot in the plane. He just rides, ensuring that everything works correctly, and he doesn't touch anything, even during take-offs and landings, unless absolutely necessary. If that is the case, there are five ways to do this, with shutting off of the generator switches being the last resort. Like the QF-4N, the QF-4S will also satisfy a secondary role in scenarios where it is not the target. These will include use of the Navy standard tow-target system (NSTTS), ECM pods (AN/ALQ-126 and AN/AST-4), threat simulators, and expendable devices, including lot testing of flares and supersonic pods to support the F-22 program. Other non-target roles include the aircraft's use as Tomahawk cruise missile chase aircraft (command, control, and destruct), airborne photo chase for weapon separation testing, threat simulation and support aircraft for the Integrated Defensive Avionics Program, Joint Supersonic Wake Infra-Red Signature Program, and NASA sonic boom signature measured test program."

Pilots who joined the QF-4 program have generally found it offered rewarding and varied flying. Among those who welcomed the chance to stick with the Phantom was CDR Al Bradford, who flew the F-4 with VF-96 in 1970.

"I returned to VF-96 in 1973 (after a period as F-4 Tactical Manual Coordinator), and remained until the squadron was decommissioned in 1975. I happened to fly the last VF-96 F-4J (BuNo 155580, *Showtime* 100), with LCDR Mike Malchiodi, for transfer. That was a sad day for the F-4, as VF-96 had the most MiG kills of the Vietnam War and the only Navy aces, as well as the distinction of winning the Admiral Clifton Award for the best Navy fighter squadron twice in its recent past. In 1975 I was ordered to the Pacific Missile Test Center at Point Mugu where I flew the F-4. There I was assigned as Project Officer for the BAe Sky Flash AAM [a British development of the AIM-7E-2]. My backseater was LT Tim Miles of the Royal Navy, and we fired eight or nine of the missiles, later to be adopted for the British Tornado and Phantom. In 1980 I

EF-4B BuNo 153070 with the markings of the Fleet Electronic Warfare Support Group. (USN via Peter B. Mersky)

F-4J BuNo 153852 was typical of many operated by NATC and its associated test organizations throughout the F-4's naval career. (Boeing/McDonnell Douglas)

returned to Point Mugu and flew the QF-4B, retiring from the service in 1985 and working as a contractor at the base. At the present time they are still flying the QF-4N/S aircraft, and they still sound as wonderful as ever."

"FMS" F-4J

Although the Phantom enjoyed large-scale export success with eleven other nations' air forces, the variants chosen for foreign military sales (FMS) were all versions of USAF models, though the F-4K (FG.1) and F-4M (FGR.2) for the UK's Royal Navy and Royal Air Force were based on the F-4J and its AWG-10 radar. However, following the Falklands conflict in the South Atlantic the British Government found itself with insufficient fighter resources to maintain an air defense posture over the Falklands "exclusion zone." It was decided to use RAF Phantom FGR.2s to provide that cover, but a number of surplus USN F-4Js was sought to fill the consequent gap in UK air defense. Fifteen airframes were selected from stored USN stock and put through NARF Noris for re-work and modification, though one was accidentally dropped into the sea from a helicopter en route from AMARC to San Diego and replaced. Modifications did not include slats and VTAS, but the AWG-10 was updated to digital AWG-10B standard. Wiring was incorporated for the Skyflash missile and the GE SUU-23/A gunpod (which was already in use by the RAF). Sets of "zero-hour" J79-GE-10B smokeless engines were fitted, and the ALQ-126 DECM equipment was removed, though its external "bulges" remained. All the life-extending strengthening and anti-corrosion measures used for the F-4S SLEP were included. The RAF decided to tolerate the lack of anti-skid braking (a handicap on rain-soaked British runways) and the air-impingement engine start method.

RAF crews received training with VMFAT-101 using the F-4S, and roll-out of the batch of F-4J (UK) aircraft began at NARF on 10 August 1989, the last of many F-4s to pass through the North Island facility. The re-lifed Phantoms made the trans-Atlantic flight to RAF Wattisham in Suffolk by 4 January 1985 carrying souvenirs of their time in the San Diego area in the form of nicknames, such as *Brigantine Bomber, Avenida Arrow*, and *Mulvaney's Missile*; tributes to some of the local late-night bars. They formed 74 "Tiger" squadron and immediately proved popular. Backseaters found many differences from the F-4M (FGR.2) cockpit, including a more user-friendly radio system, reduced instrument "clutter," and a cleaner radar picture than the one produced by the FGR.2's AWG-12 unit. Pilots appreciated the superior high-altitude performance and comparative lack of g-restrictions in maneuvering after their aging Rolls-Royce powered Phantoms. USN seats and crew personal equipment were retained until they were replaced by standard UK items stripped from time-expired Phantom FG.1 airframes shortly before the Phantom's withdrawal. In 1990 the disbandment of the RAF's Phantom training unit made more FGR.2s available, and it was decided to dispose of the F-4J(UK)s in favor of type standardization and spares economies within a shrinking defense budget. On the eve of the Gulf War, when RAF jets were once again called into action, the F-4Js were withdrawn and scrapped, used as battle-damage repair or fire-training hulks, or donated to museums. One (BuNo155529) was repainted in the VF-74 *Be-Devilers* markings, which it wore when it was sent out to the Vietnam War zone as a replacement aircraft around 25 October 1972, having previously flown with VF-33. It was then displayed in the American Air Museum, Duxford, UK.

One of the test programs using F-4s was for the XAIM-54A Phoenix in 1968. Here F-4B BuNo 148412 carries the missile, semi-recessed into a converted centerline tank which contained its support and guidance equipment. (Boeing/McDonnell Douglas via A. Thornborough)

Dandy Vandies

Many of the RAF's "born again" F-4Js had distinguished service histories in Navy and Marine squadrons, including BuNo 153783 (ZE 352 in 74 Squadron service), which had flown with VMFA-333 and later became VX-4's famous "Black Bunny" before being replaced as *Vandy* 1 by F-4S BuNo 155539. In that role the aircraft (originally *Vandy* 9 with the squadron) had tested an all-black color scheme for night visual experiments and anti-corrosion tests. Later, it test-flew a low-visibility gray scheme. Its F-4S successor also carried a spectacular blue and gold livery for the 1986 74th Anniversary of Naval Aviation, but was in turn replaced by F-4S 158358 until 1989, and then by BuNo 158360 as the final *Vandy* 1 flagship Phantom. In the Bi-centennial year VX-4 also fielded *Vandy* 76, an F-4J (BuNo 153088) which NARF resprayed in a striking multi-colored Screaming Eagle paint job (less politely known as the Turkey Bird), ostensibly to make it more visible in ACM. 1977 saw the VX-4 color range also including F-4J BuNo 158350 in overall gloss white, with its other markings in light gray.

White was also the favored color for YF-4J BuNo 151473 (a converted F-4B used by Westinghouse to test the AWG-10 radar in the late 1960s). This unique aircraft, known as "The Ghost," spent most of its career at NAS Lakehurst, NARF, El Centro, and China Lake. It was used for ejector seat trials before moving to Point Mugu in January 1997 to test new parachutes on the SJU-5 seat for the F/A-18 Hornet.

With QF-4S Phantoms still going through conversion in 1999, some of the Navy's Phantoms remain in daily use for tasks where their performance and versatility are still outstanding. Structurally, they were not substantially different from their F-4H grandparents four decades ago. McDonnell never really saw any need to change a successful format, though the St Louis designers moved with the trends and drafted in outline a TF30 turbofan-engined F-4B, the

swing-wing F-4J(VS), and even a V/STOL variant. The main objective at St Louis was to keep the production line rolling and concentrate the updates on "plug in" items, like avionics and armament. Phantoms and their crews set the standards for all subsequent Navy fighters. They were far safer than their predecessors, and pioneered the concept of BVR Fleet air defense, which was more fully realized in the F-14A. The Phantom also far exceeded the performance of most of its contemporaries in the attack role, and its versatility in this respect created the requirement for its natural successor from the St Louis production lines, the F/A-18 Hornet.

NATC tested twin SUU-11 gun pods mounted on a Shrike Dual Launch Adapter (DLA) in 1968. (via James E. Rotramel)

Glossary

AAA anti-aircraft artillery
AAC airborne air controller
ACLS automatic carrier landing system
ACM air combat maneuvering (dogfighting)
ADF automatic direction finding (radio navigation aid)
ADIZ air defense identification zone
AGL above ground level
Alpha Strike: combined strike by one or more Carrier Air Groups
AMCS airborne missile control system
Arc Light: strike by B-52s.
ARVN Army of Republic of Vietnam
AQ radar technician
Atoll: Soviet AA-2 missile, based on AIM-9B Sidewinder
bandit: confirmed enemy aircraft
BARCAP barrier combat air patrol (protecting Fleet at sea)
BDA bomb damage assessment
BIS Board of Inspection and Survey
BIT built-in test (within a piece of equipment, e.g. radar)
BLC boundary layer control (of air over a flying surface of an aircraft)
Blue Tree: escort mission for reconnaissance aircraft
bogie: suspected but unidentified enemy aircraft
BuAir Bureau of Aeronautics
BUFF B-52
BVR beyond visual range
CADC central air data computer
CAG commander of air group/carrier air group
CAINS carrier alignment inertial navigation system
CAP combat air patrol
CAS close air support
CBU canisterized/cluster bomb unit
CCIP continuously computed impact point
CIC combat information center
CILOP conversion in lieu of procurement
CNO Chief of Naval Operations
CO commanding officer /'skipper'
COD carrier on-board delivery
COMNAVAIRLANT Commander Naval air Forces, US, Atlantic

CSDU constant-speed drive unit
CVW carrier air wing
CW continuous wave radar emission
DACM dissimilar air combat maneuvering (dogfighting training using different aircraft types)
DAS direct air support
DECM deceptive electronic counter measures
Dixie station: Task Force 77 station approx. 70 nm SE of Saigon in Gulf of Tonkin
DMZ demilitarized zone
ECCM electronic counter-counter measures
ECM electronic counter measures
ECP engineering change proposal
ELINT electronic intelligence
EOGB electro-optically guided bomb
EW electronic warfare
FAC /FAC(A) forward air controller (airborne)
FCLP field carrier landing practice (on land base)
FR Fighter Regiment (VPAF)
GCA ground controlled approach
GCI ground controlled interception
ICS intercommunications system between front and rear cockpit, using 'hot mike' in oxygen mask
IFF identification, friend or foe
IFR instrument flight rules
INS inertial navigation system
IOC initial operating capability
IP initial point (on attack route to target, probably a radar offset)
IR infra-red
JCS Joint Chiefs of Staff
KIA killed in action
Kts: knots
LGB laser guided bomb
LSO landing signals officer
MAC McDonnell Aircraft Company (McDonnell Douglas Corporation after April, 1967)
MAG Marine Aircraft Group
MAW Marine Aircraft Wing

MAWTU Marine Air Weapons Training Unit
MCAS Marine Corps air station
MER multiple ejector rack (for ordnance)
MIA missing in action
MiGCAP combat air patrol to deter MiGs from attacking naval strike force
mil: gunsight setting measured in fractions of a degree (6400 mil = 360 degrees)
MOREST mobile arresting gear
NARF Naval Air Rework Facility
NAS Naval Air Station
NATOPS Naval Air Training and Operating Procedures Standard-ization program
NFO Naval Flight Officer (RIO in F-4)
NFWS Naval Fighter Weapons School (Top Gun)
NOLO no live operator aboard (QF-4)
NORAD North American Air Defense
NORDO aircraft without working radio
NPE Naval preliminary evaluation
NVA North Vietnamese Army
NVN North Vietnam
ORI operational readiness inspection
PIRAZ off-shore radar controller destroyer
POL petrol, oil, lubricants (stored)
POW prisoner of war
RAG replacement air group
Red Crown: off-shore vessel controlling air operations via radar
RHAWS radar homing and warning system
RIO radar intercept officer
RO radar/recce systems operator (RF-4B)

ROE Rules of Engagement
ROK Republic of Korea
SAM surface to air missile
SAR search and rescue (of downed aircrew)
SATS short airfield for tactical support
SEAM Sidewinder expanded acquisition mode
SLAR/SLR sideways-looking (airborne) radar
Steel Tiger: strikes on the Ho Chi Minh trails network in Laos.
SURE sensor update and refurbishment program (RF-4B)
SVAF South Vietnamese Air Force
TACAN tactical aid to navigation (UHF)
TARCAP combat air patrol over target area
TER triple ejector rack (for ordnance)
TFW Tactical Fighter Wing (USAF)
TIC troops in contact
TOT time on target
TPQ bombing mission using radar guidance to signal drop
Transpac: trans-Pacific crossing by air
UDP unit deployment program
VC Viet Cong
VF naval fighter squadron (FITRON)
VFR visual flight rules
VID visual identification (of potential target)
video: radar imagery on cockpit screen
VMFA Marine fighter-attack squadron
VPAF Vietnamese Peoples' Air Force (North Vietnam)
VTAS visual target acquisition system
Westpac: deployment to West Pacific area
WSO weapons systems operator
XO Executive Officer (usually became CO of squadron)

Appendix 1:
Dimensions and Performance Statistics

Overall Dimensions (these figures are generally applicable to the F-4B, RF-4B, F-4J, F-4N and F-4S).

Wingspan:	38 ft. 4.9 inches. 27 ft 7 inches (folded).
Length:	58ft 3 inches (RF-4B 62 ft 11 inches).
Height:	16 ft 5.4 inches (F-4J 16 ft 3 inches).
Landing gear wheelbase (between centers of wheels):	23 ft 3 inches.
Nose-gear steering limits:	70 degrees each side of center-line.
Height of wingtip above ground (folded):	11 ft 5.5 inches.
Ground clearance beneath tanks: centerline	1ft 1 inch, wing 1ft 9inches.
Distance from ground to top of center fuselage:	10 ft 8.5 inches.
Wing dihedral:	nil (main section), 12 degrees (outer, folding sections).
Wing area:	530 square feet. Sweepback at quarter-chord: 45 degrees.
Wing incidence:	1 degree. Aspect ratio: 2.82.
Wing sections:	NACA (modified) 006.4-64 (root), 004-64 (wing-fold point), 003-64 (tip).

Performance

n.b. these statistics are approximations as performance is subject to numerous variables such as temperature, external load, condition of aircraft etc. The F-4 NATOPS pilot's manual includes 138 pages covering performance statistics and graphs.

	F-4B	F-4J	RF-4B
Weights (lbs)			
Empty:	27,897	30,778	31,200
Typical combat	38,505	41,673	40,267
Maximum:	55,950	56,000	54,800
Arrested landing:	-	40,000	-
Field landing:	32,192	46,000	33,598
Wing loading (lbs.sq. ft.	72.7	78.6	84.1 (combat)
Speed:			
Max at 40,000 ft (mph):	1,490	1,430	1,407
Max. at sea level:	845	889	890
Rate of climb (clean):	40,800	41,250	47,500 (ft. per minute)
Combat ceiling:	56,850	55,000	60,000
Ferry range (miles):	2,300	1,956	1,750
Fuel (max, US galls:	3,326	3,338	3,309

Appendix 2:
U.S. Navy F-4 Units

VF-11 *Red Rippers*.
From F-8E to F-4B mid-1966. One combat cruise on USS *Forrestal* with CVW-17 (June-September, 1967: curtailed by fire). To F-4J September, 1972 and F-14A May 1980. With CVW-17 1966-80.

VF-14 *Top Hatters*.
Converted from F3H-2N to F-4B May, 1963 and deployed with CVW-1 on USS *F.D.Roosevelt*, 1964-66. One combat cruise (June 1966-February 1967). Returned to 6th Fleet for five more CVW-1 deployments on USS *Roosevelt* (1967) and *J.F.Kennedy* (1969-73) To F-14A mid 1974.

VF-21 *Freelancers*.
From F3H Demon to F-4B late 1962. Deployed on USS *Midway*. War cruises with CVW-2 on *Midway* (March-November, 1965) and USS *Coral Sea* (July 1966-February 1967) and five war cruises on USS *Ranger* (November 1967-June 1973). To F-4J July, 1968 and F-4S January 1980. With CVW-14 1981-82. Final Phantom cruise in F-4N (USS *Coral Sea* 1982-83). To F-14A November 1983.

VF-22L1 (Code 7L)
Formed with F-4B January, 1969 as first USNR F-4 unit. De-commissioned October, 1970. To F-8.

VF-31 *Tomcatters*.
From F3H to F-4B October, 1963. Deployed on USS *Saratoga* with CVW-3 1964-late 1980. One war cruise (April 1972-February 1973). To F-4J January, 1968 and F-14A January, 1981.

VF-32 *Swordsmen*
From F-8D to F-4B late 1965. First cruise was war cruise with VF-14 on USS *F.D.Roosevelt* (June 1966-February 1967) with CVW-1. With 6th Fleet on USS *Roosevelt* (1967) and *Kennedy* May 1968-1973. To F-14A early 1974.

VF-33 *Tarsiers*
From F-8E to F-4B November, 1964 with 6th Fleet. War cruise on USS *America* (April-December, 1968) after transition to F-4J in October, 1967. Returned to 6th Fleet (CVW-7) on USS *Independence*. To F-14A early 1981. Flew F-4J longer than any other unit.

VF-41 *Black Aces*
From F3H to F-4B February, 1962. To Key West, 1962. Combat cruise on USS *Independence* May-December 1965. To F-4J February 1967 (first deployable F-4J unit) and joined 6th Fleet on USS *F.D.Roosevelt*. It was the sole fighter unit on the ship's 1972-73 cruise). To F-4B/N 1973-74 and F-14A June, 1976.

VF-51 *Screaming Eagles*
From F-8J to F-4B mid-1971. Two CVW-15 war cruises on USS *Coral Sea* (November 1971-July 1972 and March-November 1973). To F-4N early 1974 until April 1977. Final Westpac cruise 1974-75 on *Coral Sea*. To 6th Fleet and *F.D.Roosevelt* (May 1976-April 1977) with CVW-19. To F-14A mid 1978.

VF-74 *Be-devilers*
From F4D-1 to F4H-1 by July, 1961 (first deployable Phantom squadron). Three 6th Fleet cruises (1963-66) on USS *Forrestal* (CVW-8). and Westpac CVW-17 war cruise on USS *Forrestal* (June-September 1967). Remained with CVW-17 in 6th Fleet until 1971, then second war cruise on USS *America*, CVW-8 (June 1972-March 1973). To F-4J early 1972. Returned to 6th Fleet with CVW-17 (1974-87) on USS *Forrestal*. To F-4S early 1982 and F-14A May, 1983.

VF-84 *Jolly Rogers*
Converted from F-8C to F-4B July, 1964. War cruise on USS *Independence* with CVW-7 (May-December, 1965). To 6th Fleet on USS *Independence*, (1966-68; two cruises, CVW-7) and USS *F.D.Roosevelt* (1970-75, CVW-6). To F-4J, February 1967, F-4B in 1973, F-4N in 1974 and F-14A June , 1976.

VF-92 *Silver Kings*
From F3H to F4B November, 1963. Eight CVW-9 war cruises on USS *Ranger, Enterprise, Constellation* and *America* (August 1964 to October 1973, CVW-9). Final Westpac cruise on USS *Constel-*

lation (June-December, CVW-9,1974). Stood down, June 1975 and disestablished, November, 1975.

VF-96 *Fighting Falcons*

A re-designation of VF-142 on 1 June, 1962 produced VF-96, using VF-142's F-4B/Gs after VF-142 had used them for only four months. Deployed on USS *Ranger* June and September 1962 for short cruises and returned to *Ranger* for Westpac cruise (November 1962-June 1963, CVG-9). Eight war cruises (as for VF-92, above) and final CVW-9 cruise on *Constellation* (June-December, 1974). Disestablished, November 1975.

VF-101 *Grim Reapers*

Operated F3H-2 and F4D-1 until end of 1962 and became East Coast RAG (Det A at Oceana) June,1960. Flew F4H-1 and F4H-1F/F-4B from early 1961. At Key West 1963-66, returning to Oceana as Det A, May 1966. Received F-4J December, 1966 and retained them until transition to F-14A, January 1976. Key West Det remained in operation for F-4 training until mid-1977 when assets passed to VF-171. Deployed on USS *America* with CVW-8, 1971.

VF-102 *Diamondbacks*

From F4D-1 to F-4B September, 1961 as second East Coast F-4 unit. Deployed on USS *Enterprise* shakedown cruise (5-8 April, 1962) coded AF, with VF-62 (F8U-1). Cruised on USS *Enterprise* with CVW-6 (August-October 1962, February-September 1963, February-October 1964) with VF-33 (F-8E). With CVW-6 on USS *America* (1965-67) and one combat cruise with CVW-6, USS *America* (April-December, 1968). To F-4J early 1968. Returned to 6th Fleet 1969-81. With CVW-7 (1969-78) and CVW-6 (1979-81), then to F-14A.

VF-103 *Sluggers*

From F-8E to F-4B, March 1965. To 6th Fleet (CVW-3) on USS *Saratoga* for five cruises, 1966-71. To F-4J late 1968. One war cruise on USS *Saratoga* (April 1972-February 1973). Returned to 6th Fleet (CVW-8) on USS *America*, 1974. To F-4S, 1981. Last cruise 1982 (CVW-17) on USS *Forrestal* as last operational East Coast F-4 squadron (with VF-74). To F-14A, January 1983.

VF-111 *Sundowners*

From F-8H to F-4B early 1971. Two CVW-15 war cruises on USS *Coral Sea* (November 1971-November 1973). Remained with Pacific Fleet and USS *Coral Sea* until 1875. Began conversion to F-4J, 1976 but reverted to F-4N for final cruise with CVW-19 on USS *F.D.Roosevelt* (October 1976-August 1977). To F-14A by end of 1978.

VF-114 *Aardvarks*

From F3H to F4H-1, late 1961. Deployed with CVW-11 on USS *Kitty Hawk's* first cruise (September 1962-April 1963) with VF-111 (F-8D) as first deployed West Coast F-4 unit. Nine further CVW-11 cruises on USS *Kitty Hawk*, eight of them teamed with VF-213 and six of them war cruises (October 1965-November 1972). To F-4J October, 1969 and F-14A May, 1976.

VF-116

Brief re-designation of VF-213 which was not carried through, though some F-4Bs were re-marked in anticipation.

VF-121 *Pacemakers.*

West Coast RAG from mid-1968 at Miramar. First USN F-4 operator. Received F4H-1 from December, 1960. F-4A/B until 1967 when first of F-4Js received. Some F-4S received, 1980. Dis-established September, 1980.

VF-142 *Ghostriders*

Re-established October, 1963 at NAS Miramar after original VF-142 re-designated VF-96. From F-3H to F-4A January, 1962 and F-4B late 1963. Made seven CVW-14 war cruises on USS *Constellation, Ranger, Enterprise* (May 1964-June 1973) To F-4J mid 1969. Final cruise with CVW-8, 6th Fleet on USS *America* (January-June 1974). To F-14A late 1974.

VF-143 *Pukin' Dogs.*

Formerly VF-53 with F3H, re-designated VF-143 mid-1962 and equipped with F-4B and deployed on USS *Constellation* February, 1963. Operational history the same as VF-142 (above) after this.

VF-151 *Vigilantes*

From F3H to F-4B early 1964. Made seven war cruises with CVW-15 on USS *Coral Sea* and *Constellation* (December 1964-July 1970) and with CVW-5 on USS *Midway* (April 1971-March 1973). To F-4N March, 1973. From September, 1973 to March, 1986 homeported at Yokosuka, with CVW-5 cruises on USS *Midway*. To F-4J mid-1978. Three Indian Ocean deployments 1979-80. To F-4S February, 1981. With VF-161, it was the last deployed USN F-4 unit. To F/A-18A June, 1986.

VF-154 *Black Knights*

From F-8D to F-4B November, 1965. Made six CVW-2 war cruises on USS *Coral Sea* and *Ranger* (July, 1966- June, 1973. To F-4J August, 1968 and F-4S December, 1979. To F-4N mid-1980 for final two cruises with CVW-14 on USS *Coral Sea*. To F-14A, January, 1984.

VF-161 *Chargers*

From F3H (last unit to transition) to F-4B November, 1964. Deployments and sub-type changes then same as for VF-151. Disestablished April, 1988.

VF-171 *Aces*

Re-commissioned 8 August, 1977 at Oceana as Atlantic Fleet F-4B/N RAG, with a Det at Key West for ACM training. To F-4J, November 1978 and F-4S April, 1981. Disestablished June, 1984.

VF-191 *Satan's Kittens*

From F-8J to F-4J March, 1976. Deployed on USS *Coral Sea* with CVW-15 June 1976-March, 1978. Disestablished 1 March, 1978 and re-established December, 1986 on F-14A.

VF-194 *Red Lightnings*
From F-8J to F-4J May, 1976 and CVW-15 with VF-191. Disestablished March, 1978. Re-established December, 1986 on F-14A.

VF-201 *Renegades* (Code AF)
From F-8H to F-4N April, 1976 as USNR unit at Dallas, Tx. with CVWR-20. To F-4S early 1984 and F-14A December, 1986.

VF-202 *Superheats* (Code AF)
From F-8H to F-4N April, 1976 as USNR unit at Dallas, Tx with CVWR-20. To F-4S April, 1987 shortly before transition to F-14A.

VF-213 *Black Lions*
From F3H to F-4B/G July, 1964. Deployed with CVW-11 on USS *Kitty Hawk*, October, 1965-June, 1966 war cruise (F-4G) and November 1966-November, 1972 (F-4B and F-4J). To F-4J August, 1969. Two more cruises on USS *Kitty Hawk* (November 1973-December, 1975) and final short cruise with CVW-8 (retaining NH codes) on USS *America* in North Atlantic. To F-14A late 1976.

VF-301 *Devil's Disciples* (Code ND)
From F-8L to F-4B February-June, 1974 as USNR squadron at Miramar with CVWR-30. To F-4N September, 1975 and F-4S early 1981 (aircraft from VF-21/VF-154).

VF-302 *Stallions* (Code ND)
From F-8K to F-4B December 1974 as USNR squadron at Miramar with CVWR-30. To F-4N September,1975 and F-4S late 1980. To F-14A March, 1985. Flew last Miramar-based F-4.

VAQ-33 *Firebirds* (Code GD)
Used several F-4B (February, 1970-January, 1981). Designation of only remaining F-4B, BuNo 153070 changed to EF-4B, 1976 (their first F-4B lost mid-1970). Two EF-4J BuNo 153076, 153084 (De-cember, 1976-January, 1981). Assigned to Fleet Electronic Warfare Support Group providing electronic training for defense systems radar operators by simulating various high-speed airborne threats and targets. Specialized DECM/IFF equipment fitted for this purpose. Based at Oceana (1978-80) and NAS Norfolk.

VX-4 *Evaluators* (Code XF)
F-4H-1 from 1961 and F-4A, F-4B (until May, 1981), F-4G (1966-67), F-4J (1968-83 and 1990), F-4N (1973-75), F-4S (1982-86).

VX-5 *Vampires* (Code XE)
F-4B and several F-4N (1963-70).

VC-7 *Tally-Hoers* (Code UH)
F-4A e.g. BuNo 148258 (1971-72), F-4J (January,1969-August, 1973).

U. S. Navy Flight Demonstration Squadron, *Blue Angels*
F-4J (January, 1969-August, 1973) at NAS Pensacola.

Naval Air Test Center (Code 7T/SD)
Various F-4A-S up to 1988.

Naval Air Weapons Center and Pacific Missile Test Center (until January, 1992), *Bloodhounds*
QF-4B (from April, 1972), QF-4N (from February, 1983), QF-4S (from July,1995).

Other Users.
Naval Air Test Facility (NAS Lakehurst). Naval Test Pilot School (NAS Patuxent River). Naval Ordnance Test Station and Naval Air Weapons Center, (NAS China Lake).
Naval Weapons Evaluation Facility (Kirtland AFB). Naval Air Development Center (Johnsville, Pennsylvania). Aerospace Recovery Facility/National Parachute Test Range (NAF El Centro).

Appendix 3:
U.S. Marine Corps F-4 Units

VMFA-112 *Cowboys* (Code MA)
From F-8A (last USMC F-8 unit) to F4N December, 1975 at Dallas. To F-4J July, 1983 and F-4S January, 1985-January, 1992. Last USMC F-4 unit. To F/A-18 September, 1992.

VMFA-115 *Silver Eagles* (Code VE)
From F4D-1 to F-4B September, 1963 with MAG-24, 2nd MAW at Cherry Point. To 1st MAW, Iwakuni June, 1965-August, 1977 with six deployments to MAG-11 at Da Nang and Nam Phong, MAG-13 at Chu Lai and Naha, Japan. To 2nd MAW Beaufort July, 1977-July, 1985 with two cruises on USS *Forrestal* (1980, 1981).To F-4J August, 1975, F-4S October 1981 until December 1984 and F/A-18 July, 1985.

VMFA-122 *Crusaders* (Code DC)
From F-8E to F-4B July, 1965 at El Toro with 3rd MAW. To 1st MAW at Iwakuni and MAG-24 Kaneohe Bay August, 1967-August, 1974. Made one deployment to MAG-11 at Da Nang (1967-68), one to MAG-13 at Chu Lai (1969-70). Re-established with MAG-31, 2nd MAW at Beaufort December, 1975 until February, 1986 with F-4J (to F-4J 4 December, 1975). To F-4S October, 1982. Was designated as F-14A unit but received F/A-18 March, 1986.

VMFA-134 *Smokes* (Code MF)
From A-4F to F-4N March, 1984 as USMCR unit at El Toro. To F-4S November, 1986 and F/A-18 May, 1989

VMFA-212 *Lancers* (Code WD)
From F-8E to F-4B March, 1968 with MAG-24, 1st MAW at Kaneohe Bay with rotations to Iwakuni. Deployed to MAG-15 at Da Nang (April-June, 1972). To F-4J December, 1971. To Kaneohe Bay June, 1972. To F-4S February, 1983 and F/A-18C April, 1989.

VMFA-232 *Red Devils* (Code WT)
From F-8E to F-4J 19 September, 1967 with MAG-33, 3rd MAW, El Toro. To 1st MAW, Iwakuni (March, 1969-October 1977) with single deployments to Chu Lai (31 March, 1969-7 September, 1969), Da Nang (6 April-1 September 1972), Nam Phong with MAG-15

(1972-73) and Naha (1973-74). To Kaneohe Bay with 1st MAW (October 1977-October, 1988). To F-4S mid-1979 and F/A-18C April, 1989.

VMFA-235 *Death Angels* (Code DB)
From F-8E to F-4J September, 1968 at Kaneohe Bay with MAG-24, 1st MAW. Rotations to Iwakuni. To F-4S January, 1982 until June, 1984. Last active-duty USMC F-4 unit. Was scheduled for F-14A but transitioned to F/A-18C August, 1989.

VMFA-251 *Thunderbolts* (Code DW)
From F-8J to F-4B October, 1964 with 1st MAW, MAG-32 (MAG-31 after March, 1971) at Beaufort. To F-4J June, 1971, F-4S August, 1981 and F/A-18 March, 1988.

VMFA-312 *Checkerboards* (Code DR)
From F-8E to F-4B February, 1966 with MAG-32, Beaufort. To MAG-14 at Cherry point (February 1971-August 1974) and back to Beaufort with MAG-31 (August, 1974-July, 1987). To F-4J from February, 1973, F-4S July, 1981 (until July, 1987) and F/A-18A April, 1988.

VMFA-314 *Black Knights* (Code VW)
From F4D-1 to F4H-1 June, 1962 at El Toro. To MAG-15, Iwakuni (January, 1965-September,1970) with deployments to MAG-11 at Da Nang (13 January, 1966-14 April, 1966), Chu Lai (1 August, 1966-mid-1967 and 17 November, 1967-September, 1967). To MAG-11 at El Toro 1970-June 1983. To F-4N October, 1973 until May, 1982. Became first USMC F/A-18 unit 15 December, 1982.

VMFA-321 *Black Barons / Hell's Angels* (Code MG)
From F-8K to F-4B December, 1973 as USMCR unit at Andrews AFB. To F-4N June, 1977 and F-4S November, 1984 until 13 July, 1991. To F/A-18A November, 1992.

VMFA-323 *Death Rattlers* (Code WS)
From F-8E to F-4B August, 1964 with 3rd MAW, El Toro. To 1st MAW December, 1965- March, 1969 deployed to Da Nang with

MAG-11 (1 December, 1965-1 March, 1966 and 15 July 1966-15 May, 1967); MAG-13 at Chu Lai (15 August, 1967-25 March, 1969) with a Det in Taiwan (July-December, 1966). To El Toro with MAG-11, 3rd MAW (March 1969-September, 1982). To F-4N 1974 and F/A-18A 21 March, 1983.
Cruise on USS *Coral Sea* with CVW-14, (November, 1979-June, 1980).

VMFA-333 *Shamrocks* (Code DN)
From F-8E to F-4J March, 1968 at MCAS Beaufort with MAG-31. Deployed on USS *America* with CVW-8 for a 6th Fleet Mediterranean cruise (6 July-16 December, 1971), the first USMC F-4 Squadron to join a CVW. Deployed on USS *America* with CVW-8 and transferred to 7th Fleet for war cruise (5 June, 1972-24 March, 1973). To CVW-8 on USS *Nimitz* (June, 1975-February, 1977) and NATO deployments to Europe, 1979. To F-4S mid-1981 until October, 1987 and F/A-18 December, 1987

VMFA-334 *Falcons* (Code WU)
From F-8C to F-4J July, 1967 (first USMC F-4J unit) with MAG-33, El Toro. To MAG-11 at Da Nang (August, 1968-January, 1969) and MAG-13 at Chu Lai (January-November, 1969). To MAG-15 at Iwakuni (November, 1969-March, 1971) and El Toro (March-December, 1971) Disestablished 30 December, 1971.

VMFA-351 (Code MC)
From F-8K to F-4B at NAF Atlanta November, 1975 as USMCR squadron. To F-4N mid-1977. Disestablished 22 May, 1976.

VMFA-451 *Warlords* (Code VM)
From F-8E to F-4J January, 1968 with 2nd MAW, Beaufort. To F-4S June, 1978 (first F-4S unit, but with un-slatted aircraft) and F/A-18 (April, 1987). Scheduled deployment with CVW-17 in 1976 canceled.

VMFA-513 *Flying Nightmares* (Code WF)
From F4D-1 to F-4B January, 1963 with 3rd MAW, El Toro. To MAG-11 at NAS Atsugi (November, 1964) and Da Nang (15 June-14 October, 1965). With 2nd MAW. Cherry Point with MAG-24 (October 1965-June, 1970). To cadre status and then to AV-8A 15 April, 1971.

VMFA-531 *Gray Ghosts* (Code EC)
From F4D-1 to F4H-1 November, 1962 at Cherry Point (MAG-24). To Atsugi (1st MAW) June, 1964 with MAG-11 and Da Nang (10 April-6 June, 1965). To Cherry Point with 2nd MAW (July-April, 1968). To El Toro April, 1968-November, 1982 with MAG-33, MAG-13, MAG-11. Deployed with CVW-14 on USS *Coral Sea* (November, 1979-June, 1980). To F-4N August, 1975. F-14A transition begun 1975 but canceled. To F/A-18 January, 1983.

VMFA-542 *Bengals* (Code WH)
From F4D-1 to F-4B (November, 1963) at El Toro with 3rd MAW. To MAG-11, 1st MAW Iwakuni (April, 1965-January, 1970) with deployments to Da Nang (10 July-3 December, 1965; 1 March-1 August, 1966; 10 October-31 December, 1966; 10 July, 1968-31 January, 1970) and a Det to Chu Lai (May, 1967-July, 1968). To El Toro, February, 1970 and dis-established June, 1970. To AV-8A November, 1972.

VMFAT-101 *Sharpshooters* (Code SB from 1969-73 and SH from 1973-87)
Established on F-4J with some F-4B January, 1969 at El Toro as Marine F-4 training squadron. Moved to Yuma, mid-1970 with code-change to SH in 1973. Absorbed assets of VMFAT-201 in July, 1974. Included F-4N (from 1976), F-4S (from 1983). F/A-Trained USN F-4 crews after VF-171 RAG was dis-established in June, 1984. To 18 (October, 1987) at El Toro.

VMFAT-201 *Hawks* (Code KB)
Formed from VMFA-531 on F-4B at Beaufort (31 March, 1968) and transferred to Cherry Point. To F-4J from 1969. Dis-established (30 July, 1974) and assets passed to VMFAT-101.

VMCJ-1 *Golden Hawks* (Code RM)
Replaced RF-8A (October, 1966) and operated alongside EF-10B (replaced by EA-6A) with MAG-11 at Iwakuni. To Da Nang (28 October, 1966-1 July, 1970).To Iwakuni (July, 1970-July, 1975) with MAG-15. Deployed on USS *Midway* (July, 1974-August, 1975). Dis-established September, 1975 and assets passed to VMFP-3.

VMCJ-2 *Playboys* (Code CY)
From RF-8A to RF-4B during 1965 at Cherry Point with 2nd MAW. Carquals on USS *Independence* (January, 1966) and Dets deployed with CVW-17 (January-July, 1971) on USS *Forrestal* and CVW-3 (June 1971-February, 1973) on USS *Saratoga*. Dis-established 30 July, 1975 and assets to VMFP-3.

VMCJ-3 (Code TN)
First RF-4B unit (May, 1965), at El Toro with 3rd MAW until July, 1975 when it was re-designated VMFP-3 and absorbed RF-4Bs from VMCJ-1 and VMCJ-2

VMFP-3 *Eyes of the Corps* (Code RF)
Formed in July, 1975 on RF-4B from the three VMCJs. Dets at Kaneohe Bay (MAG-24), with CVW-5 in Japan and in the USA. Dets deployed on USS Midway 1975-80 (CVW-5). Re-named *Rhinos* (1987). Dis-established 30 September, 1990.

Appendix 4:
USN and USMC Aerial Victories in SE Asia

09/04/65	MiG-17	F-4B	151403	VF-96	Murphy/Fegan	Showtime 602/NG	CVW-9	AIM-7
17/06/65	MiG-17	F-4B	151488	VF-21	Page/Smith	Sundown 101/NE	CVW-2	AIM-7
17/06/65	MiG-17	F-4B	152219	VF-21	Batson/Doremus	Sundown 102/NE	CVW-2	AIM-7

(n.b. a third MiG-17 was later confirmed as destroyed, probably by debris from Batson/Doremus's victim. See Chapter 3).

06/10/65	MiG-17	F-4B	150634	VF-151	MacIntyre/Johnson	Switchbox 107/NL	CVW-15	AIM-7
13/07/66	MiG-17	F-4B	151500	VF-161	McGuigan/Fowler	Rock River 216/NL	CVW-15	AIM-9
20/12/66	An-2	F-4B	152022	VF-114	Wisely/Jordan	Linfield 215/NH	CVW-11	AIM-7
20/12/66	An-2	F-4B	153019	VF-213	McCrea/Nichols	Black Lion 110	CVW-11	AIM-7
24/04/67	MiG-17	F-4B	153000	VF-114	Southwick/Laing	Linfield 210/NH	CVW-11	AIM-9
24/04/67	MiG-17	F-4B	153077	VF-114	Wisely/Anderson	Linfield 2—/NH	CVW-11	AIM-9
10/08/67	MiG-21	F-4B	152247	VF-142	Freeborn/Elliot	Dakota 202/NK	CVW-14	AIM-9
10/08/67	MiG-21	F-4B	150431	VF-142	Davis/Elie	Dakota 2—/NK	CVW-14	AIM-9
26/10/67	MiG-21	F-4B	149411	VF-143	Hickey/Morris	Tap Room 1—/NK	CVW-14	AIM-7
30/10/67	MiG-17	F-4B	150629	VF-142	Lund/Borst	Dakota 203/NK	CVW-14	AIM-7
09/05/68	MiG-21	F-4B	153036	VF-96	Hefferman/Schumacher	Showtime 1—/NG	CVW-9	AIM-7

(n.b. Major Jack Hefferman was a USAF exchange officer)

10/07/68	MiG-21	F-4J	155553	VF-33	Cash/Kain	Rootbeer 212/AE	CVW-6	AIM-9
28/03/70	MiG-21	F-4J	155875	VF-142	Beaulier/Barkley	Dakota 201/NK	CVW-14	AIM-9
19/01/72	MiG-21	F-4J	157267	VF-96	Cunningham/Driscoll	Showtime 112/NG	CVW-9	AIM-9
06/03/72	MiG-17	F-4B	153019	VF-111	Weigand/Freckleton	Old Nick201/NL	CVW-15	AIM-9
06/05/72	MiG-17	F-4B	150456	VF-51	Houston/Moore	Screaming Eagle 100/NL	CVW-15	AIM-9
06/05/72	MiG-21	F-4J	157249	VF-114	Hughes/Cruz	Linfield 206/NH	CVW-11	AIM-9
06/05/72	MiG-21	F-4J	157245	VF-114	Pettigrew/McCabe	Linfield 201/NH	CVW-11	AIM-9
08/05/72	MiG-17	F-4J	157267	VF-96	Cunningham/Driscoll	Showtime 112/NG	CVW-9	AIM-9
10/05/72	MiG-21	F-4J	157269	VF-92	Dosé/McDevitt	Silver Kite 211/NG	CVW-9	AIM-9
10/05/72	MiG-17	F-4J	155769	VF-96	Connelly/Blonski	Showtime 106/NG	CVW-9	AIM-9
10/05/72	MiG-17	F-4J	155769	VF-96	Connelly/Blonski	Showtime 106/NG	CVW-9	AIM-9
10/05/72	MiG-17	F-4B	151398	VF-51	Cannon/Morris	Screaming Eagle 111/NL	CVW-15	AIM-9
10/05/72	MiG-17	F-4J	155749	VF-96	Shoemaker/Crenshaw	Showtime 111/NG	CVW-9	AIM-9
10/05/72	MiG-17	F-4J	155800	VF-96	Cunningham/Driscoll	Showtime 100/NG	CVW-9	AIM-9
10/05/72	MiG-17	F-4J	155800	VF-96	Cunningham/Driscoll	Showtime 100/NG	CVW-9	AIM-9
10/05/72	MiG-17	F-4J	155800	VF-96	Cunningham/Driscoll	Showtime 100/NG	CVW-9	AIM-9
18/05/72	MiG-19	F-4B	153068	VF-161	Bartholomay/Brown	Rock River 110/NF	CVW-5	AIM-9
18/05/72	MiG-17	F-4B	153915	VF-161	Arwood/Bell	Rock River 105/NF	CVW-5	AIM-9
23/05/72	MiG-17	F-4B	153020	VF-161	McKeown/Ensch	Rock River 100/NF	CVW-5	AIM-9
23/05/72	MiG-17	F-4B	153020	VF-161	McKeown/Ensch	Rock River 100/NF	CVW-5	AIM-9
11/06/72	MiG-17	F-4B	149473	VF-51	Teague/Howell	Screaming Eagle 114/NL	CVW-15	AIM-9
11/06/72	MiG-17	F-4B	149457	VF-51	Copeland/Bouchoux	Screaming Eagle 113	CVW-15	AIM-9
21/06/72	MiG-21	F-4J	157307	VF-31	Flynn/John	Bandwagon 106/AC	CVW-3	AIM-9
10/08/72	MiG-21	F-4J	157299	VF-103	Tucker/Edens	Clubleaf 206/AC	CVW-3	AIM-7
11/09/72	MiG-21	F-4J	155526	VMFA-333	Lasseter/Cummings	Shamrock 201/AJ	CVW-8	AIM-9
28/12/72	MiG-21	F-4J	155846	VF-142	Davis/Ulrich	Dakota 214/NK	CVW-14	AIM-9
12/1/73	MiG-17	F-4B	153045	VF-161	Kovaleski/Wise	Rock River 102/NF	CVW-5	AIM-9

There were two other kills by USN/USMC on exchange tours with USAF F-4 squadrons:

17/12/67	MiG-17	F-4D	66-8709	13 TFS	Capt. D. Baker (USMC)/1LT Ryan (USAF)	AIM-4	
12/08/72	MiG-21	F-4E	67-0239	58 TFS	Capt.L.Richard (USMC)/ LCDR M.Ettel (USN)	AIM-7	

Appendix 5: USN/USMC F-4 Production

Sub-type	Block	BuNos	Quantity
YF4H-1	1a	142259-142260	2
F4H-1/F-4A	1	143388-143392	5
F-4A	2b	145307-145317	11
F-4A	3c	146817-146821	5
F-4A	4d	148252-148261	10
F-4A	5e	148262-148275	14
(Total 45)			
F4H-1F/F-4B	6f	148363-148386	24
F-4B	7g	148387-148410	24
F-4B	8h	148411-148434	24
F-4B	9i	149403-149426	24
F-4B	10j	149427-149450	24
F-4B	11k	149451-149474	24
F-4B	12l	150406-150435	30
F-4B	13m	150436-150479	44
F-4B	14n	150480-150493	14
F-4B	14n	150624-150651	28
F-4B	15o	150652-150653	2
F-4B	15o	150993-151021	29
F-4B	15o	151397-151398	2
F-4B	16p	151399-151426	28
F-4B	17q	151427-151447	21
F-4B	18r	151448-151472	25
F-4B	19s	151473-151497	25
F-4B	20t	151498-151519	22
F-4B	20t	152207-152215	9
F-4B	21u	152216-152243	28
F-4B	22v	152244-152272	29
F-4B	23w	152273-152304	32
F-4B	24x	152305-152331	27
F-4B	25y	152965-152994	30
F-4B	26z	152995-153029	35
F-4B	27aa	153030-153056	27
F-4B	28ab	153057-153070	14
F-4B	28ab	153912-153915	4
(Total F-4B: 649)			

n.b. Block 14n included the following which were built as F-4G and surviving aircraft reverted to F-4B standard: 150481, 150484, 150487, 150489, 150492, 150625, 150629, 150633, 150636, 150639, 150642 and 150645 (total 12).

RF-4B	20	151975-151977	3
RF-4B	21	151978-151979	2
RF-4B	22	151980-151981	2
RF-4B	23	252982-151983	2
RF-4B	24	153089-153094	6
RF-4B	25	153095-153100	6
RF-4B	26	153101-153107	7
RF-4B	27	153108-153115	8
RF-4B	41	157342-157346	5
RF-4B	43	157347-157351	5
(Total RF-4B: 46)			
F-4J	26z	153071-153075	5
F-4J	27aa	153076-153088	13
F-4J	28ab	153768-153779	12
F-4J	29ac	153780-153799	20
F-4J	30ad	153800-153839	40
F-4J	31ae	153840-153876	37
F-4J	32af	153877-153911	35
F-4J	32af	154781-154785	5
F-4J	33ag	154786-154788	3
F-4J	33ag	155504-155569	66
F-4J	34ah	155570-155580	11
F-4J	34ah	155731-155784	54
F-4J	35ai	155785-155843	59
F-4J	36aj	155844-155866	23
F-4J	37ak	155867-155874	8
F-4J	38al	155875-155889	15
F-4J	39am	155890-155902	13
F-4J	40an	157242-157260	19
F-4J	41ao	157261-157273	13
F-4J	42ap	155903	1
F-4J	42ap	157274-157285	12
F-4J	43aq	157286-157297	12
F-4J	44ar	157298-157309	12
F-4J	45as	158346-158354	9
F-4J	46at	158355-158365	11
F-4J	47au	158366-158379	14
(Total F4J: 522)			

F-4N *Beeline* ECP aircraft. The R&D aircraft was BuNo 153034 (FP02) followed by 227 conversions (P001-P226 including 153034 as a 'production' F-4N). BuNos were retained and conversion numbers allocated in conversion sequence.

Block 1 (P001-005) 150430, 150652, 150491, 150452, 150635
Block 2 (P006-010) 150444, 151398, 150460, 151424, 150407
Block 3 (P011-015) 150634, 150441, 151016, 151451, 150422
Block 4 (P016-020) 151491, 150996, 151015, 150445, 150472
Block 5 (P021-025) 151433, 151442, 151434, 151006, 151400
Block 6 (P026-030) 150425, 150479, 150640, 150627, 150450
Block 7 (P031-035) 150485, 150625, 152235, 152267, 150466
Block 8 (P036-040) 151439, 152241, 151480, 150412, 151004
Block 9 (P041-045) 152291, 151431, 150630, 152280, 150411
Block10 (P046-050) 152230, 150651, 150468, 151513, 151468
Block 11 (P051-055) 151435, 150648, 150429, 151459, 151417

Block 12 (P056-060) 151476, 151463, 150492, 152278, 151519
Block 13 (P061-065) 150482, 150475, 151484, 150476, 152227
Block 14 (P066-070) 150465, 151430, 151469, 150642, 151456
Block 15 (P071-075) 151436, 152306, 151000, 151444, 151448
Block 16 (P076-080) 150436, 152259, 152258, 152272, 152253
Block 17 (P081-085) 151489, 152229, 150419, 152288, 150643
Block 18 (P086-090) 150438, 151413, 150456, 150448, 150632
Block 19 (P091-095) 150423, 151406, 152254, 152236, 150415
Block 20 (P096-100) 150489, 151003, 152294, 152302, 151471
Block 21 (P101-105) 152991, 152223, 152275, 150442, 151487
Block 22 (P106-110) 150426, 150638, 152969, 153024, 152967
Block 23 (P111-115) 152227, 151502, 151498, 151422, 150478
Block 24 (P116-120) 152237, 152210, 153059, 153047, 151464
Block 25 (P121-125) 150432, 150481, 153026, 151401, 153045
Block 26 (P126-130) 153065, 150464, 152977, 151475, 152318
Block 27 (P131-135) 152281, 153053, 152295, 153039, 153050
Block 28 (P136-140) 153023, 152975, 153914, 152981, 153017
Block 29 (P141-145) 152252, 152313, 152323, 151514, 151452
Block 30 (P146-150) 151477, 150480, 150440, 153008, 152982
Block 31 (P151-155) 151446, 153016, 152293, 150639, 151008
Block 32 (P156-160) 152243, 153915, 152226, 151510, 152221
Block 33 (P161-165) 151440, 152222, 152326, 153058, 153012
Block 34 (P166-170) 152968, 152965, 151504, 153067, 152225
Block 35 (P171-175) 152310, 152214, 152317, 152970, 153010
Block 36 (P176-180) 152212, 152250, 150435, 152208, 153056
Block 37 (P181-185) 152269, 152996, 152282, 151415, 150993
Block 38 (P186-190) 151465, 152307, 152300, 152263, 152983
Block 39 (P191-195) 153036, 153057, 152270, 152990, 152298
Block 40 (P196-200) 152244, 153019, 153027, 152986, 153062
Block 41 (P201-205) 153030, 153011, 151503, 151461, 152321
Block 42 (P206-210) 151002, 152290, 152246, 151449, 151455
Block 43 (P211-215) 152971, 153064, 150490, 152992, 152303
Block 44 (P216-220) 152279, 150628, 151007, 152217, 151011
Block 45 (P221-225) 150484, 153006, 151482, 152327, 151511
Block 46 (P226-) 152284, (153034).

F-4S ECP/ SLEP aircraft (in Block and production order). Production aircraft up to J047 received STRAPS (AFC 601 Part 1) structural and other updates and AFC 601 Part II (SLATS), the retro-fitting of maneuvering slats was done from kits later. BuNo 155892 (JX35) was given the SLATS mod and then returned to St Louis for structural tests. BuNo 155855 (J048) was the first production aircraft to receive STRAPS and SLATS Validation/verification airframes: 157286 and 158360 (SLEP only).

Block 1 (J001-005) 155565, 153791, 153845, 153853, 153818
Block 2 (J006-010) 155822, 153784, 153787, 155736, 155783
Block 3 (J011-015) 153902, 153779, 155805, 155765, 153909
Block 4 (J016-020) 155821, 155786, 155562, 153825, 153828
Block 5 (J021-025) 155858, 155555, 153833, 155521, 155732
Block 6 (J026-030) 155549, 155836, 155828, 155847, 155839
Block 7 (J031-JX35) 155840, 155848, 155561, 153810, 155892
Block 8 (J036-040) 153826, 155901, 155572, 153860, 155559
Block 9 (J041-045) 155887, 153805, 154788, 155560, 153859
Block 10 (J046-050) 153809, 155845, 155855, 153821, 155899
Block 11 (J051-055) 155872, 155575, 153820, 155527, 155893

Block 12 (J056-060) 153874, 155869, 153884, 153823, 153824
Block 13 (J061-065) 153832, 155749, 155573, 155878, 153847
Block 14 (J066-070) 155568, 155541, 155897, 155570, 153882
Block 15 (J071-075) 153856, 155900, 153780, 155531, 153843
Block 16 (J076-080) 155864, 153904, 153891, 153842, 153907
Block 17 (J081-085) 157267, 155801, 155779, 155896, 153819
Block 18 (J086-090) 153911, 155579, 157269, 153858, 153910
Block 19 (J091-095) 153880, 153868, 155757, 153835, 153869
Block 20 (J096-100) 153900, 153808, 155532, 155552, 157298
Block 21 (J101-105) 157297, 154781, 153879, 153840, 155745
Block 22 (J106-110) 153881, 153827, 153896, 153857, 153800
Block 23 (J111-115) 153814, 153851, 155542, 155522, 155834
Block 24 (J116-120) 155769, 155833, 153873, 153862, 155767
Block 25 (J121-125) 153855, 153872, 155812, 153877, 153887
Block 26 (J126-130) 155539, 155829, 155735, 155747, 157245
Block 27 (J131-135) 157257, 155784, 155519, 153908, 153798
Block 28 (J136-140) 155792, 155759, 153792, 155854, 155733
Block 29 (J141-145) 158348, 155741, 157279, 157309, 155517
Block 30 (J146-150) 157293, 155863, 155772, 155550, 153889
Block 31 (J151-155) 155544, 155883, 153893, 157276, 158370
Block 32 (J156-160) 155794, 154786, 157283, 153890, 158346
Block 33 (J161-165) 153903, 153899, 155518, 155528, 155773
Block 34 (J166-170) 157249, 155753, 155761, 155766, 155746
Block 35 (J171-175) 155743, 157250, 157259, 157308, 157290
Block 36 (J176-180) 157291, 157248, 155820, 155525, 153864
Block 37 (J181-185) 153898, 155731, 155871, 155806, 155890
Block 38 (J186-190) 155830, 158350, 155530, 155818, 155543
Block 39 (J191-195) 155859, 155808, 157292, 157296, 155524
Block 40 (J196-200) 155781, 155827, 155515, 155862, 155838
Block 41 (J201-205) 154782, 157301, 158354, 155891, 157268
Block 42 (J206-210) 155807, 155810, 158374, 155545, 155874
Block 43 (J211-215) 155813, 155876, 155754, 155567, 157304
Block 44 (J216-220) 155881, 157242, 157243, 155739, 157272
Block 45 (J221-225) 155764, 157260, 155879, 155849, 157281
Block 46 (J226-230) 158376, 158372, 157278, 158352, 157282
Block 47 (J231-235) 155566, 155888, 157255, 155851, 158353
Block 48 (J236-240) 155898, 157246, 158362, 155547, 155787
Block 49 (J241-245) 155740, 157307, 155823, 157254, 157287
Block 50 (J246-250) 155558, 155825, 157305, 158351, 155852
Block 51 (J251-255) 157286, 158358, 155504, 157264, 157300
Block 52 (J256-260) 155880, 158359, 158373, 157251, 155844
Block 53 (J261-262) 158369, 158360.

PROJECT SURE RF-4B ECP

R&D/ pre-production aircraft: BuNo 151983 VAL/VER airframes: 157348, 157351

Production sequence (FR01-30):
Block 1 153105, 157345, 151980, 153107, 157347
Block 2 151977, 151981, 151983, 151978, 157348
Block 3 151975, 153096, 153094, 151979, 153095
Block 4 153103, 153106, 153109, 153108, 153091
Block 5 153110, 153093, 153092, 152101, 153102
Block 6 157351, 153102, 157346, 157350, 157349
(Data courtesy John Harty/McDonnell Douglas)

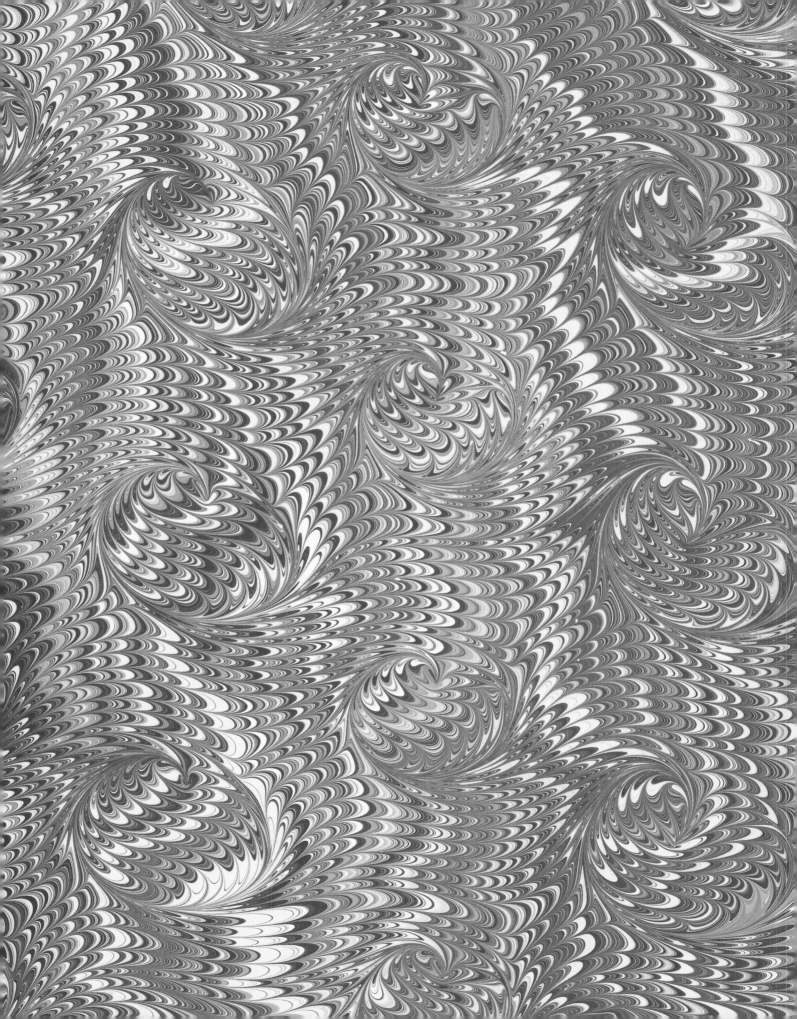